Praise for Charged

"*Charged* is the book I seriously wish had been available before I entered engineering school many years ago. It is a marvelously readable tour through the underpinnings of how our reality appears to work — that often flies in the face of convention… But always makes great sense. Absolutely brilliant!"

—**John L. Petersen, Founder and President of the Arlington Institute**

"Gerald Pollack is one of my favorite scientists. He asks penetrating questions about everyday phenomena, like clouds, the rotation of the earth and the flight of birds and comes up with radical yet simple suggestions for a new scientific vision. He reveals that our current scientific understanding, which we usually take for granted, is often a misunderstanding that lulls us into a false sense of certainty. His book is a breath of fresh air, and indeed gives a new way of thinking about fresh air itself. *Charged* is clearly written with delightful diagrams and drawings which make concepts easier to grasp. It is rare to encounter such a brilliant, original and stimulating book that changes the way we see the world around us. I am very grateful to Pollack for writing it."

—**Rupert Sheldrake, PhD, biologist and author of *Science Set Free***

"A lot of my culinary creativity and insight is inspired by my motto: question everything. Gerald Pollack is an explorer and visionary who figures out the fundamental questions we all need answers to (even if we haven't realized it yet) and explains them with lucidity and clarity. *Charged* will open your eyes, your mind and your understanding."

—**Heston Blumenthal, award-winning chef, cookbook author and restaurateur, including 3-Michelin-Star The Fat Duck**

"Delightfully illustrated and perspicuously written, *Charged* uncovers and untangles spellbinding anomalies of everyday life — from the unexpected origin of clouds, winds and raindrops to that of falling and flying of objects, animals, and machines — through the pervasive role of electrical attraction and repulsion in the natural world. Pollack's proposal is profoundly simple, bold, original, democratic, relevant, and grounded. This book shall captivate children and Nobel laureates alike."

—**Alex Gómez-Marín, professor of the Spanish Research Council and director of the Pari Center in Italy**

"Prof. Pollack is extrapolating upon his great discoveries. This is the kind of thinking needed to advance our paradigms scientifically. Reading this book changed how I look at…Everything!"

—**George Wiseman, Alternative Energy Innovator and President of Eagle-Research**

"Another foundational book from Gerald Pollack. In this new work, Pollack builds on his discoveries of charge separation in water to extend straight out to the edge of the cosmos. Understanding electrical charges as a veritable counterforce to gravity gives us a new future to shape. This is the next textbook for all our science. I read it with glee. Every new revelation about how Nature works came with a satisfying "click" that made sense. Read it in little "sips" to hold on to the unveiling of wonder."

—Gina Bria, Founder of Hydration Foundation

"A reminder that the true purpose of science is to understand and appreciate the true workings of nature. *Charged* continues this exploration. It redirects focus to matters that should have been the center of attention long ago."

—Ari Pentilla, CEO of Prorink

"Another proof of Pollack's open mind to look at common phenomena with fresh eyes and question the status quo. And again a very pleasant read, with a very didactic and fun writing style. A must read for students, who should know that what they are taught may not yet be the ultimate truth, and to urge them to explore."

—Everine van de Kraats, PhD, Founder and Director of
World Water Lab and World Water Community

"I was positively astounded by this book. There are countless positive provocations for our sleepy, not to say clinically dead, 'incremental' scientists, especially in physics. In fact, I fear the electric shocks stemming from this book will probably not suffice to rouse most scientists from the dark tomb in which they have happily shut themselves. However, I do hope it will help to awaken those who have not yet been fully gripped by the whirlpool of modern scientific religion. It is a book that brings new light, new challenges, and new perspectives on how to deal with nature in a much simpler way."

—Igor Jerman, PhD, Director of Bion Institute, Slovenia

"Pollack's book explodes the bubble: electrical charge is a super power. He illustrates — quite literally with great drawings — that if one electron is removed from each of two vertically stacked bodies, the repulsive force will be so high that in order to prevent the rise of the upper body, you would need to pile 50,000 trucks on top. This revelation has dramatically changed my own perception of the material realm. Schools must include it right away. So many things are governed by charge and charge collapse — but how can charge collapse? Pollack explains all of this, and how it directly governs the natural world around us. And he presents it in an intriguing, readable way that cannot help but draw you in."

—Ralf Otterpohl, Dr.-Ing, Professor Emeritus,
Hamburg University of Technology

"The same scientist who revolutionized our understanding of water a decade ago, is about to do the same to the role of electric charge in the world… I almost fell off my chair several times when reading the Introduction, but chapter after chapter convinced me that he had very solid evidence for every claim. And Prof. Pollack's courage for challenging the status quo — rarely popular — is astounding and inspiring!!"

—Yonah Alexandre Bronstein, PhD, President and CTO of ASTAV, Inc.

"In the current century, the electroneutrality paradigm is being challenged by experimental results in areas ranging from atmospheric science to synthetic chemistry. This book is an excellent contribution, presenting a wealth of information illuminated by the original thinking and provocative style of a leading contributor to knowledge on Earth's surface (and beyond)."

—Fernando Galembeck, Emeritus Professor, Unicamp, Brazil

"This book demystifies the pivotal role of electrical charge in nature, to be understood by everyone. By applying 'Occam's razor' of simplicity, Pollack enables the scientist as well as the novice to understand and value the profound connections between electricity, water, and all life."

—Stuart Rudick, EverBlue Ventures

"Here, science is practiced as it should be. A researcher makes his own observations and experiments in nature and then begins — wide-eyed, curious and impartial like a child — to investigate whether there is already evidence from others that supports what he has concluded. It takes a lot of courage for a scientist to question common narratives. At the same time, it takes perseverance not to be swayed by ignoramuses and nay-sayers. I hope that the book — as it did for me — will also encourage other researchers to "stick at it.""

—Dr. Cornelia Renate Gottwald, Magnetic Field Therapist, Germany

"Many anomalies don't fit the currently accepted paradigm. Through these anomalies, a new paradigm may emerge. Pollack follows this path, wielding Occam's razor (the principle of simplicity) to reveal multiple new paradigms of understanding, all centered on electrical charge. He demonstrates the centrality of electrical charge in the understanding of so many familiar phenomena involving the Earth and the entire universe. The result is mind-blowing."

—Won H. Kim, Professor, Yonsei University, Korea

CHARGED

The Unexpected Role of Electricity
in the Workings of Nature

GERALD H. POLLACK

ILLUSTRATED BY ETHAN POLLACK

E&S
EBNER AND SONS

Ebner & Sons Publishers
ebnerandsons.com
Seattle WA, USA

ElS

Published by Ebner and Sons Publishers, Seattle WA

info@ebnerandsons.com

www.ebnerandsons.com

Written by Gerald Pollack

Illustrated by Ethan Pollack

Charged: The Unexpected Role of Electricity in the Workings of Nature

ISBN 9780988778900

Other formats:

    hardcover ISBN 9780988778917

    ebook (epub) ISBN 9780988778924

Library of Congress Control Number 2025906167

Printed in South Korea

1st printing

Book layout design by Elizabeth Mullaly

Book cover by Ethan Pollack

Cover woodcut image from *The Philosophy of Storms* by James Pollard Espy (1841)

This book is typset in Iowan Old Style BT and Azo Sans.

To the memory of Emily Freedman,
my life-partner for a quarter century,
who unselfishly supported all aspects
of this work and lent gentle words of
encouragement along the way.

Contents

Preface

Imagine a sixth-grade science class — maybe even yours.

The teacher begins by showing the students a hand-drawn sketch. The sketch depicts the ancients, our long-forgotten ancestors, staring in awe at a vast field of erupting volcanoes. You can imagine the scene. With a superimposed background of dark, ominous-looking clouds spewing terrible flashes of lightning, what else could those ancients think but that "these awesome phenomena must be created by some kind of god." Poor, unfortunate ancients, laments the teacher. They know not what they think.

Then comes the inevitable rescue: "It was not until the advent of modern science that we came to understand these things."

But do we really understand them? Do modern scientific paradigms improve our grasp of everyday phenomena? Do they help us understand why clouds float? Or why spiders can sometimes fly? Or why the earth relentlessly turns on its axis?

This book seeks answers to such questions. It does so by identifying fresh beginnings — simple scientific foundations with appreciable explanatory power. It acknowledges that phenomena not readily understandable to early humans might remain obscure even to people of today, *i.e.*, that at least some prevailing paradigms might be missing that simple, reassuring, ring of truth. To be clear, I don't profess *a priori* that most current paradigms are necessarily misguided; that would be arrogant. But the complexity of many of those paradigms does lead us to wonder whether simpler foundational concepts might be worthy of exploration.

In considering the prevailing paradigms, we immediately think of the towering figures of science who brought us there: Albert Einstein, Max Planck, Richard Feynman, and others. Even though most of us might not fully grasp the entirety of their theoretical constructs, we defer to them almost reverentially. We accept what they profess, even if we don't get it all. Hence, a question to you, the reader: Have their arguably arcane theories brought you to a satisfying understanding of the workings of nature? Or, have you reluctantly concluded that the fundamentals lie beyond your comprehension, leaving ownership of ground truth to those scientific sentinels? Have modern scientific deities replaced those early imagined gods?

For many, today's science can seem practically impenetrable. We may easily follow the superficial descriptions, only to lose our way when trying to comprehend the detailed underpinnings. For example, some of you might have heard about the Higgs boson (sounding suspiciously like the latest cool dance). Touted as a fundamental particle of nature, confirmation of this so-called "God" particle had seemed significant enough to earn Peter Higgs, an English theorist, the 2013 Nobel prize. But what does the Higgs boson mean? How can we understand the implications of its existence for our daily lives? If you feel at a loss over how such sub-atomic particles fit into any context relevant to your world, you're not alone. We remain collectively bewildered, much as we imagine our distant ancestors.

Lacking comprehension of many of the foundational underpinnings of natural science, we can easily feel helpless. We don't know what to think or do about the critical issues that now plague our society. You know the issues: pollution, global warming, electromagnetic smog, chronic disease, and others. Absent any intuitive grasp of first principles, we're forced to rely on the presumed wisdom of those scientific "experts" out there. Yet, the problems grow progressively more serious with each passing year. "Expert wisdom" has thus far failed to help solve our problems.

Thus we ask: How much responsibility lies in the increasing complexity of today's scientific foundations? Have we progressed as far from the ancients as we'd like to think?

Have Scientific Principles Always Been So Obscure?

Science has not always been so daunting. Beginning six centuries ago, the world's pre-eminent scientific philosophers had continually professed the importance of *simplicity* (or ontological parsimony). Newton set a prominent example. Had that temperamental soul managed to develop a sense of humor, he might have found occasion to express his position by invoking the principle now known as KISS — "Keep it simple, stupid!"

In fact, Newton went on to formalize the principle of scientific simplicity. He knew well of the philosophical principle set forth earlier by the 14th century English friar William of Ockham (sometimes spelled "Occam"). Occam asserted that, "It is useless to do with more what can be done with less." In other words, keep your arguments simple. Newton built on that precept. He extended Occam's principle to the realm of science, arguing (in language fashionable at that time), "We are to admit no more causes of natural things than such as are both true and sufficient to explain their appearances." True, and sufficient — that's it. Out of the mouth of Newton.

Today, we call that principle "Occam's razor." We might better call it "Newton's razor," or even "Aristotle's razor" for his similar idea, expressed two millennia ago: "Nature operates in the shortest way possible."

What's meant by "short"? Or "simple"? How do those concepts help you understand the world around you, or even exploit that understanding for creating a kinder, gentler, more accessible world?

The terms "short" and "simple" mean adhering to a logical cause-and-effect sequence — *A* causes *B*; *B* causes *C*, *etc.* Consider, for example, the explanation for the common whistling teapot. Applying heat to water in a teapot causes the water to boil; boiling water creates steam; steam in a closed vessel produces pressure, which forces the vapor to flow through a narrow nozzle. Flow through that narrow nozzle creates vibration, which your ears detect, and your brain interprets as the familiar whistle. Simple cause-and-effect linkages create an easily grasped logical progression. We can fathom those linkages without very much difficulty; each one is short, and simple.

Alternative scenarios may lack any such straightforward logic. Consider the following (admittedly silly) option: When water boils, sound comes from gremlins lodged in the water. Annoyed by heat, those gremlins spring to life. As the water begins boiling, vibrational forces cause the gremlins to experience pain. They scramble to escape, screaming as they pass through the narrow nozzle. We hear that screaming as a whistle.

It goes without saying that this scenario will strike you as preposterous. But why? We reject the hypothesis not only because of the absence of a clear, logical sequence, but also from the unfamiliar notion of observable gremlins scrambling about. How does the very existence of those gremlins link to our life's experience? Why should those creatures preferentially settle in water? What mechanism endows them with the capacity to feel pain? And, if the gremlins' departure from the kettle occurs contemporaneously with the kettle's diminishing water level, might that imply that gremlins could in fact be *made* of water? Ouch!

Unanswered "how" and "why" questions leave us feeling uneasy. Even if we had imagination enough to consider a gremlin-centered reality, we sense that the required sequence of events will prove so cumbersome as to make nonsense of that explanation. The correct explanation, we intuitively feel, is likely to be a lot simpler.

What we intuitively sense was formalized by Occam and others after him: The most likely option has the fewest missing links in the chain of logic. That's what we mean by simplicity. Simplicity, in turn, implies elegance — a theme that Mother Nature likely employs in good measure. We are ineffably attracted to simple, elegant explanations. They carry the ring of truth.

Of course, simplicity need not *necessarily* equate with truth; no assurances exist that the simplest explanation will always turn out to be the correct one. Nevertheless, the pursuit of simplicity offers hope that science could, after all, make the workings of our universe understandable even to mere mortals like ourselves. Simplicity democratizes science. It opens the possibility that nature's first principles might be universally accessible.

Has Simplicity Vanished from the Scientific Scene?

The principle of simplicity seems so sensible that you may well wonder why I bother to dwell on its origins.

Oddly, this principle has practically disappeared from the contemporary scientific scene. My academic colleagues often revel in presenting their work as complex — as though the ability to sort through intricate details places them in an esteemed position. The trend toward complexity may have begun a century ago with a statement famously attributed to Einstein: *"Everything should be made as simple as possible, but not simpler."* Einstein understood the virtue of simplicity; yet, seeing evidence for such counter-intuitive phenomena as relativity, he had to admit that science might prove not so simple after all. He began harboring some doubts.

With each passing decade, simplicity has progressively given way to complexity. Recently, seeking information on the emission of beta particles, I went to the handiest source — Wikipedia (the free online encyclopedia). Here's what I found: "This process is mediated by the weak interaction. The neutron turns into a proton through the emission of a virtual W^- boson. At the quark level, W^- emission turns a down-type quark into an up-type quark, turning a neutron (one up quark and two down quarks) into a proton (two up quarks and one down quark). The virtual W^- boson then decays into an electron and an antineutrino."

Atomic physicists likely have no problem with this explanation, but God help the rest of us poor souls. It's a foreign language — yet, conspicuously found in an encyclopedia designed for common understanding. Is such complexity really necessary for the description of a simple particle? Perhaps. But, must we then conclude that Occam's razor has begun losing its edge?

My physicist friends assure me that science has *not* gone astray — merely that Occam's razor has become as outdated as the steam engine. Physics, they claim, *is* non-intuitive. That notion gains grand support from Richard Feynman, the late 20th century's preeminent physicist. In the introduction to his book on quantum electrodynamics, *QED*, Feynman reassures the reader about their prospective immersion into the complex material to follow: "It is my task to convince you not to turn away because you don't understand it. You see, my physics students don't understand it either. That's because I don't understand it. Nobody does."

Was Feynman's provocative statement designed merely to humor his colleagues? His charm was definitely endearing to those who knew the man. Just read his book, *"Surely You're Joking, Mr. Feynman!"*, and you'll see why. It was practically impossible to not love this captivating guy. I, myself, succumbed. I still cherish the man's memory, relishing the elegantly simple way he could explain so many difficult concepts in physics. With regard to his own most notable scientific contribution, however, one might wonder: Were those introductory words merely offered as a readership draw? Or, did Feynman feel constrained to admit that his seemingly monumental contribution was beset with daunting complexity?

Does nature really operate in a way that perplexes even the most brilliant of minds? Niels Bohr, who gave us the original planetary model of the atom, thought otherwise. He believed that, if a principle had genuine merit, you could explain it to your grandmother. (Few grandmothers were scientists in Bohr's day.) But today, we have come to accept that scientific fundamentals are not necessarily accessible to non-experts, let alone to their grandmothers.

Given this increasingly pervasive complexity, I've come to believe that the scientific enterprise may have lost its way. We think of the enterprise as ancient, and hence by now well developed. Indeed, judging from the written record you might conclude that we've progressed from *NO* science understanding to *SOME* science understanding. But does that necessarily mean we have *THE* science understanding? Scientific practice generally assumes not: It operates formally by challenging prevailing theories that seem incomplete, or at odds with experimental observation; it then attempts to replace them using a minimum set of postulates to approach natural truths. Most of current science, I'm convinced, fails to operate in that mode.

Nevertheless, much of science still seems young and enthusiastic, like a child excited to learn to ride a bicycle. Children may stumble many times before finally getting the hang of it. Science, likewise, seems to me to be functioning in a stumbling mode. Yet, the youthful exuberance of many engaged in the scientific enterprise cannot help but motivate fresh approaches for seeking out reality. Many scientists and non-scientists alike hunger for fundamental scientific truths. In our hearts and minds, we crave firm anchors to reality.

I'm full of hope that the flowers of scientific youth may blossom into something meaningful, enduring, and perhaps even beautiful.

Restoring Occam?

In seeking simplicity, we need to recognize that even the simple can sometimes become complex when one probes deeply enough. Lying beneath an explanation that might conform to Occam's razor sit questions that can become vexing to answer. For example, the sun supplies much of the earth's energy, but how did the sun come to acquire that deliverable energy? I'm not sure the answer will be as simple as we might hope. Simplicity may have its limits. Notwithstanding, we begin with the presumption that a search for simple foundational concepts may provide the keys to the kingdom.

This book begins by recognizing a critical problem with the currently accepted scientific models: anomalies. Anomalies are observations brushed aside because they don't fit into a prevailing paradigm. But that raises a question: If the paradigm is adequate, then why shouldn't those observations fit?

Once I realized that many of the dominant paradigms of modern science came with well-recognized anomalies, I began feeling impelled to search for more reliable foundational truths and rebuild from there. Matters unclear should become clear. In that pursuit, I soon realized that anomalies can serve as gifts: They offer points of focus that need to be reconciled, for without reconciliation there can be no real progress. This book represents my best effort to build an understandable model of our natural environment free of obvious anomalies — at least anomalies obvious to me.

As I began proceeding along that course, I soon noticed the same feature appearing again and again: electric charge. As you will soon see, I came to appreciate that even small amounts of positive or negative charge can produce stunning forces, easily capable of driving diverse natural processes. This revelation spurred me into action, for I came to realize that charge forces — attractions and repulsions — could lead to simple, powerful hypotheses of the sort that Occam might have anticipated.

Essentially, I began to see that much of nature might prove directly interpretable in terms of the interplay of positive and negative charges. This seemed exciting and powerful — like knocking on the mother of all doors.

The Book That Follows

Perusing the Table of Contents can offer a glimpse of the diverse areas of science that this book explores. Those areas range from the nature of gravitation to the genesis of weather. I also touch on subjects of more common interest, such as how birds fly and how fish swim. And very much more.

Current science offers explanations for all those phenomena. Often, however, those explanations barely get us past the first round of questions. Yes, gravitation arises from the attraction of masses. But what mechanism *causes* those masses to attract? And more: Everyone knows that birds fly by flapping their wings. But what about those birds that don't commonly flap — how can they stay aloft? Another conundrum: Dark clouds may release rainfall; but not always. By what wisdom do those clouds decide whether or not to oblige us to open our umbrellas? How do they figure it out?

In the material that follows, I hope to appeal to common sense. My goal is to move us toward fundamental ideas that integrate us to reality instead of isolating us from it. The ideas should resonate in our hearts as truth, more powerfully than the bits and pieces of science deemed compulsory to represent reality.

In this pursuit, I do not hold back. If we are asked to blindly accept concepts lacking an accessible foundation of first principles, I speak up. If the emperor has no clothes, I say so.

The book is written for any reader. An advanced education is not necessary — I have done my reasonable best to keep the presentation straightforward. Since so much reference material can be found on the Internet (exercising due caution), I streamlined the text by omitting references to easily accessible information. On the other hand, when facts seemed unexpected or crucial for an argument, the relevant references have been included.

I recognize that colleagues steeped in their respective fields will likely come down hard on some of the proposed explanations, for few of us welcome challenges to our firmly held belief systems. History teaches us that only rarely do challengers escape the wrath of the fields' elders. No exception is anticipated here. After reading what follows, those colleagues may decide whether my efforts have provided some modicum of reassurance or have merely muddied the waters. I hope those colleagues will appreciate that this effort is driven solely by a compulsion to understand nature at its core, a pursuit that cannot always be accomplished without a substantial dose of disruptive thinking.

Like any scientist with even a modicum of humility, I don't regard what follows as a statement of ground truth. As new evidence comes my way, I find myself continually refining my views. In that spirit, the ideas presented here are offered for rent, rather than for long-term purchase. My intent is to use electric charge as a vehicle to illustrate that we may be ripe for a return to simplification, a return to the deepest roots, with a new path forward.

I have nevertheless attempted to build an edifice of understanding based on simple observations and straightforward logic. While hoping to hone Occam's razor, my greater aspiration is to catalyze readers to question, and even reconsider, aspects of the scientific *status quo* that make little sense to them.

In so doing, I hope to inspire a more fruitful approach to understanding how the world works.

GHP

Seattle, WA, December, 2024

It ain't what you don't know that gets you into trouble.
It's what you know for sure that just ain't so.

— Mark Twain

Science is the belief in the ignorance of the experts.

— Richard Feynman

A ROADMAP

Questions You Might Never Have Thought to Ask

This Roadmap will help orient you to the route that this book will follow. Please consider it as "Chapter Zero."

A Shortcut to the Road Ahead

The key to this book lies in the unexpectedly central role of electrical charge throughout nature. The book will take you on an electrical journey. It will explore how an appreciation of the role of electrical charge might offer a way forward to a simpler and more straightforward understanding of the science of everyday life — the science that lets us understand how the world works.

You might presume you already *know* how the world works. After all, you regularly think about explanations for natural phenomena. It's raining — so you surmise that those droplets must be pulled to the earth by gravitation. Suppositions like that can feel satisfying; you're comforted by your apparent understanding of nature. Who needs any further comprehension?

Yet, even in that seemingly straightforward realm, matters are not so straightforward. A scientific study challenges that simple gravitational interpretation (see Chapter 8): In drawing those raindrops toward the earth's surface, at work is something beyond just one mass pulling another. To explain the observed high speed of descent, another force must be at play.

Like the falling raindrop scenario, presumptions exist about the mechanisms underlying many everyday phenomena. Those mechanisms may seem reasonable on the surface; we're inclined to accept them as valid, and most of them probably are. Nevertheless, we cannot be sure, especially if said mechanisms are burdened by embarrassing anomalies, or contradictions. Determining whether alternative explanations may better fit the evidence requires us to dig more deeply. Such excavation could potentially uncover fresh interpretations that prove less clumsy, more elegant, and ultimately more appealing. We can't know until we look.

To begin the search, I invite you to accompany me along that journey of exploration. I will first offer you a capsule preview of the subjects covered in the Roadmap — a preview of the preview if you will. As you will immediately see, the focus lies squarely in the proposed central role of electrical charge. Following that teaser, we will explore the subject matter more substantively.

• **Earth Physics:** Most of us envision electrical charge forces as puny. Is lightning an exception? Could electrical charges plausibly generate colossal amounts of force?

• **Earth-Centered Dynamics:** Wind gusts are natural, but how are they generated? If the answer is not immediately obvious, then here's another question to think about: What makes the earth spin? Could both of those phenomena possibly involve electrical charge forces?

• **Weather:** In generating those thunderous cloudbursts, might electrical charge play a more central role than the usual suspects, temperature, and pressure? Could electrical discharge be more than just a side effect?

• **Gravitation:** We think of gravitation as arising purely from the attraction of masses. That paradigm mostly works; but sometimes it fails. Is it outlandish to consider whether electrical charge forces may contribute in some unsuspected way?

• **Learning to Fly:** Yes, birds fly by flapping their wings. But some species rarely flap. Could electrical charge improve our understanding of how birds remain aloft? Or, how Frisbees float? Or, how spiders can make their way from land to distant ships, as Darwin once observed? Indeed, could spiders really *fly*?

- **The End of the Road(map):** Could our brief journey of exploration lead to a set of foundational principles, providing solid ground for trekking into the scientific future?

- **Summing Up: Unlocking Nature's Mysteries:** Guidelines for effective probing.

Next, I will launch into a more detailed overview of how this book develops a fresh approach to issues that have frequently bedeviled both lay people and students of science alike. It will raise questions you might never have thought to ask.

But First, How Can We Understand Electrical Charge?

The term "charge" may seem as remote from common experience as Pluto. Even though we live in a pervasively electrical culture, the concept of electrical charge would seem for most to lie well beyond our sphere of understanding.

Let's begin with simple elementary-school science. On some winter's day, you may have used a comb and wondered how the static electrical charge gave you that flyaway hair. Charge seemed involved; yet, somehow, the phenomenon remained shrouded in mystery. Perhaps it still does. Even the very definition of charge may seem elusive: Protons constitute the unitary positive charge, electrons the unitary negative charge — but what exactly *is* charge? Can we really define it? Or, is it merely a postulate?

In this context, I'm reminded of the plight of my colleague, Bill, who began his career at an East Coast children's science museum. His duties included telling kids about electricity. To his surprise, he found that he couldn't. Despite his rigorous grounding in electrical science, children asked questions about those moving charges that he couldn't honestly answer. Frustrated, he decided to leave his position and came to work at my university. Bill continued working to understand the basis of electrical phenomena as a sideline, eventually arriving at what he felt to be a more nuanced view of how those charges might really flow.

Like Bill, you might (or might not) have some inkling of how electricity is supposed to work. However, you may still wonder whether a book about a phenomenon as seemingly elusive as electrical charge could really matter.

Charge could matter more than you think. First, the *physics* of electrical charge matters because even relatively small variations of charge can yield surprisingly powerful effects. Try stuffing a bunch of negative charges together — they strongly resist. Most undergraduate science students can calculate the repulsive force among those charges, which turns out to be unexpectedly huge. I know I'm jumping the gun here, but I can tell you that its magnitude will probably astonish you (see next section). Given such power, charge-based forces could have impressive potential as drivers of natural phenomena.

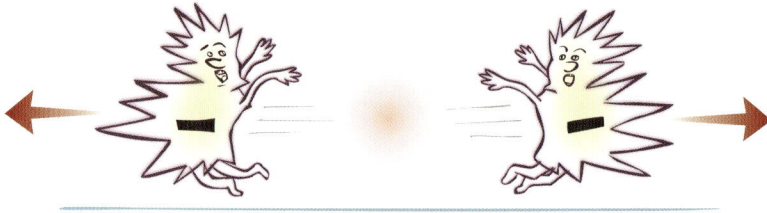

Second, the *concept* of charge matters because it may offer potentially useful ways to explain natural phenomena for which satisfying explanations do not yet exist. I'm talking about phenomena that may range from the genesis of weather all the way to the flight of gliders. Charges could turn out to be central protagonists.

Those two attributes, concept and underlying physics, motivate us to forge ahead. The potential for fresh interpretations could open new pathways to understanding — not only satisfying our innate curiosity, but also providing avenues for practical exploitation. The discovery of X-rays made possible the field of diagnostic imaging, in the same way as the discovery of microbes enabled the development and production of life-saving antibiotics. Fresh understandings inevitably lead to practical applications, some of which could never have been conceived in advance of the discovery that spawned them.

So, we press on with our journey of exploration. We admit that electrical charge has no easy "definition." It is nothing more than a postulate — a simple proposition that has gained acceptance by proving its worth and may yet prove useful for gaining further understanding.

To guide you along this quest for fresh understanding, I offer the following roadmap. As we head toward unexplored territories, it should help provide orientation.

SECTION I: The Charged Earth

Charges lurk practically everywhere. Even the Earth, itself, bears a net negative charge, with complementary positive charge residing in the atmosphere. This fact will come as news to many, but not to geophysicists and atmospheric scientists, who mostly recognize these separated charges as a confirmed fact. On the other hand, even some of those specialists may not fully appreciate the force-generating potential carried by Earth's charge, or by charges generally.

To illustrate their vast strength, imagine the following scenario. You are lying on the ground, with a friend suspended one meter above you. (The suspension method is irrelevant.) Go ahead and imagine — I'll wait.

Now suppose further that you could remove one percent of the electrons from each of those two bodies. Eliminating those negative charges leaves both of you positively charged, so you repel one another. How large is that repulsive force? Stated a different way, how much weight would you need to place atop your friend to prevent her from being thrust further upward due to the repulsive force?

Calculated by physics Nobelist Richard Feynman, the answer beggars the imagination: Under these stated conditions, you'd need the weight of the entire earth!

Charge forces are surprisingly strong, and therefore full of potential to do work.

Continuing in that same vein, consider the concept of air pressure. According to the conventional view, air pressure comes from the weight of the atmosphere pressing on the earth's surface — much like a lead block pressing on the earth. Air, however, is nothing like a lead block — it's a gas. Unlike solids and liquids, whose constituent molecules interact with, or even cling to one another, gas molecules tend to keep their distance. That's the very definition of a gas. If those air molecules remain separated, then how can their *collective* weight press on the earth? The lowest molecules surely can press, but what about the upper ones? If they're mainly disconnected, then how can they press collectively?

For resolving this enigma, explanations (mostly based on molecular collisions) can be found in the scientific literature, although those explanations raise questions that need answering. It's a cascade of queries. An alternative path comes from the presence of charge: Since the atmosphere bears positive charge and the earth bears negative charge, those two entities will attract. *Could the resulting attraction suffice for keeping the atmosphere clinging to the earth, thereby manifesting as pressure?*

SECTION II: Earth-Centered Dynamics

Have you ever wondered what creates the wind? Or what force propels a wind gust? Questions of that sort often stymie atmospheric scientists. Expedient responses invoke pressure gradients — but what establishes those gradients? Few scientists seem to exhibit very much interest in answering that simple question.

Once again, a candidate for explaining matters could reside in electrical charge. Since positive charge pervades the atmosphere, any difference in the amounts of charge between two nearby atmospheric regions will impel the excess charges to flow into the area with fewer charges. *Could such airflow underlie what we experience as wind? If not, then what creates a wind gust? And how can those gusts sometimes be so localized?*

Here's a related question that you have probably not spent much time worrying about: What force keeps the earth spinning on its own axis? Did the earth get a kick-start long ago, and continue to spin because of its immense inertia? Inertia could certainly maintain spin — but for a few *billion* years?

If that seems questionable, then consider an alternative hypothesis: A continuous supply of energy propels the spin. Could any such energy come from our benevolent sun? *Can you imagine how the sun's energy might create localized gradients of atmospheric charge, which propel airflow over the earth's surface (wind), shearing past the Earth to keep it spinning?*

If you cannot, then please do read on to see whether a proposed mechanism makes sense to you.

SECTION III: Weather

Forecasters know a lot about weather, yet they have often failed to predict monumental events such as New England's massive snowfalls in 2014 and 2015. Even forecasts for rain the next day sometimes come with only a 50 percent probability. What additional variables might help us improve those forecasts? How can we *understand* weather?

Consider our emerging protagonist: electrical charge. Lightning reveals that clouds may bear charges. Hence, charge could conceivably serve as a useful lens through which we can examine the genesis of weather. Current weather hypotheses remain vague. They leave unresolved questions such as the following: How did those charges get into the clouds in the first place? And how might the clouds' charges interact with the earth's charges?

Seemingly ignored by many mainstream scientists, evidence originating from decades ago shows that it's not just thunderclouds that bear charge, but all clouds. That includes even those gentle-looking, puffy, white, fair-weather clouds. Evidence from our own laboratory concurs: A cloud bears negative charge because each one of the tiny water droplets that make up the cloud bears a net negative charge.

But, wait a minute! We know that negative charges repel one another, leaving us with a paradox: How could these like-charged droplets draw together to form a cloud? How do they "condense"?

And once formed, why do clouds remain suspended? Suspension seems entirely illogical: Clouds mainly comprise water droplets, but garden-variety water droplets routinely fall to the earth. Yet clouds don't fall; they stay aloft. It's been calculated that the water contained in a large cumulonimbus cloud may weigh up to a hundred million pounds. Nevertheless, those hefty clouds rarely plummet to the earth. Could the cloud's negative charge repel the earth's negative charge strongly enough to keep those clouds aloft? Is that why clouds float?

And if so, then why does it rain? You'd think that gravitation might do the trick. But, if charge repulsion keeps the cloud droplets suspended, then how would gravitation suddenly take over and pull those drops down to the earth?

And further, a dark cloud will hold its water sometimes, while at other times releasing its contents in a downpour. What turns the faucet on?

Finally, when the rain does fall, why does its intensity vary so much? We may experience anything from a gentle sprinkle to torrents of rain accompanied by driving winds that leave you chasing your hat down the street. How does this happen? And why do rainclouds sometimes grow into typhoons, or spawn tornadoes that can twist to such catastrophic effect?

Could some of the answers lie in an unrecognized role of electrical charge?

SECTION IV: Gravitation

Many of us have learned that masses attract. Your mass attracts the earth's mass, and vice versa. Such mutual attraction grounds you to the planet's surface. This understanding of gravitation seems entirely reasonable, but where does that leave you when you begin to wonder *why* masses attract?

When I asked this blunt question to several physicists, they became frustrated — seemingly embarrassed by their inability to respond in a simple, intuitive way. Consistently, they invoked the recondite subject of Einsteinian space-time geometry. That kind of response appears to satisfy most physicists, but the rest of us find that our efforts to reach a straightforward account of gravitation come to nothing. Plainly put, we are left with no clear sense of how this familiar force of the universe works. We resign ourselves to taking the physicists' word for it because we can't easily articulate the reason why masses attract.

Thinking within the framework of charge raises another possibility. If the earth is charged, then could the *charge* of the earth contribute to the gravitational pull? This idea will surely seem foreign — I know. But please appreciate that standard gravitational theory is beset with multiple anomalies, more than commonly thought. Especially at short mass-to-mass separations, and also at long separations, those anomalies become serious, requiring large "correction factors" to fit the theory. We may opt to sweep those anomalies under the carpet and press on; or we can explore contributions that might be less burdened by pesky anomalies, too many of which inevitably raise questions.

Regarding the prospect of charge-related contributions, it may surprise you to learn that towering scientific figures including Einstein, Faraday, and Feynman, along with some present-day scientists, have previously considered electrical or electromagnetic origins of gravitation. The concept is less radical than you might think.

Consider the following potential mechanistic contribution, involving charge forces.

According to Faraday's law, a charged body induces opposite charge on the closest surface of anything nearby. Thus, Earth's massive negative

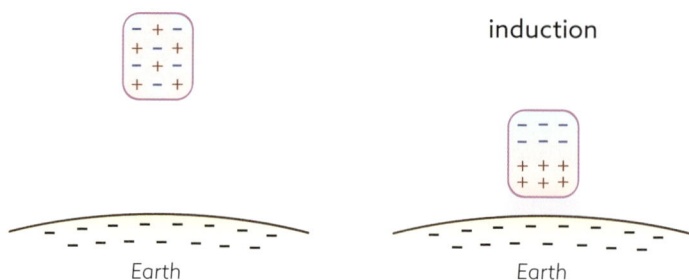

induction

charge will induce positive charge on the bottom of any mass situated on the earth or just above it. The top of that mass, meanwhile, will contain equal negative charge.

The earth's negative charge will pull on the positive charge situated on the lower surface, while pushing on the negative charge on the far surface. The pulling force will always win out — it will be marginally stronger than the pushing force because the positive charges lie closer to the earth's surface than the negative charges. Hence, the net force will be weakly attractive. But then again, physicists characterize the gravitational force as "extremely weak," some 10^{40} times weaker than the electromagnetic forces presumed to hold atoms together.

So, it's worth asking: *Could any such charge-based attraction contribute to earthly gravitation? If so, then by how much? And could such attraction help account for cosmic gravitation as well?*

SECTION V: Learning to Fly

After considering the weighty problems of cosmic gravitation, we turn our attention to objects somewhat closer to Earth — namely birds and airplanes. Many of us feel satisfied with our understanding of how those objects fly, but this section raises some troubling questions.

Take birds, for example. Everyone "knows" that birds fly by flapping their wings. Yet, peering through my living-room window, I routinely see eagles gliding substantial distances with hardly a flap. A seemingly effortless flight might include directions south, north, east, and, also west, in which the eagle may variously lose and even gain altitude. Eagles seem capable of traveling wherever they wish with no obvious difficulty — with practically no wing flapping.

Scientists have offered a variety of simple explanations, among them: eagles find updrafts; they're lightweight; they're aerodynamically

constructed; *etc*. I have even heard some attempts to "explain" this phenomenon by suggesting that eagles "soar," as though that constitutes an adequate explanation. Those accounts come almost reflexively; yet rarely does anyone question whether they suffice. *Do we really understand how birds can maneuver through the air without flapping their wings?*

I'm certainly aware of arguments based on wing contour. You may be surprised to learn, however, that like planes, birds can fly upside down (Chapter 15). If airfoil shape keeps birds suspended, then why wouldn't upside down flight bring embarrassing plummets to the earth?

Without fully divulging my proffered explanation, let me just tease you with the potential applicability of one well-recognized phenomenon: the triboelectric effect. Any object passing through the air will acquire negative charge. Similarly, a fixed object with a stream of air flowing past, will likewise acquire negative charge. (Think of your blow dryer, fluffing out your just-dried hair by charging each strand.) Hence, a soaring bird will inevitably acquire negative charge. The presence of that charge raises a question: Could the soaring bird remain aloft at least in part because the negatively charged bird repels the negatively charged earth?

In short, could electrical forces be relevant for flying? Might this phenomenon hold some relevance for airplanes as well?

SECTION VI: Moving Ahead

Similar questions arise for the issue of how fish swim. What, exactly, propels fish forward? You may feel as though responding with "tail swinging" easily answers that question. Though mainly pushing sideways, to some extent a swinging fin also pushes water backwards, thereby propelling the fish forward. The concept seems simple enough — but in ordinary viscous water, could any such tail-swinging mechanism generate enough propulsive power to accelerate black marlins to speeds in excess of 60 miles per hour?

If not, then how do fish swim?

And, slightly tongue-in-cheek, if nature has settled on tail swinging as the optimal means for achieving aquatic locomotion, then why haven't marine engineers even *thought* to emulate that process? Given their ingenuity, you'd think that implementation ought to be straightforward. So, how come the Queen Mary II doesn't propel itself forward using the swinging-fin mechanism?

Finally, consider sailboats. How those boats move downwind seems like a no-brainer, to be disposed of in a single thought: The wind pushes on the sails, and off you go. However, sailboats can also move upwind. Modern sailboats manage to advance as close as 40 degrees off the wind. Racing boats can sail closer to 30 degrees. Iceboats, by comparison, can head practically dead into the wind. Those cold weather sailing vehicles move *against* the wind, sometimes reaching speeds up to three times the wind velocity. Imagine advancing into a gale-force wind blowing fiercely on your face. How can that happen?

To explain this anomaly, sailors with a scientific bent may reflexively invoke "Bernoulli's principle," a well-known principle of fluid dynamics. But as we shall see, the argument is far from satisfying. So, the question arises: Could sailboats plausibly head into the wind in the absence of something extraneous pushing them or pulling them in said direction? And if they cannot, then what's the nature of the force that might propel them?

Could unsuspected charge forces possibly extricate us from the horns of this dilemma?

SECTION VII: Summing Up: Unlocking Nature's Mysteries

Numerous questions about how things work, both here on Earth and in the broader universe, remain unanswered. This book attempts to find answers. As I wrote this book, I tried to measure up to the earlier-mentioned standard (see Preface): "You do not really understand something unless you can explain it to your grandmother." The grandmother I was blessed to know had no scientific training. Nevertheless, in spite of a few intricacies sprinkled throughout the book, and notwithstanding some immigrant-language challenges, I would hope that she could have understood most of what follows in this book.

I can condense the book's message down to a single pervasive theme: the overarching role of electrical charge. We find charge everywhere. And, it generates forces far beyond our intuitive sense of things. That makes electrical charge a natural candidate to consider as a key driver of many physical phenomena, as the rest of this book will attempt to demonstrate.

As you've undoubtedly noted from this Roadmap, the material in this book will stray far from well-trodden scientific pathways. Mainly, we will be exploring uncharted terrain. I do not profess that all of what follows necessarily represents ground truth; I'm confident that it won't. Nevertheless, I hope this adventure will open your eyes to operative mechanisms that you might not have thought to consider, some of which, I hope, may illumine the workings of our natural world.

SECTION I

The Charged Earth

Is Earth charged? Or is it merely a rotating blob with lots of mass?

In this section, we deal with the Earth's net negative charge — long known, but not widely appreciated. That charge has important consequences, whose nature we begin to explore in the three chapters that follow.

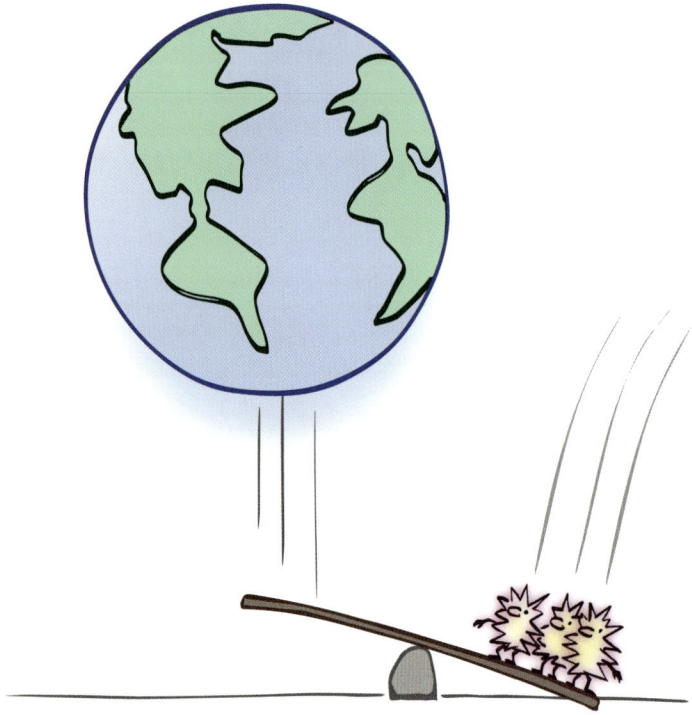

CHAPTER 1

Can a Few Hidden Charges Move the World?

You'd think that somebody like myself, with a solid grounding in electrical engineering, would immediately see the potential for keeping electrical charge central to their intellectual universe. But no — at least not at the outset. I certainly understood that electrical charges in motion created *current*, but beyond that simple formulation, electrical charge seemed to lack very much practical significance for my work or my thinking. It hardly factored into my understanding of how the world worked.

That was until a casual remark by a Russian colleague turned this view upside down. Andrey Klimov worked in my laboratory at the University of Washington well over a decade ago. He was older than most lab members. His long experience, together with a combination of abundant practical knowledge, unending curiosity, and exceptional creativity, coalesced into a single not-so-modest package of pure genius. Andrey could hold forth eloquently on many arcane topics. Sometimes I found myself too preoccupied to listen, but one remark definitely caught my attention.

Andrey claimed that even a few closely spaced charges could create forces of *enormous* magnitude. Wondering about this bold assertion, I began thinking about nuclear explosions, where the energy held in infinitesimal atomic nuclei could have sufficient power to obliterate a city. While I understood that nuclear explosions might not work exactly that way, my mind buzzed with excitement as I considered the potential for those closely packed charges to develop immense forces.

I thought: Just how much force could a few closely packed charges generate? Curiosity drove me to the usual sources, which yielded information on force magnitudes that dumbfounded me. Comprehending such vast forces seemed beyond my imagination. How could I have failed to recognize this? I couldn't help wondering what kinds of actions those huge forces might produce: Hurricanes? Turning of the earth? Keeping clouds suspended? As I thought about all of this, I felt myself sufficiently drawn to consider opening a new line of scientific inquiry — on the possible role of electrical charge forces in everyday life. Thus began the journey I share with you in these chapters.

To begin, let's look at a few examples of those forces. The Roadmap already provided one illustration: the body-levitation scenario. Flabbergasting as that thought experiment may seem, I will offer additional examples here, which I believe may impress you equally. I will then go on to show that the charges underlying those forces may exist throughout nature. We merely need to seek them out and acknowledge their potential to help us understand the world around us.

How Much Work Can Be Done by an Accumulation of Charge?

To begin, consider an old-fashioned 120-watt incandescent light bulb, operating at 110 volts. Electrons passing through its filament generate the familiar radiant glow. Now, imagine a miniscule container capable of holding all the electrons passing through that filament during one second. That's one Coulomb (C), equal to the quantity of charge conveyed in one second by a steady current of one Ampere. Concentrating anything to a point is of course impossible, but we can ignore such practical problems in this thought experiment. Now place that tiny container of concentrated charge at a point sitting somewhere on the surface of the earth.

Then, repeat this process. Place this second cluster of negatively charged electrons one meter above the first. Since like-charges repel, the upper cluster will rise farther into the air above. To prevent that rise, can you guess how much weight you'd have to place atop the upper cluster?

When I posed this question to an undergraduate class, the students who didn't respond with blank stares came up with guesses ranging up to about one kilogram (2.2 pounds). Like myself prior to that pivotal conversation with Andrey, the undergraduates hadn't a clue.

I calculated the answer initially in Newtons. Since few of us have much experience with that unit of force, let me translate it into more familiar units: jumbo jets. According to my arithmetic (see box), it would take some 5,000 Boeing 747s stacked atop the upper charge cluster to keep that cluster from rising.

CALCULATING FORCE IN JUMBO-JET UNITS

To calculate the force (F) between two like charge clusters, as above, we can use Coulomb's Law,

$$F = k \frac{q_1 q_2}{r^2}$$

Here, q_1 represents the first charge; q_2, the second charge; r, the distance between those two charges; and k, Coulomb's constant ($8.99 \times 10^9 \ Nm^2/C^2$).

For our purpose, we set q_1 and q_2 to equal 1 Coulomb, and r equal to 1 meter.

Those values give us $F = 8.99 \times 10^9$ Newtons, or, as more commonly expressed, 9.17×10^8 kg.

If the average empty Boeing 747 weighs a mere 183,523 kg, then it would take the weight of 4,994.9 of those jumbo-jets pressing on the upper charge cluster to keep that cluster from rising. That's close to 5,000 jumbo jets — an awesome number to contemplate. All from a bunch of charges running through some lightbulbs.

Illustrating those stacked jumbo jets in an esthetically appealing way seemed something of a challenge, so my artist-son and I decided to replace those weighty objects with ones less unwieldly: garbage trucks. To keep the upper cluster of charges from rising, it would take about 50,000 garbage trucks (**Fig. 1.1**).

Figure 1.1. One-coulomb blobs of like-charge, spaced one meter apart, exert enough force to lift unexpectedly heavy weights.

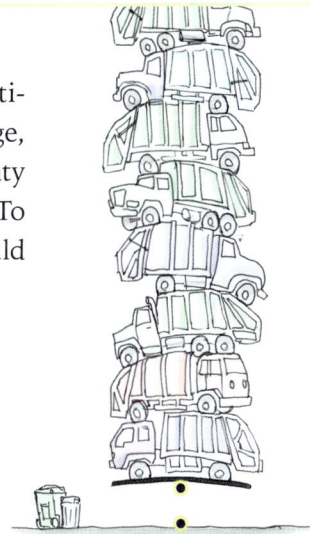

It amazed me to think that the charges gathered for one second from an ordinary lightbulb could provide enough force to lift a stack of trucks more than *one hundred miles high*. Andrey was right: Charges can exert enormous forces.

The Force of Electrical Charge Overwhelms the Force of Gravity

Perhaps you've never compared the energy from light bulb filaments to the weight of jumbo jets, or even garbage trucks. So, let's consider another example involving more familiar items: metal spheres. Imagine positioning one charge-neutral, marble-sized copper ball atop another and asking the following question: How many like charges must those two spheres acquire (or lose) before repulsion drives the upper one to rise against the pull of gravity (**Fig. 1.2**)?

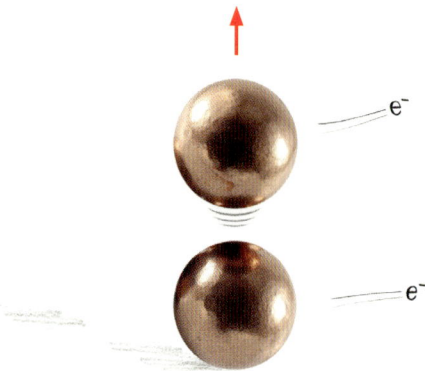

Figure 1.2. Removing one out of every several million electrons from each sphere can produce levitation.

The answer may surprise you. If you could remove only one out of every 5.7 million electrons (0.000017 percent) from each sphere, then the residual net positive charges in each one would create enough repulsive force to just barely lift the upper sphere. (More substantial levitation could come from additional charge removal.) Clearly, *trivial amounts of charge can create electrostatic forces sufficient to counter the pull of gravity*. You don't need very much.

The quantitative difference between charge-based force and gravitation-based force becomes even clearer in Donald E. Scott's provocative book, *The Electric Sky*.[1] Scott presents the following simple analysis to demonstrate the power of charges.

Place a proton and an electron some distance apart from one another. The two particles have opposite charge and will, therefore, attract. But the two particles also have mass, creating a second attractive force based on gravitational attraction.

You might guess that the charge-based force will exceed the gravitational force, and simple computations bear that out. But, by how much? Any such comparison is facilitated by a simplifying factor: Both forces vary as the inverse square of the separation distance — which means that the extent of proton-electron separation doesn't matter. Whether it's one micrometer or one kilometer, the ratio of the two attractive forces remains the same.

As it turns out, Scott and others have already calculated that, whatever the distance that separates the two objects, the electrostatic force exceeds the gravitational force by roughly 10^{38}. That's 1 followed by 38 zeros. To get a feel for the magnitude of this astronomical number, consider the diameter of our solar system relative to the diameter of a proton — roughly 10^{19} depending on how you calculate. But that's not even close: To get it right, you'd need to compare the proton's diameter to the span between the Earth and some distant galaxy (**Fig. 1.3**), an almost incomprehensible ratio.

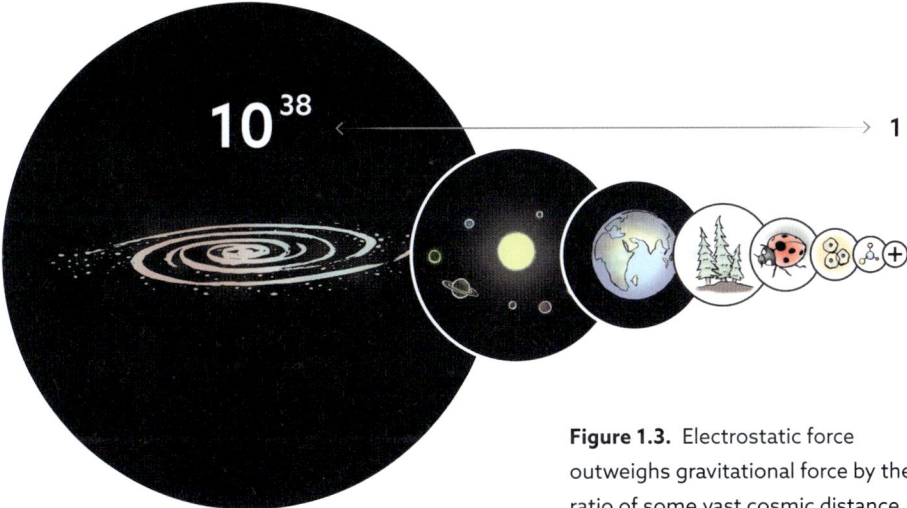

Figure 1.3. Electrostatic force outweighs gravitational force by the ratio of some vast cosmic distance to the diameter of a proton, $10^{38} : 1$.

When comparing the forces of gravity and charge, therefore, *only charge counts*. Charge forces can be huge beyond imagination. Gravitation not. In later chapters, I'll show how such charge forces could be exploited by objects of relatively large size to overcome gravity and levitate from the earth. Such feats could be more common than you might think. Indeed, given their immense magnitude, charge forces could plausibly rule the day.

Does Any of This Really Matter?

I've used these illustrations to demonstrate that charges can exert mammoth forces. In the real world, those forces can matter only if materials can retain the charges; otherwise, the considerations above hold little significance beyond perhaps providing some theoretical meat for specialists to chew on.

Can materials sustain any such net charge?

An example familiar to engineers and physicists alike is the dielectric, a material known exactly for that feature — its capacity to keep charges separated. The capacitor is another example. Capacitors hold positive charge on one plate, negative charge on the other. Those separations of charge can be sustained over appreciable periods of time. Thus, charges can indeed accumulate in distinct compartments.

While the capacity for charge accumulation seems clear, some chemists nevertheless challenge the notion that materials can retain that charge. That notion may be valid, but perhaps only partially. Think, for example, of a rechargeable battery. The battery holds charge, effectively separated at its respective poles. If you leave it idle, the battery will eventually run down. Explained in terms of the "principle of electroneutrality," this rundown feature illustrates the natural tendency for separated charges to recombine over time. Hence, chemists may well be accurate in asserting that separated charges do tend to recombine. However, a simple fix can prevent any such recombination.

That fix is energy. By supplying enough energy, those charges can remain fully separated. In the example at hand the energy comes from the electrical mains: Merely plug the battery's recharger into the receptacle and maintain full charge separation. With energy input, charges can accumulate in distinct compartments.

To put this energy principle in a more general context, what we are really considering is the distinction between open and closed systems. Closed systems do not allow penetration of any external energy; hence separated positive and negative charges will inevitably recombine over time; separation cannot be sustained. For such systems, therefore, the prevailing wisdom of electroneutrality would seem indisputably at play.

However, few systems are closed.

Commonly infiltrating the scene is the radiant energy of the sun. Many commonplace, yet underappreciated, examples of the ceaseless influx of solar energy sustaining persistent charge separation appear in my previous book, *The Fourth Phase of Water*.[2] Even modest amounts of the sun's energy can suffice. That book establishes that *because of this energy*, substances in nature can — and typically do — maintain separated charge. It might not surprise you, given that book's title, to learn that one such substance is water's fourth phase, also known as exclusion-zone (EZ) water. Simply put, EZ water contains charges that are separated from opposite charges, the separation maintained by the input of solar energy.

For sustaining the separation of opposite charges, then, we need energy. That energy could come from the open system of our environment, commonly from that glowing yellow sphere in the sky that we know as our life-giving Sun. We will deal extensively with that energy in the chapters that follow, largely in terms of its potential for building charge-based forces.

The bottom line: Given the incoming energy of the sun, clusters of like charge may well accumulate in many natural settings. Considering their considerable force-generating potential, those charges may well matter.

Kelvin and Friends

A stunning example of sustained charge separation may be seen in the clever demonstration known as the Kelvin water dropper (**Fig. 1.4**). It's so much fun, and teaches us so much, that I can't resist telling you about it.

Conceived originally by Lord Kelvin in 1867, this setup yields a dazzling climax. But before I describe it, I want to relate an anecdote about the distinguished physical chemist whom we honor for this invention. Lord

Kelvin famously remarked one day in a huff that nothing heavier than air could ever fly. Or, more precisely: "I can state flatly that heavier-than-air flying machines are impossible." Coming from so illustrious a scientist (after whom a temperature scale was named), that less-than-humble remark quickly gained fame when, only a few years later, the Wright brothers flew their first airplane at Kitty Hawk.

Drat! Even luminaries can err.

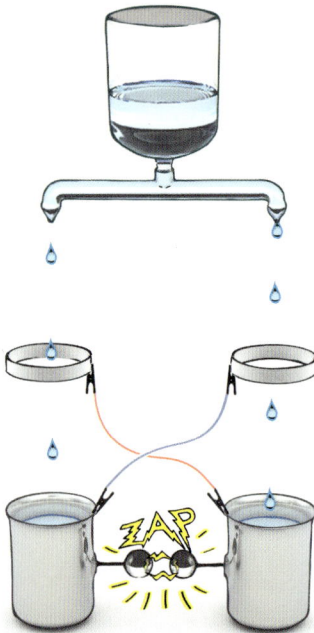

Figure 1.4. In the Kelvin water dropper demonstration, two cups of water acquire opposite charges. When the electrical potential difference between them grows sufficiently high, the two vessels discharge onto one another.

Kelvin's apparatus works by separating water from a common source into two side-by-side metal buckets, which quickly develop opposite charges (**Fig. 1.4**). When those separated charges build to a critical level, a spark discharges across the gap between them with a flash of light and an audible crack. Arcing through a centimeter or more of air requires the separated charges to build in huge amounts, corresponding to an electrical potential difference of tens of thousands of volts. Without question, this apparatus separates charges.

This description may tempt you to construct one of these devices at home. However, it's much simpler to watch a video. Look on YouTube for a fine one produced by Professor Walter Lewin.[w1] The video shows Professor Lewin demonstrating the effect, without explanation, to awe-struck Massachusetts Institute of Technology (MIT) freshmen. He then challenges them to explain the phenomenon as a homework assignment. If you wish, have a look before reading further.

How Does Kelvin's Apparatus Work?

If the principles underlying the workings of the Kelvin discharge apparatus interest you, this section offers a five-step attempt at explanation. If not, please feel free to skip to the next section.

As **Figure 1.4** shows, this experimental setup uses gravity to feed water from a container to the center of a narrow horizontal pipe. This pipe lets water fall from both of its open ends, through two conductive metal rings, each situated below one end of the pipe. Beneath those rings sit two conductive metal buckets, into which the falling droplets settle. Wires cross connect each bucket to the ring above the other bucket.

With this apparatus in mind, we can now trace the few steps that lead from steadily dripping water to a spectacular electrical discharge.

- Let us suppose that the first droplet falls from the left end of the pipe; and let us further suppose (as is common) that it contains some trace of charge, say negative. If so, then, when the droplet falls, the left bucket would acquire that same trace of negative charge.

- Because a metallic wire connects the left bucket with the right ring, the left bucket's negative charge will be shared with that ring.

- A negatively charged right ring would now electrostatically attract positive charges in the hanging water above, while repelling its negative charges. The incipient droplet will therefore contain positive charge. When it lands, the drop adds a measure of positive charge to the right bucket.

- The right bucket's positive charge will conduct through its wire to the left ring. The left ring's now-positive charge will induce negative charge on the developing droplet above, attracting it to fall. Once this process begins, all drops falling to the left will increase the left bucket's negative charge, while every drop falling into the right will add to that bucket's positive charge. Opposite charges build in the respective buckets, each droplet incrementing that charge.

- Eventually, the charges in the respective buckets build high enough that said buckets cannot help but discharge onto one another.

Variants work in much the same way. If the first negatively charged droplet happened to fall into the right bucket instead of the left, or if that droplet continued to fall into the left bucket but had a positive charge instead of negative, then everything would work similarly; only the polarities would switch.

In any of those scenarios, electrical discharge is practically inevitable. Regardless of which bucket accumulates positive charge and which accumulates negative charge, the opposite charges in the respective buckets will continue to build, as will the ring charges. As those ring charges grow more intense, the induction effect on the water above will keep strengthening. Each falling droplet will contain more charge than the one prior, as will the collecting buckets. Eventually, those bucket charges must grow so strong that the air between the two containers ceases to function as an insulator, allowing a spark to arc across the gap.

More From Mr. Kelvin: Some Take-home Messages

Prior to the grand finale of electrical discharge, we can enjoy a subtle sideshow that drives home another point: charge's immense strength.

For simplicity, please consider only the left side of the setup, and suppose that it begins accumulating negative charge. As the bucket becomes more highly charged, that sizable accumulation will begin to matter. The bucket's high charge may then repel newly falling droplets of similar charge, even forcing them to reverse course and rise upward. That does happen (**Fig. 1.5**). This striking observation graphically illustrates that *electrical charges can create forces strong enough to easily defy gravity.* We surmised that potential earlier. Here, we see evidence for it. Imagine!

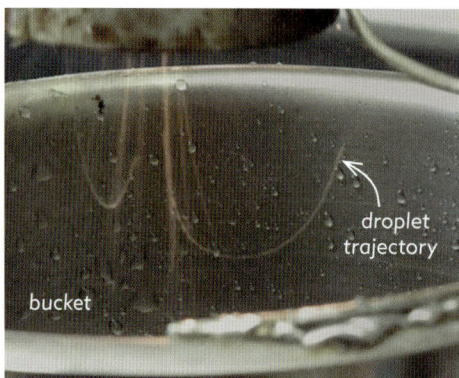

Figure 1.5. Droplets, illuminated with red light, falling into a Kelvin water-dropper bucket and photographed with long exposure time. Red tracks show the droplet trajectories. When water in the bucket acquires sufficient charge, the droplets approaching the water get repelled upward (curved red tracks), defying gravity.

The water dropper also shows that substances can sustain net charge for appreciable periods of time. When students in our laboratory restricted the supply so that droplets fell less frequently, the buckets still accumulated positive and negative charges. Notwithstanding the extended time course, still, the discharge occurred. Such slow buildup and sustenance validate water's capacity for charge retention over the long term.

The Kelvin water dropper's demonstration of charge retention is by no means unique. Undergraduate students in my laboratory at the University of Washington discovered residual charge not only in those buckets, but also in fresh fruits. Using pairs of identical electrodes, they found that fruits sustain a negative electrical potential of approximately 80 to 100 mV with respect to ground. When those fruits were squeezed, the juice retained most of that charge.

Similarly, when we used an electrode pair to pass charges into a bath of water, the regions near the electrodes sustained opposite charge for hours after the electrodes had been withdrawn.[3,4] Once again, we find sustained accumulations of charge, even in unanticipated circumstances.

Indeed, one day I was literally stunned by an electrical shock, coming from merely *touching* water. A colleague had prepared a type of alkaline drinking water with claimed therapeutic features. Though it had been sitting in its plastic bottle for weeks, sufficient charge evidently remained in that water to create that "shocking" experience. Zap!

I hope this discussion makes it clear that substances can maintain net charge over long periods of time. Water is merely one example, though an important one, because that substance exists in abundance throughout our environment. Later, we'll encounter multiple examples.

The Consequences of Charge

The potential impact of any such net charge is rarely considered. It's there, but who cares? Recognizing both the presence of charges throughout our environment, and the enormous forces they can generate, opens the possibility of fresh understandings about how the world may work.

Perhaps the most obvious example is weather. From observing lightning, we know that some clouds must contain charge. In fact, it's not just thunderclouds — fair-weather clouds contain charge as well.[5] Yet, despite the pervasiveness of electrical charge in the sky above, how often have you heard your weatherperson pontificate on the status of atmospheric charge? Charge is certainly there, but few seem to care very much about its presence.

The following chapters will consider a variety of charge-based forces all around us (including those involved in weather). I will attempt to show how those forces point toward new ways to deepen our understanding of phenomena ranging from the origin of wind all the way to the mechanisms underlying flight.

I ask you to try to keep an open mind as we reconsider whether we really understand the basis underlying some of those everyday phenomena. In many cases I believe we don't. While we may easily reiterate the standard tropes, penetrating beneath the surface often reveals a profound absence of genuine understanding of what's going on. That's been my experience. When we encounter any such seemingly dead-end obstacles, I will pose alternatives for you to consider, mainly based on electrical charge forces. Please pause to consider whether those explanations make sense to you, and also whether they offer a more satisfying understanding of the phenomena at hand. Building simple truth is the obvious goal.

The deeper goal, however, is more profound: identifying a unique physical fount from which bountiful understanding might spring. We will be searching for that foundational principle. If we can properly identify it, then the resulting insights should help account for everyday phenomena even well beyond those phenomena considered in this book. Identifying that principle — should one genuinely exist — could deeply enrich our body of understanding. That's the goal.

I ask you, therefore, to keep your mind open. Staying receptive to any such possibility offers the prospect of striking gold: Imagine the wealth of insights gained from uncovering one of nature's most deeply held secrets. It's happened before; it could happen again.

I suggest to you that foundational principles may lie in the unexpectedly profound role of electrical charge.

Summary

Most of us cannot sense the presence of charges. Therefore, electrical charges routinely escape our notice, particularly in situations in which we don't expect them. Yet, evidence reveals substantial bodies of charge quartered throughout our environment. Multiple examples exist. And, in situations with adequate levels of environmentally derived energy, those charges may persist indefinitely.

Less obvious is the fact that charge-based forces can reach magnitudes that would impress even the most skeptical of us. Demonstrations show huge forces, easily overwhelming the pull of gravity, and thereby allowing heavy objects to levitate. Even garbage trucks and jumbo jets can get lifted. Hence, in the real world, charges would appear to count, perhaps in a big way. Any failure to account for their presence, therefore, comes at our own peril.

Plausibly, those charge forces could underlie phenomena commonly ascribed to other mechanisms. Seeking out and judging the extent to which those charge forces may pervade the workings of nature sets us on the course ahead. I invite you to accompany me on this journey to explore how far charge-based explanations can take us.

CHAPTER 2

Is the Earth Negatively Charged?

Having dealt with the astonishing forces that charges can develop, we ask the obvious question: Where might those charges reside? Is there a dwelling place on Earth for those formidable actors?

Most of us think of Earth as electrically neutral. When we connect an electrical appliance to a ground socket, we feel assured that the appliance is securely connected to an ocean of bland neutrality; the appliance becomes safe and well grounded. So, you can imagine my reaction when I learned that the earth was not neutral at all — it had a net negative charge.

That jarring revelation came to me initially from Andrey Klimov (**Fig. 2.1**), the same Russian colleague who alerted me (Chapter 1) to the massive force that can arise between charged bodies. Now he was telling me that the earth was charged. He embarrassed me further by asserting that the earth's negative charge had become so broadly recognized in Russia that even his friends from the former USSR (Union of Soviet Socialist Republics) knew it since their middle-school days. Shame on the rest of us!

Figure 2.1.
Andrey Klimov. 1946 –

Andrey's negative-charge assertion came as a shock (if you'll excuse the pun). My undergraduate training in electrical engineering had provided no hint that a connection to the ground meant a connection to a vast sea of negativity. But a check of the scientific literature quickly revealed that Andrey was right; the earth's negative charge was by now a classical observation, known since the early experiments of Jean Peltier in the middle of the 19th century. Evidently, some of us had missed out.

This chapter deals with some of those missed revelations. It considers the origin of the earth's net negative charge and begins touching on its possible implications. It also considers the complementary positive charge. To get there, I must first deal with a few technical underpinnings, so please bear with me. I believe that this pathway will lead to something fundamental, a framework within which earthly charges matter. Indeed, that framework may be key for unlocking the workings of many everyday phenomena — all familiar, but few understood.

NEGATIVE CHARGE ORIGIN

Evidence for the Earth's Net Negative Charge

Figure 2.2. Richard Feynman (1918–1988), prominent Nobel physicist whose work dominated physics during the latter portion of the 20th century.

A negatively charged Earth? Who knew (apart from the Russians)? Well, apparently, the great physicist Richard Feynman did (**Fig 2.2**). Following the revelation by my colleague Andrey Klimov, a student of mine handed me a copy of Feynman's three-volume set, *Lectures in Physics*,[1] which many physics graduate students assiduously study. Once I read what the legendary physicist had to say about the earth's electrical charge, my skepticism began to melt away. Volume 2 of his *Lectures* devotes a full chapter to evidence for Earth's negative charge, and also to the atmospheric electric field that develops as a consequence of that charge.

The term, "electric field," may sound esoteric to some, but it's a useful formalism for describing the region between separated charges (**Fig 2.3**). The field is commonly represented by a series of lines running from plus to minus. Places where those field lines are spaced more closely denote a relatively stronger field. The lines commonly contain arrowheads, indicating the direction the field would push on a positive test charge. A negative test charge would get pushed equally strongly in the opposite direction.

Formation of electric fields needn't necessarily require separated charges. More generally, a single charge, or a cluster of like charges, can also create a field. We often consign such unipolar fields to the realm of theory because nature does not commonly build isolated charges of one polarity. Rather, it uses energy to *separate* charges, as Chapter 1 showed, and as more than a few examples below will likewise demonstrate.

Hence, we presume, as have others, that the well-recognized atmospheric electric field arises from separated charges. A useful analog is the capacitor. Capacitors produce electric fields between their charged plates (**Fig. 2.3a**). Typically, the field lines are evenly spaced (except for edge effects). So, the field will exert the same strength in pushing charges no matter where in the field those charges may reside. Uniformity prevails throughout the capacitor's field.

By contrast, the earth's electric field lines are not parallel; they lie closer together near the earth's curved surface than at higher altitudes (**Fig. 2.3b**). At those altitudes, the field strength is lower. Hence, uniformity does not prevail. On the other hand, the effect

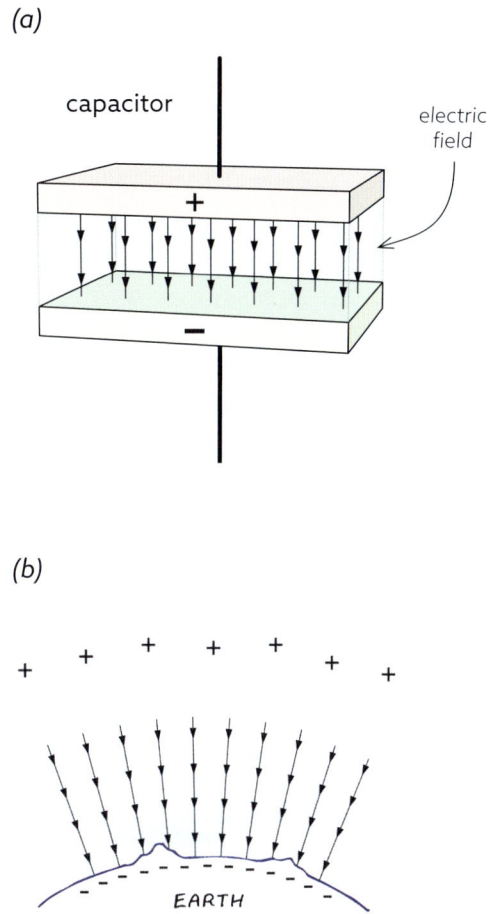

(a)

(b)

Figure 2.3. (*a*) Capacitor field. Separated charges create an electric field running from the positive plate (pink) to the negative plate (yellow). (*b*) The earth's electric field arises from positive charges above and negative charges on or beneath the earth's surface.

should be quantitatively modest since the atmospheric shell is so thin relative to the earth's radius that the field lines diverge only modestly. So, the capacitor model stands as a reasonable though imperfect representation of the atmospheric electrical field. No good reason exists to discard that model.

If the atmospheric electric field comes from separated charges, and the negative component resides on the earth, then an obvious question arises: Where might lie the corresponding positive component? To create the atmospheric field, those positive charges must reside well above the earth's surface, and we'll attempt to pinpoint their location in a moment.

Between that upper positive charge repository and the earth's negative charges lie the electric field lines, the object of our focus. Unlike the better-known magnetic field lines, which run more or less parallel to the earth's surface, the electric field lines intersect the earth perpendicularly (**Fig. 2.3b**). The two entities should not be confused.

The strength of the earth's electric field has been measured — it comes to about 100 volts per meter at the earth's surface.[2] As Feynman pointed out, probably with a chuckle, this means that if you were a two-meter-tall person (6.5 feet) standing on the ground, your nose would be roughly 200 volts more positive than your toes. Think of it!

While the existence of the earth's net negative charge might strike you as implausible (the Russians notwithstanding), numerous experiments carried out during the first half of the 20th century have confirmed the existence of that charge (for review, see Aplin *et al.*[3]). Geophysicists are keenly aware of it. In fact, geophysics textbooks routinely consider earth negativity, although failing to make much of its implications beyond global electric-circuit aspects, *i.e.*, beyond the trace currents that flow through the atmosphere to the earth. Few of us concern ourselves with that seemingly arcane subject. On the other hand, because the earth's negative charge has so much potential to explain what we see every day, we will find good reason to care deeply about its nature.

But first, let us return to the earthly capacitor and consider the origins of the negative and positive charges that make up that capacitor. I will deal first with the negative charge, then the positive. Both are critical for what follows.

Possible Origin of the Earth's Negative Charge

For the better part of a century, scientists have believed that the earth's negativity came from electrical discharges — *i.e.*, from lightning striking the earth. Such belief has seemed natural, for in some locales, intense discharges occur daily. Over the entire earth, the lightning count comes to approximately five *million* strikes per day.[w1] Since those flashes transfer negative charge to the earth, some physicists have argued, substantial negativity could accumulate on the earth's surface.

Even if originating from localized strikes, any such earthly negativity should quickly spread. Because of the earth's known high sub-surface conductivity, any locally concentrated negative charge should quickly disperse to flanking regions.[1] In this way, charges should even out. So, if scattered lightning discharges occur frequently enough and their charges can spread rapidly enough, those strikes could potentially sustain negative charge over the entire surface of the earth.

The lightning-derived negative-charge thesis would seem to make sense also because the lightning comes from clouds, and clouds bear net negative charge.[4] That negativity could discharge onto the earth. Inside storm clouds in particular, which ordinarily unleash gobs of lightning, the bottoms of those clouds (except for localized pockets of positivity) are confirmed to be highly negative.[w2] Lightning could transfer some of that negative charge to the earth.

So, while the lightning thesis seems tenable, some issues nevertheless raise questions. A few worth pondering: Why should negatively charged clouds discharge onto a negatively charged earth? And if such transfer were somehow possible, then why shouldn't that earthly excess immediately disperse into the atmosphere above, leaving the earth with little or no net negativity?

An altogether different source of earthly negativity comes to light with the advent of recent discoveries regarding the earth's water.[4] Oceans cover 70 percent of the earth's surface, while most of the remaining 30 percent contains water in other contexts, including lakes, rivers, and vegetation. Effectively, water blankets the earth. That water should be typically neutral. But, if some fraction of earthly water were to bear net negative charge — I'll explain forthwith — then a straightforward explanation for the earth's negative charge could prevail, independent of lightning.

The possibility that water *could* contain net negative charge seems borne out. Chapter 1 provided the example of the Kelvin water dropper, where a single source of neutral water separates into negative and positive components. So, water molecules can demonstrably split into oppositely charged components. Whether those molecules ordinarily *do* split is the question at hand.

The answer is affirmative. Recent discoveries to which I just alluded[4] (also see *What is EZ Water?* on page 22) reveal that splitting H_2O water molecules into OH^- and H^+ occurs whenever water interfaces with certain material surfaces (**Fig 2.4**). Those surfaces serve as templates; they can nucleate the growth of the negatively charged component, whose OH^- components pack like Legos to form an ordered phase of liquid-crystalline water. That negatively charged phase can build extensively. Since that crystalline zone excludes most solutes, I have earlier referred to it as the "exclusion-zone," or "EZ" for short.[4] The water inside that zone is then called EZ water, and typically bears negative charge (**Fig. 2.4**).

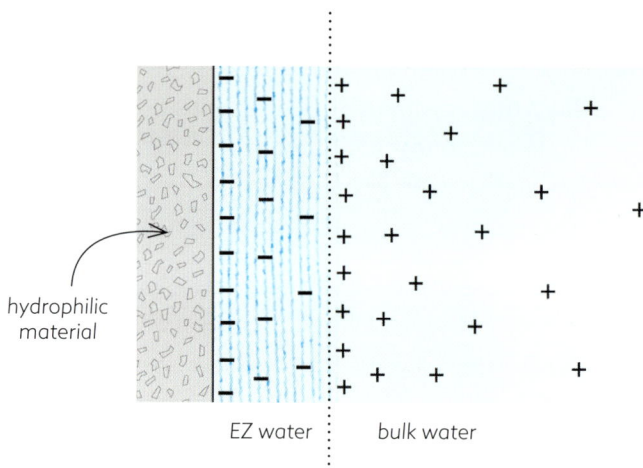

Figure 2.4. Building next to hydrophilic surfaces, EZ water typically contains concentrated negative charge, while complementary positive charges distribute themselves throughout the bulk water beyond.

In contrast to the packed OH^- moiety, the positively charged H^+ (*i.e.*, protons, which latch onto water molecules to form hydronium ions, H_3O^+) reside freely in the bulk water beyond the EZ (**Fig. 2.4**). Those positive ions disperse widely.

The respective charges in those two zones can be substantial: Point electrodes immersed, respectively, in EZ and bulk waters easily carry current sufficient to light an LED.[w3]

Before thinking that you can get something for nothing, let me say up front that the battery-like potential energy associated with this separated charge does not come from nowhere. It comes from light energy. Evidence has shown that incident light, *i.e.*, radiant energy, supplies the power required for breaking water molecules into their charged components.[4] Energy of that kind is needed for separating charge.

This light-mediated splitting of water into separated compartments constitutes a paradigm that has only recently emerged. Evidence has shown how radiant energy, coming both from the sun above the earth and from the heated magma within the earth, can split the earth's water molecules into positively charged (bulk) and negatively charged (EZ) components.[4] That feature will be central to much that follows throughout this book. That is why I spend time detailing its features.

So, we return to the question raised earlier but still not yet answered: Could the earth's negative charge come from its water, particularly from a negatively charged component of that water? Could certain earthly surfaces nucleate the buildup of EZ water?

And, to tickle your grey cells further, consider the hydronium ions, the hydrated protons that build as EZs build. Could those ions, free and positively charged, repel one another strongly enough that some of them get pushed out of the water and into the air above? Could any such evaporated hydronium ions constitute the positive terminal of earth's electric field?

As we address these questions, please bear in mind the strength of charge forces — enough, practically, to move mountains.

Earthly Negativity from EZ Water?

Regarding the potential for earth negativity to reside in water, consider a hypothetical example. Suppose that, on the earth, the concentration of negatively charged EZ water were to exceed the concentration of positively charged hydronium ions. Any such imbalance would cause the earth to bear net negative charge.

What is EZ Water?

EZ water, also known as fourth-phase water, was discovered in the early part of the 21st century in the author's laboratory and confirmed by others.[5-8] In early experiments, tiny particles were suspended in a water bath. When any of various gels were immersed into that aqueous suspension, the particles would migrate out of the zone adjacent to the gel, leaving a particle-free region next to the gel. Migration was not the result of any charge repulsion — the same zone of exclusion could be observed with particles of either polarity, and, also, with gels that were either charged or neutral. That particle-free region was aptly called the "exclusion zone" or simply "EZ." And, the water contained within that zone was labeled "EZ water."

Subsequent studies showed that EZ water's physicochemical properties differed from those of ordinary water. All features were at variance (for summary, see: Pollack, 2013[4]). From that body of evidence, we could eventually deduce that EZ water consists of hexagonal sheets stacked one upon another (see figure). Those sheets grow from the hydrophilic (nucleator) surface outward, into the liquid water.

The stacked molecular sheets typically bear negative charge, in amounts up to one elemental charge per unit hexagon. That's a substantial charge density. The corresponding positive charges, which break off from water molecules to allow those hexagonal sheets to form, reside in the bulk water, beyond the EZ (see **Fig. 2.4**). Since pure water alone is involved in demonstrating this charge-splitting process, the charge separation logically results from the splitting of H_2O, into OH^- and H^+. The former units serve as building blocks for EZ lattice buildup, while the positive ions fill the water spaces beyond, free to dance.

Some may recall that splitting of water molecules into their oppositely charged components happens in plants all the time. Driven by light, that process constitutes the first step of photosynthesis. The recognized existence of natural charge-splitting provided an early hint that in the present context as well, we might be dealing with a light-driven separation of charge. Experiments amply confirmed that supposition.[9] We found that incident light drove the water-splitting process — just as it does in photosynthesis.

So, light was certainly playing a key role in EZ buildup. While various optical wavelengths were effective, the most effective wavelength turned out to lie in the infrared region of the spectrum.[9] It

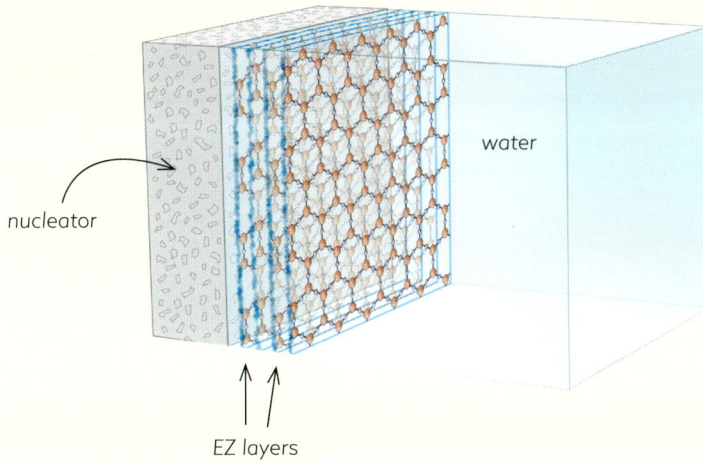

Hexagonal EZ lattice builds adjacent to hydrophilic surface ("nucleator"), using water as raw material.

was 3.0 μm, the same wavelength that water absorbs the most. That was a pleasing correspondence.

By then we began feeling confident. Recognizing a similar light-induced water-splitting mechanism operating in both EZ buildup and photosynthesis lent some measure of reassurance, for nature, after all, arguably works in the "shortest" way possible. Simplicity prevails — one mechanism operating in multiple contexts.

This light-driven feature works well in nature because of the abundance of infrared-light energy in the environment.

It's all over the place. And it's free for the taking. To imagine how that energy might impact the larger world should not be much of a challenge, for as you may appreciate by now, charges can exert extraordinary force. In human biology particularly, amassing that force can come from energy derived not just from the environment outside, but also from metabolic processes occurring inside. Both supply infrared (heat) energy, which builds charge.

For much that happens throughout nature, then, EZ-based charge could potentially play a central role.

How realistic is such an option? What about the complementary hydronium ions? Wouldn't they neutralize any excess EZ's negative charge? Before we attempt to answer those questions, let's first examine four places where EZ water might accumulate.

1. The first is seawater. The ocean contains mainly water and salt. Scientists typically think of dissolved salt as breaking into separated Na^+ and Cl^- ions, which then spread throughout the water. While that supposition comes almost reflexively, it raises a question that not many have thought to ask: Why would positive and negative ions want to separate from one another? Opposite charges attract. Do we really understand the true nature of sea water?

    An entirely different understanding of salt water comes from light-scattering experiments, which reveal the surprising presence of many large clusters populating those salt solutions.[10] The clusters presumably build from the salt and water molecules, since those are the only two species present. However, the water molecules contained in the cluster would likely differ from ordinary liquid water; otherwise, those clusters would not remain distinct from the bath. Indeed, light-absorption experiments show that some or all the cluster water exists in the EZ state.[11] Hence, the salt water of the oceans could well contain copious amounts of EZ water, replete with negative charge.

2. A second possible source of negatively charged EZ water lies in the biological organisms that inhabit the land and the sea. Every living cell contains abundant EZ water. That water builds next to the many hydrophilic surfaces comprising the cell's dense macromolecular matrix. So, cells are full of EZ water. While the buildup of that negative charge should produce an equal number of complementary positive charges (H^+), those free positive charges repel one another; hence, they should get driven out of the cell, leaving residual net negative EZ charge.[12] In fact, the cell's net negative charge is so pervasive that it has been touted as a hallmark of life.[13]

    Those EZ-containing cells should be present throughout the land and sea, in both plant and animal kingdoms. They should appear not only in the context of the many known unicellular organisms, but also as multicellular species, some dwelling on the earth's surface and many more lying near the bottoms of the oceans, where infrared energy issuing from thermal vents may fuel those species' existence. It's crowded down there, with substantial concentrations of EZ water conferring negative charge.

3. A third source of EZ negativity could be the water lying near the bottom of the sea. The earth is primarily composed of silicate minerals, which are hydrophilic in nature. Since hydrophilic interfaces tend to grow EZs,[14] the sea-ground interface could theoretically generate abundant negatively charged EZs — particularly in places where nearby thermal vents supply infrared energy. Hence, the seas' bottoms may well harbor abundant negative charge.

4. Negatively charged EZs may extend even farther down beneath the earth's solid surface. Scientists have recently uncovered massive bodies of water captured deep within the bowels of the earth. Because of their presumably ancient origin, these ultra-deep aquifers have been dubbed "primary" waters.[w4] Many of them connect to the earth's surface through fissures or springs.[w5] Rocky interfaces at those depths could provide nucleating surfaces for EZ growth, while radiant infrared energy from the heated magma beneath could supply the energy required for buildup. Additionally, the extreme pressure exerted from waters lying above those deep sites should expand those EZs, pressure being a feature that demonstrably enhances EZ buildup.[15] Hence, earthly negativity arising from EZ water may well extend rather deeply downward, perhaps to as much as tens of kilometers beneath the earth's surface. While that's still a small fraction of the earth's radius, it's deeper than just the "surface."

Thus, the earth's peripheral shell could contain negatively charged EZ water arising from at least four potential sources: sea water; living organisms; ocean bottoms; and possibly deep primary waters. Together, those sources could provide *a thick blanket of negative charge that envelops the entire earth* (**Fig. 2.5**).

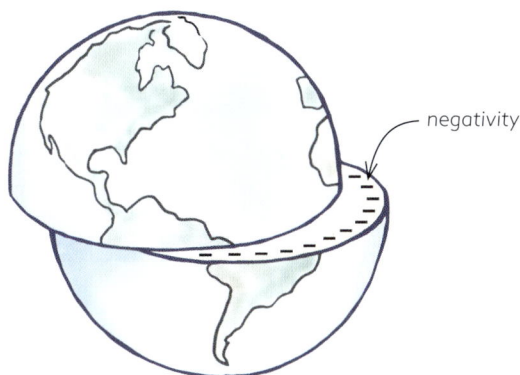

Figure 2.5. Negative charge envelops the earth's sub-surface.

THE POSITIVE CHARGE COMPLEMENT

Neutralization by Positive Charge?

But wait! What about the complementary hydronium ions that build as EZs build? Wouldn't those positive ions balance any negatively charged EZs? Shouldn't they confer neutrality?

Not necessarily. Unlike EZs, which cling strongly to hydrophilic or charged surfaces and thereby remain in place, the hydronium ions suffer less constraint. While some may draw close to the negatively charged EZs by natural attraction (**Fig. 2.4**), the ones more distant from the EZ's outer boundary should remain relatively immune to the EZ's influence. They should freely disperse — which they demonstrably do.[16] They also repel one another; hence, repulsive spreading should occur in all directions, including upward. That repulsive pressure may push hydronium ions across the air-water interface and out. In that way positive charges could be delivered to the atmosphere by evaporation (see Chapter 7).

Readers familiar with my previous book[4] may recall the presence of an EZ layer at the surface of the water, potentially thwarting any such hydronium-ion exit. However, that coverage is generally incomplete; it is commonly grid-like. Thus, regularly disposed regions of EZ will intersperse with bulk-water regions containing abundant positive charges. It is the latter regions through which positive charges may forcibly exit upward into the atmosphere.

Any such upward positive-charge transfer (evaporation) would leave the EZ-filled water beneath with net negative charge. This phenomenon could occur wherever EZs build. It should happen most actively during daytime, when the sunlight driving EZ buildup and charge separation is most intense. During that period, relatively more positive charges should rise to populate the atmosphere. In fact, atmospheric positive charge does increase appreciably during the daytime (see Chapter 4), increasing the plausibility of this explanation.

I emphasize that the negative charge of the earth appears to be more than just at surface-level; it may penetrate rather deeply. This follows from the anticipated contributions of ocean bottoms and the deeper "primary" (earth-associated) waters, the latter amounting to several times the water contained in all the earth's oceans, lakes, and rivers combined.[w6] Those waters lie principally within the 5 – 30 km (3 to 18 mile) thickness of the earth's crust, and may even penetrate into the mantle beneath.[w7] Indeed, the mantle has been confirmed to contain ice-like (EZ?) water, embedded next to deep natural diamonds.[17] All of this implies substantial thickness to the shell of negativity that encapsulates the earth.

While the above considerations do not definitively resolve the question of origin of the earth's negativity, my purpose was to introduce an alternative possibility, beyond lightning, that has arisen from recent work. The explanatory power of this mechanism will be probed briefly here and in more detail in the next several chapters. By demonstrating its ability to explain familiar phenomena in seemingly straightforward ways, I will attempt to establish the mechanism's reasonableness.

Do the Numbers Work?

The question: How much earthly charge is required for creating an atmospheric field of known magnitude? Whether from earthly EZ water or from lightning discharge, could the magnitude of the earth's negative charge suffice? Do the numbers match?

Scientists addressed this question almost a century ago.[18] Creating an electric field of 100 volts per meter, they estimated, requires an earth-surface charge density on the order of 100 pC (pico-Coulombs) per square meter. (A Coulomb, remember, is a unit of charge.) Newer data puts the estimates higher by a factor of up to ten times. Even with the larger, more conservative number these values calculate to something less than 10^{10} electrons per square meter of earth surface, which in turn translates to a single extra electron in an area 10 micrometers square. If micrometer units are unfamiliar, you can think of this charge density as one extra electron over the surface area of a typical biological cell (**Fig. 2.6**).

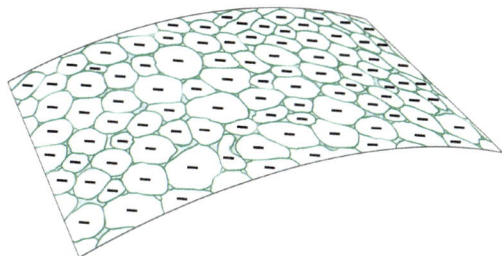

Figure 2.6. In a sheet of cells, one extra electron per cell is sufficient to create an electric field similar in magnitude to that in the lower atmosphere.

For context, that electron density amounts to only one thousandth the electron density in a single EZ sheet — a drop in a million buckets.

In fact, that calculation overestimates the extremely modest charge requirement. The analysis had presumed that all the required charges lie in a single plane. More realistically, the required charges distribute over the many stacked planes that compose the EZ. That translates to fewer charges needed per plane. So, the per-plane requirement should be considerably less, even, than the figure calculated above — realistic enough to qualify for explaining the measured electric field magnitude. The numbers seem reasonable.

A related question concerns the spread of that negative charge over the earth's surface. The water-based hypothesis implies that the negative charges might exist only in regions rich with water. Arid regions, in theory, should accumulate little charge, leading to a patchy negative charge distribution around the earth's surface rather than the uniform one implied above. However, the earth's shell is conductive throughout: Sub-surface aquifers of mineralized and primary waters lying beneath those arid zones should confer plenty of conductivity; so, locally generated charges ought to have little problem spreading abundantly over the earth's sub-surface shell. We should anticipate relatively uniform negativity throughout (**Fig. 2.7**).

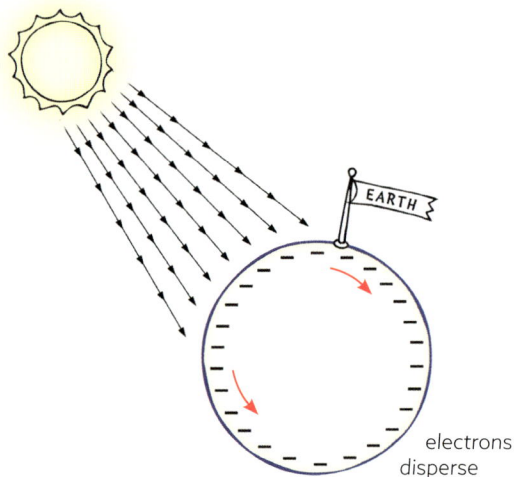

Figure 2.7. Negative charges should disperse rapidly around the earth because of the earth's high sub-surface conductivity. Rapid dispersal ensures a reasonably uniform distribution of sub-surface negativity.

electrons disperse

One exception to the general rule of earth-surface negativity seems worthy of mention because it will come up again: The earth's near-surface charge can sometimes turn *locally* positive. Positively charged patches may appear just beneath low-altitude clouds. When clouds are low enough (because of diminished negativity), their residual charges may nevertheless come close enough to the earth's charges to exert influence. In theory, the clouds' residual negative charges should drive the earth's negative charges downward, while simultaneously drawing its positive charges upward. Such "inductive" action can create local patches of earth-surface positivity just beneath the respective clouds.

We'll consider the induction phenomenon later when we deal with weather; but here we maintain our focus on ordinary fair-weather conditions, where, as a general rule a uniformly charged earth should prevail. The situation is straightforward. We will see that those earthly negative charges matter: We'll look at the substantial force that those charges can generate, ultimately considering their possible role in phenomena ranging from the origin of gravitation to the genesis of wind and the spinning of the earth.

In sum, the numbers appear to work. The earth could easily harbor the requisite amount of negative charge required for creating the known electric field.

Where Lie the Complementary Positive Charges?

While the earth's net negative charges appear numerically reasonable for sustaining the atmospheric electric field, that's not the end of the story. A second, positively charged, terminus should exist somewhere above the field. It's the separated bodies of charge that bear responsibility for creating the electric field in between. Where might those positive charges lie? How strong are they? And why should we care?

We should care because those charges lodged up in the atmosphere may matter. Later, I will show how they could play central roles in multiple phenomena: hurricane formation; the flights of wingless creatures; the penetration of sailboats running on ice almost directly into the wind; *etc*. So, let's proceed. Where are those positive charges situated?

The contemporary view puts those charges way up in the ionosphere, in a conductive region some 50 to 100 km (31 to 62 miles) above the earth. At that altitude, scientists presume that incoming solar energy collides with gas molecules to create the positive ions. Collision must, of course, simultaneously dislodge complementary negative charges, and how those electrons may get segregated to leave net positivity in said zone has not, to my knowledge, been well articulated.

While the presumption of ionospheric origin may bring some explanatory power, one critical observation does not fit at all. Let me explain.

Earthing: Capitalizing on the Earth's Negative Charge

What happens when you ground yourself to the earth? Will you get zapped by the earth's negative charge?

Well, kind of. Some time ago, while intentionally placing my bare feet on an aluminum pad, connected by a stout wire to an iron rod driven downward into the earth's negatively charged bowels, I immediately felt a mild tingle. Current must have flowed between the conducting pad and my feet.

I have since learned that many people ground themselves regularly. The reason people subject themselves to such seemingly bizarre treatment: health benefits. Just as a barefoot walk on the beach may feel pleasant, "earthing" (grounding) is said to confer a similar sense of wellness. Studies report numerous health benefits.[20]

A straightforward mechanism may underlie any such health benefit. The cells in your body bear negative charge: Electrodes stuck into cells routinely report negative electrical potentials on the order of 50 mV to 100 mV. According to textbook views, membrane pumps and channels bear responsibility for separating ions to create this intracellular negativity. Negatively charged proteins inside the cell may also contribute.[21] But there is another option: It now seems clear that the massive amounts of negatively charged interfacial (EZ) water enveloping the proteins and other macromolecules inside the cell (accompanied by the expulsion of positive ions) could well contribute the lion's share of the cells' negative charge.[4,12]

While healthy cells maintain that high negativity, pathological cells may not. Cancer cells, for example, show negative electrical potentials of only 10 to 15 mV.[22,23] Similarly, if your liver is malfunctioning, then the liver cells' electrical negativity may well be compromised. And the same for your brain cells, or your muscle cells, etc. Connecting your charge-depleted body to an ample supply of negative charge may then return those cells toward (if you'll pardon the expression) "healthy negativity." It can happen as the earth pushes its abundant negative charges into your

The ionospheric presumption posits a body of positive charge high up, complementing the body of earthly negative charge. Those separated charges constitute the anticipated capacitor. In the case of ideal capacitors (**Fig. 2.3a**), the electric-field strength between those plates is spatially invariant — it doesn't matter whether the field is measured near the positive plate, near the negative plate, or anywhere in between. The field should have the same strength all over. On the other hand, a spherical capacitor (following the earth's shape) is not quite ideal. The field lines spread somewhat (**Fig. 2.3b**); hence, some progressive decrease in field

charge-depleted cells; or, with similar consequence if that negative earth draws positive charges out from those cells. Either way, the resulting increase of bodily negative charge should build EZ water, recharging your cells with that unique aqueous commodity required for proper function.[4,12,24]

Such charge transfer can be accomplished in simple ways: by resting your bare feet on a well-grounded metal plate; by walking barefoot on a wet, sandy beach; or, by burying yourself in wet sand or moist mud. All such actions should build EZ in your cells, thereby promoting health.

So, besides its potential for creating geophysical actions, the earth's bounty of negative charge brings health benefits. When it comes to the body's health, one might say that negativity is positive and positivity is negative.

Humanity had long enjoyed those earthing benefits without necessarily understanding their underlying basis. Now, we can at least hypothesize what's happening: Earthing builds EZ water, the latter required for proper function. On the other hand, with the advent of modern insulating shoes, we have inadvertently cut ourselves off from that earthly connection. Instead, some people resort to the odd practice of tree hugging. The trees' cells, after all, bear negative charge, much like other living cells, and the thought is that we can acquire some of that charge through lingering tree hugs.

Simpler, perhaps, is the more conventional practice of walking barefoot on wet grass. That may accomplish the same and also protect us from quizzical looks from our neighbors.

strength is anticipated as you measure upwards from the earth. But because the spread is modest, no glaring deviation from ideal should be expected. The electric field above the earth should be relatively uniform.

By contrast, measurements show anything but. At ground level, the field strength is typically 100 volts per meter, but measurements higher up reveal an exponential decrease with increasing altitude. At 10 kilometers (6 miles) high, the field has a value of only about 3% of that at the earth's surface; at 30 km, it is down to only 300 millivolts per meter; and at 85 km, near the capacitor's supposed upper "plate," the value diminishes to a paltry one *micro*volt per meter.[19] That's *a hundred millionth* of the value at ground level — way out of expectation of the capacitor model, which anticipates only modest variation at different altitudes, certainly not eight orders of magnitude.

I do appreciate that quantitative disparities of that sort tend to impress scientists more than others. Scientists love numbers. For others, I emphasize the *huge* difference between expectation and observation — nothing short of colossal. While we don't rule out the presence of some excess positive charges in the ionosphere, any such charges cannot represent the positive plate of the earth's electric field. The numbers don't fit at all. We need to look elsewhere.

The capacitor's positive charges must evidently lie somewhere. Possible locations above the ionosphere seem as doubtful as within the ionosphere, for any such positioning would lead to an inconsistency similar to that just described. The remaining option puts the positive charges someplace lower, in between the ionosphere and the earth — perhaps somewhere in our atmosphere. For reasons to come shortly, I believe that is their actual location.

Given the positive charge lying above the negative earth, you might expect to be able to hold a wire over your head, connect a lightbulb between that elevated wire and ground, and light the bulb. Effectively, this has been done, using a motor in place of the lightbulb. In a 1925 patent by Hermann Plauson[P1], one electrode was lifted by a set of balloons. Between that elevated electrode and the ground, a motor was connected. The motor demonstrably ran. That observation left little doubt that the atmosphere's positive charge had a substantial vertical extent.

While the location of those positive charges may seem like nothing more than an academic exercise, those charges weave the fabric of the next two chapters. I hope to show how the positive charges lying within the earth's atmosphere not only square with experimental evidence, but also crack open the door to understanding diverse unexplained phenomena. One example: The atmospheric electric field varies with the time of day (see Chapter 4). Such diurnal variation must tell us something about positive-charge origin; and that, in turn, could reveal answers to some questions that few of us have even thought to ask.

Even though I reserve serious consideration of atmospheric positivity for subsequent chapters, it's worth closing with a few points, some of which you already know. First, none of this separated charge comes free. Sustaining charge separation and the consequent electric field requires energy, just as charging any capacitor requires energy. During daytime, the sun can supply that energy; but, as nighttime approaches and solar energy fades, those separated charges can be expected to recombine. Negative and positive charges should inescapably draw toward one another like young lovers in the flower of romance. Powerful is the urge to coalesce.

Any such recombination could impact the system's positive and negative charges differently (**Fig. 2.8**). Positive charges could return to the earth; or, negative charges could rise into the atmosphere. If those two bodies of charge bear relevance for earthly phenomena, then you can envision the importance of identifying the prevailing option.

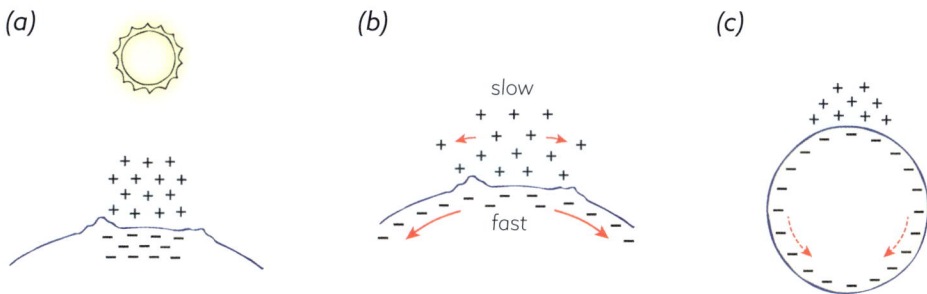

Figure 2.8. Sunlight separates charges *(a)*. The negative charges beneath the earth's surface should spread rapidly because of the sub-surface's high conductivity, whereas the positive charge should spread more slowly because of the low conductivity of the atmosphere *(b)*. Eventually, negative charges should distribute uniformly around the earth *(c)* while the positive charges should remain clustered.

Beyond such recombination is the anticipated lateral spread. Consider first the negative charges. As darkness arrives, you'd think that the earth's concentration might be seriously impacted. However, half the earth will still be receiving solar energy — there's always a bright side. With the unrelenting solar energy pressing on that bright side and continuously building negative charge there, the earth's high sub-surface conductivity should ensure that those negative charges will spread rapidly to darker regions. So, the bounty ought to be abundantly shared. Negative charge should remain roughly constant at any point on the earth's sub-surface, irrespective of local time of day.

Contrast that stability with the situation in the atmosphere above. The atmosphere doesn't enjoy the same high conductivity as the aquifer-ridden sub-surface of the earth beneath. Hence, its positive charges can't spread rapidly in the lateral direction. They can certainly spread as a consequence of positive-positive repulsion, but not nearly as rapidly as expected of the negative charges on the earth below.

Such restriction implies that atmospheric positive charges at any point above the earth ought to wax and wane pretty much in synch with the sun's energy. With morning light, that positivity should increase; and as darkness sets in, it should diminish. Hence, we may draw our conclusion: It's the atmosphere's positive charges that are most dynamic, not the earthly negative charges.

The anticipated diurnal variation of atmospheric positive charge is of such critical significance for much that follows that I will devote a subsequent chapter to its most prominent features. In short, those free positive charges are highly subject to the influence of charges arising elsewhere. Please stay with me to see how the ensuing mischief may play out.

Summary

Our beautiful planet bears net negative charge. Solid evidence exists for the presence of that earthly charge, although its existence is only variably known and widely underappreciated. The origin of this negative charge, and the electric field that builds from its presence, are matters worthy of scrutiny, for they may impact many natural phenomena.

According to traditional views, it is the relentless onslaught of lightning striking the earth that imparts earthly negativity. Lightning, after all, is an electrical phenomenon, and cloud-to-earth discharges could theoretically transfer negative charge downward to the earth. If the receiving earth were unable to dispel all that charge rapidly enough, then an earthly residual would remain, and through this dynamic, the earth would retain net negative charge.

An alternative view suggests that the earth's negativity may come from light-induced splitting of its water molecules. Such splitting, powered by light, is recognized by scientists as the first step of photosynthesis. Hence, plenty of precedent exists for the light-induced breakup of the earth's water molecules into their positive and negative components.

Since the negative components of water stick tightly to hydrophilic surfaces (a discovery coming from our laboratory[14]), arguments can be adduced to paint a picture different from what we may reflexively think. Many hydrophilic surfaces lie within natural bodies of water. Such surfaces could retain negatively charged EZ components. Those charges should thus stay put, conferring net negativity to the earth.

The complementary positive-charge components, by contrast, do not stick; they remain free. Repulsion among those positively charged ions ought to drive them in all directions — including skyward. Any such evaporative push should build atmospheric positivity, while leaving the earth with a negatively charged EZ residual. In such a way, the earth becomes negative, the atmosphere positive, features that are well established.

Any such separation cannot happen without energy. That energy should logically come from the sun. In much the same way that energy from electrical mains separates charges in your cell-phone battery, solar energy may similarly separate earth and atmospheric charges.

Interaction between those oppositely charged entities comes next. Atmospheric and earth charges attract one another. Could such attraction keep the atmosphere clinging to the earth? Is that the reason we can continue to breathe?

CHAPTER 3

Eternal Love: Does the Negative Earth Attract the Positive Atmosphere?

Focusing principally on the earth's negative charge, the previous chapter also laid the groundwork for dealing with its complement: the atmosphere's positive charge. Addressing the origin and impact of those charges begins now as we move toward a question of some significance: What keeps the atmosphere clinging to the earth?

So familiar is the atmosphere-earth linkage that we rarely stop to think about how the earth retains its precious atmosphere. Could the pull of gravity suffice? We presume it does — we implicitly suppose that the earth's solid mass pulls strongly enough on the atmosphere's gaseous mass to do the job.

But can we be certain?

Is there space for thinking that the persistent cling of the atmosphere rests at least in part on some force stronger than the decidedly weak pull of gravitation? Could simple electrostatic attraction suffice? That is the option we explore in this chapter, beginning with the simple question of where the atmosphere's positive charge originates. Where do those charges come from?

This chapter is somewhat technical; for that, I apologize.

On The Origin of Atmospheric Positivity

The more immediate question that requires answering first: Where exactly do the atmosphere's positive charges reside? Two serious options lie on the table. Scientists dealing with the earth's electrical environment place those positive charges in the ionosphere high above the earth,[wl] while atmospheric scientists tend to place them in the atmosphere closer to the earth.

The previous chapter weighed in on that difference of opinion. It provided evidence that the ionospheric assignment was out of accord — distinctly out of accord — with recognized electric-field measurements. Hence, we lay emphasis on the option preferred by most atmospheric scientists: Positive charges reside somewhere beneath the ionosphere. I believe we'll find good reason to support that option.

Now the main question: Where do those atmospheric charges come from? According to atmospheric scientists, the positive charges could arise from any of several sources. One of them is cosmic energy. Coming from remote birthplaces, cosmic energy is known to pepper the atmosphere with protons and alpha particles, both positively charged. That's one option. Another potential source of atmospheric positivity is the sun's solar wind. Containing huge concentrations of positive helium nuclei and protons, the solar wind penetrates the atmosphere, plausibly donating some of its positive charges. Finally, ordinary wind can potentially accomplish much the same (by the so-called triboelectric effect; see Chapter 14): Since swiftly moving air donates negative charges to any earthly material over which it passes, the air itself is left more positive.

While all of these sources of atmospheric positivity seem possible, even likely, I suggested in the previous chapter that the lion's share of the atmosphere's positive charge might come from the earth's water. Please recall the rationale: Solar energy separates water molecules into their charged components. The negatively charged components (hydroxide ions, OH$^-$) arguably populate water's sub-surface zones in the form of EZs clinging to diverse hydrophilic surfaces. I argued that those EZs may bear responsibility for much of the earth's net negative charge.

Meanwhile, the corresponding positively charged components (mainly hydronium ions, H$_3$O$^+$) lie initially suspended in the earth's liquid water.

Said ions repel one another. Those repulsive forces drive the ions apart, many of them driven through the water-air interface and into the atmosphere above. Such (evaporative) passage would then build atmospheric positivity. In this way, the atmosphere acquires positive charge, while the earth retains negative charge. Both bodies of charge originate from water, *i.e.*, from water supplied with solar energy.

Given the copious amounts of solar energy available for driving the separation of H_3O^+ from OH^-, I reasoned in Chapter 2 that the above-mentioned feature could be a major contributor to atmospheric positivity. Along with the three conventional sources mentioned above, such rising positivity could thus be a prodigious supplier of atmospheric charge.

Just how much positivity that mechanism can supply is an issue to be explored below. For now, I will presume it to be the prime contributor and will see how far that can take us.

How Do Atmospheric Positive Charges Distribute Themselves?

If positive charges rise from the water beneath, then, the question arises: Could those positive charges represent the positive pole of the atmospheric electric field? Identifying that pole is necessary for realizing our understanding of the overall picture. Chapter 2 narrowed the site of that pole to the zone beneath the ionosphere. The question now: Could those atmospheric positives, vertically distributed as they may be, represent the field's positive terminal?

If you have grown impatient with a too-deep immersion into electric-field esoterica, I can understand. I will do my best to ease that passage.

First, a necessary word on the form of those atmospheric positives. The charges do not necessarily exist as lone hydronium ions. Rocket-borne mass-spectrometer measurements reveal that atmospheric positive charges exist in the form of "proton hydrates," each positive ion gathering as many as 20 water molecules.[1] Thus, some combination of protons and water molecules bear responsibility for much of said positivity, although it remains convenient to speak of that positivity in terms of simple hydronium ions.

A more critical question for answering the pole question requires some consideration: How do those positive atmospheric charges distribute themselves over vertical space? Do they all hang out at the same altitude, perhaps scrunching up at some height above the earth? Or, do they spread vertically? Distribution matters: I will soon show how the vertical charge distribution could be critical for explaining various earthly features ranging from the origin of wind to the rotation of the earth. Consideration of that vertical distribution therefore seems in order.

Begin by picturing yourself (in miniature size) as one of those atmospheric positive charges. If you happen to lie relatively near the earth, then the earth's ample negativity should strongly attract you. You will linger near the earth's surface, along with many of your positive compatriots (**Fig. 3.1**).

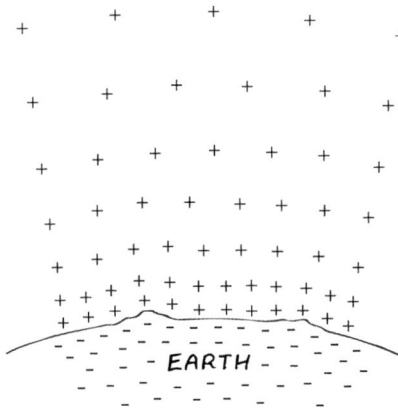

Figure 3.1. Atmospheric positive charges cluster close to the (negative) earth, while being more dispersed further up.

If you lie high above, on the other hand, beyond the strong pull of the earth's negativity, then repulsion from neighboring positive charges will dominate your senses. You may still feel the earth's negativity, but you will mainly experience the urge to distance yourself from those repulsive positives nearby. The result: an upward spread of high positive charge, with a near-earth clustering of lower positive charges. So long as those positive charges continue to get supplied from the earth's water, the expected distribution should resemble the one illustrated in **Figure 3.1**.

A charge-measuring device would confirm much the same. If you climbed a tall ladder while toting an electrometer in your backpack, here's what you would measure: Standing on the bottom rung, you would detect a lot of local positivity. You'd measure less positivity as you climbed upward. Still higher, your instrument would report still less, and if your ladder were long enough to bring you to a truly distant elevation, then your instrument would detect only a vague trace of positivity, some of it possibly coming from the ionosphere higher up. Your electrometer would confirm what we reasoned in the paragraphs above — positive charge concentrations should diminish strongly with altitude.

Now, the formal test of this charge distribution. The previous chapter showed that the measured electric field strength falls off sharply with increasing altitude. Do such experimental measurements square with the deduced vertical charge distribution?

If you're standing on the earth's surface, you'll note the earth's full negative charge beneath you. You'll also note lots of positive charges above you (**Fig. 3.2**). That profound separation should imply a strong electric field. By contrast, if you're positioned higher up, then the positive charges beneath you should mitigate some of the earth's negative charges; meanwhile, fewer positive charges will lie above you. Diminished net charge above and below implies a smaller electric field strength — as measured.

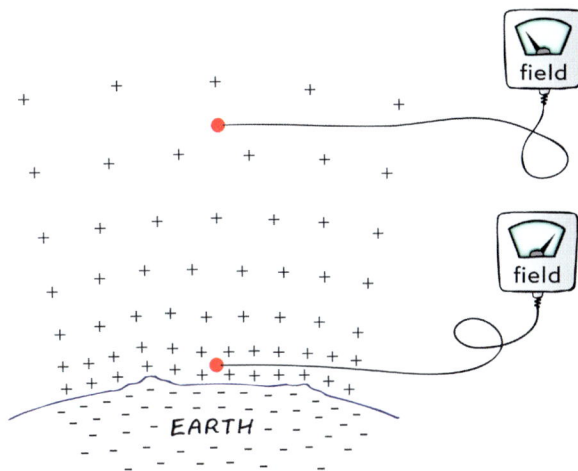

Figure 3.2. Near the earth's surface, maximum separation of positive and negative charge confers highest electric field.

So, qualitatively at least, theory fits measurements. We may persist in thinking that appreciable positive charge gets driven from the water, into the atmosphere above. That positive charge could logically represent the positive pole of the earth's electric field. The main question of this section has been answered.

Before we move on, however, we need to address another question, perhaps too obvious to ask: How could massive atmospheric positive charge lie directly adjacent to massive earthly negative charge? Wouldn't they quickly annihilate the other? Annihilation does in fact take place in another water-based context inside a water bath: When positively and negatively charged zones adjoin one another, a narrow, neutral buffer zone forms between them.[2,3] Something similar could happen where positive charge of the atmosphere meets the negative charge of the earth. Such a narrow buffer zone might in fact serve a convenient purpose: If you touch the earth, you may avoid experiencing a massive electrical shock.

Hence, the positioning of atmospheric positivity next to earthly negativity need not pose a death blow to the seemingly apparent distribution of charges. Positive atmosphere can exist near earthly negativity.

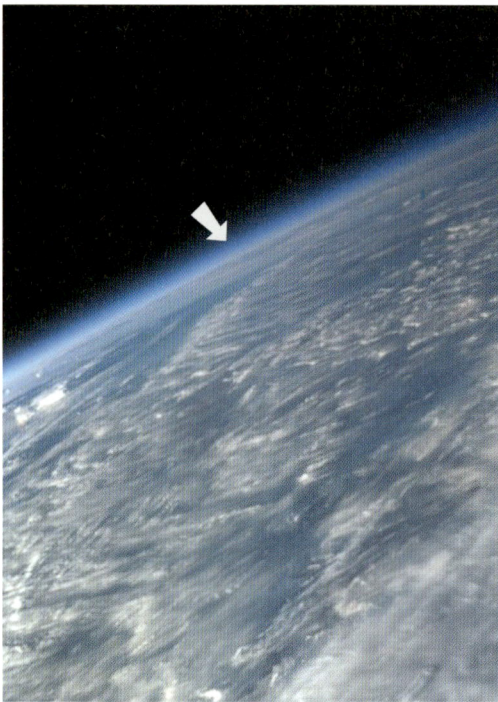

Figure 3.3. An extremely thin layer of atmosphere surrounds the earth.

While this section's focus has centered on the underlying science, we shouldn't ignore the aesthetics. If you take the trouble to climb that long ladder to track the vertical distribution of positive charge, then be sure to take notice of the awesome view from the top (**Fig. 3.3**).

Images such as the one in **Figure 3.3** are routinely sent by satellites. They depict the spherical mass of the earth surrounded by a surprisingly thin layer of atmospheric gas. That layer extends upward by perhaps only 1 percent of the earth's radius. While the splendor is evident, looking at the image raises the central question that we address in the next sections: What makes that narrow layer of gas cling so tightly to mother earth?

How Can We Understand Atmospheric Pressure?

According to common understanding, everything with mass should be pulled down by earth's gravitational attraction. The atmosphere is no exception. Nevertheless, I raise the question whether much of that pull, currently attributed to mass-based attraction, might actually come from charge-based attraction.

Why should we question the impact of gravitational attraction?

In theory, the earth's mass creates a gravitational force attracting everything on or above its surface. That includes the enveloping gas. Gravitation pulls on all air molecules, including those situated high in the atmosphere, which then press on the air molecules beneath, which in turn press on the earth. The molecular stack creates the ground-level pressure, which we call "atmospheric pressure."

Atop Mount Everest, the fewer air molecules pressing down means lower air pressure. By contrast, the pressure in the bottom of Death Valley should be higher. The pressure magnitude depends on how many air molecules lie above you.

That explanation may seem plain and simple, but it creates something of a conundrum. Pressure may be easily understandable when dealing with a pile of lead bricks: the higher the stack, the higher the pressure exerted at the bottom. But, a gas does not behave like bricks. Bricks press unrelentingly upon one another, thereby building pressure at the bottom. But gas molecules do not press in that way. By *definition* the gaseous state exists when constituent molecules remain well *separated* from one another. Molecules may occasionally collide, but most of the time they maintain their distance. Hence, the paradox: How can the pressure at the bottom depend on the pile's height if molecules maintain their distance from one another?

The conventional explanation circumvents this conundrum by invoking the effects of thermal motion. It posits that because molecules are always bouncing around, chance collisions do the job of creating pressure. Thus, collisions would exert pressure on the earth's surface in the same way that driven tennis balls exert pressure as they hit a wall. The larger the number of synchronous hits, the higher the pressure. That argument could make sense, especially when considering millions of balls.

What's hard to understand is how those implied collisions could add up in the vertical direction, as they must if the explanation is to suffice. This feature is key. Yet, even a single, random, non-colliding molecule in a representative stack would defeat that expectation. Since gas molecules (by definition) spend most of their time distanced from neighbors, you'd expect numerous gaps to be common. But with gaps in a column, how could those collisions add up vertically in synchronized fashion? And if they cannot, then how could pressure depend on the column's height, as required? Something seems missing from the sequence of logic.

I understand that this argument challenges the explanation of atmospheric pressure that we're taught; yet this deeper dig exposes what seems to me to be a logical flaw. Simply put, a column of gas molecules does not behave like a column of bricks.

This issue could resolve by pursuing a different interpretational pathway. If atmosphere and earth carry opposite charges, *then the earth must attract the atmosphere*. No way around it. Positively charged air must press unceasingly on the negative earth. Molecular collisions play no role at all. When you stand atop a high mountain, where fewer of those atmospheric positive charges press on you, you experience less pressure than if you stand on the ground, where more charges press.

This is the interpretational line we will pursue.

Why Doesn't the Atmosphere Just Blow Off?

Now the critical question: Could charge attraction have enough strength to keep the atmosphere bound to the earth?

Recall (Chapter 1) that electrostatic forces can outweigh gravitational forces by a factor of 10^{38}, a figure too large to comprehend. Thus, the charge-based forces to which we allude can be impressive, if not monumental. Indeed, if one second's worth of electrons flowing through a lightbulb filament can create force enough to raise 5,000 jumbo jets, then relatively few negative earth charges could plausibly suffice for holding that positive atmosphere in place. Just how few was demonstrated earlier

(**Fig. 2.6**): It's not many. I argue, therefore, that charge-based attraction of the atmosphere to the earth constitutes a plausible hypothesis, worthy of exploration.

But what of gravitational attraction, the more conventional interpretation? In this context, imagine viewing the earth from a (rather privileged) vantage point in space. You see the earth, together with its clinging atmosphere. But you also notice that the wind up in the atmosphere can be fierce. You might expect occasional gusts to whisk away a good deal of that weakly held air, never to return — a frightful prospect for sure (**Fig. 3.4**). Such prospect takes on a glimmer of reality when considering the recognized feebleness of the gravitational draw. Imagine the progressive erosion of our atmosphere.

Figure 3.4. The specter of atmospheric blow-off. Could the mass-based gravitational force prevent such a catastrophic event?

If the atmosphere contained sufficient charge, on the other hand, it could cling strongly to an oppositely charged earth. According to this paradigm, the atmosphere should remain in place indefinitely, allowing us the luxury of continued breathing.

Attentive readers may notice a potential hiccup in my argument. Yes, the positively charged hydronium ions in the atmosphere may cling to the negative earth, but why also the molecules of air that surround those charged species? Aren't those gases mostly neutral? Why should they attract to the earth?

We think of those gas molecules as independent of one another. However, simple arguments can be adduced to support the view that many of those atmospheric molecules may be weakly attracted to one another — loosely enough to avoid compromising their gaseous character, but strongly enough to impart some element of collective behavior.[2] If so, then the atmosphere in its entirety ought to be attracted to the earth's negative charges. We need not worry unduly over the prospect of suffocation.

Three Paradoxes

Continuing in this electrostatic vein, I take some time to illustrate the power of this charge-centered paradigm. To do so, I begin by supposing the alternative: that the conventional gravitational attraction paradigm applies. Let us see where that takes us.

Within that framework, three atmospheric parameters predominate: pressure, temperature, and density. From the behavior of each one of those familiar parameters, we should expect ready interpretations of commonly observed atmospheric phenomena. We will see, however, that such exercises lead us down rabbit holes: They lead to confounding paradoxes, interpretational ambiguities, and even downright contradictions.

I will then show how the electrostatic framework, at least provisionally, offers the prospect of more straightforward interpretations.

Paradox 1: Atmospheric Pressure. Can Pressure Determine Our Weather?

Looking down on our planet from a cosmic vantage point, you might notice patterns of weather. Cloud masses continually shift across the surface of the earth. Those weather patterns supposedly arise from local pressure variations, arising out of a complexity of factors. Weather reporters wax eloquent about high-pressure ridges and centers of intense low pressure moving hither and yon, creating your local weather. We've heard such pronouncements starting from a young age, and therefore trust that the weather reporters grasp how the weather works, even though we may not.

As you continue to marvel at the majesty of all those ever-shifting clouds below, some unleashing lightning flashes, you may recall some of what you've read up to now. Charges can be important. They are clearly at work in at least some kinds of weather: Lightning flashes may discharge electrical potentials on the order of a half million volts.

Relative to those impressive potential differences, differences of pressure are miniscule. Standard atmospheric pressure is 1 bar, or 1,000 mbar. The *record* low pressure, observed near Guam during Super Typhoon Tip on October 12, 1979, was 870 mbar. The record high, measured in the Siberian city of Agata on December 31, 1968, was 1,083 mbar. While even those record extremes differ by only about +/- 10 percent of the mean, everyday regional pressure differences are much smaller; they commonly amount to less than 1 percent of the mean. Ordinary weather-associated pressure variations are therefore extremely modest.

Meanwhile, we are told, paradoxically, that any such small pressure differences can determine our weather. A butterfly flapping somewhere in Mexico is said to trigger a hurricane in the Atlantic. It's possible — a catalytic effect, perhaps? On the other hand, one does wonder whether pressure differences so modest could consistently determine something as profound as cataclysmic weather events, or even a rainy day — or whether those pressure differences might perhaps arise as the *consequence* of weather rather than a determinant. It's hard to know (see Chapter 10).

But compare those miniscule pressure variations to the large variations of electrical potential and charge that routinely pervade the atmosphere, even over the course of a day (Chapter 4). Would a relevant parameter commonly varying by 50% likely have more influence than another relevant one varying by only 1%?

Evidence will be adduced (Chapters 7-9) to show that rather than pressure, electrical charge could well be the primary determinant of weather. Please stay tuned.

Paradox 2: Atmospheric Temperature. Does Warmer Air Always Rise?

Everyone knows that hot air rises. If you fill a balloon with hot air, that balloon can take you high into the atmosphere. So long as the balloon's air remains hot you ought to remain aloft; but should that air cool down, you may descend back to the earth. By suggesting that hot air rises, I am telling you nothing new.

However, this common assertion leads to a paradox. If hot air always rises, then that rising air should heat the region above. But the air temperature atop your local mountain peak usually remains *lower* than the temperature beneath it at lower elevations, *i.e.*, temperature tends to diminish with increasing altitude. Generally, within the troposphere (lower atmosphere), the higher you go, the colder it gets (**Fig. 3.5**).

ATMOSPHERIC TEMPERATURES

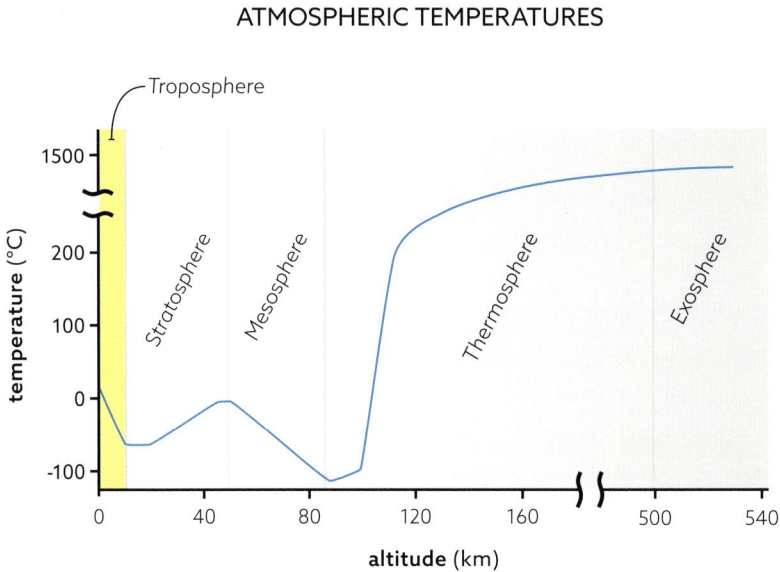

Figure 3.5. Temperatures vary depending on altitude above the earth. At certain altitudes, the temperature increases with altitude.

You might argue that this conundrum resolves because hot air cools as it rises. Think about the implications of that defense: As the rising air cools down, it becomes denser; the denser air should begin sinking through the warmer air beneath. This up-and-down pattern should create vertical air currents practically everywhere. On mountainsides for example, naturally cool air from above should slip down the slopes, replacing the warm air that had risen elsewhere. But anyone frequenting mountain slopes knows that this is not always so: Mountain air can flow upward, sideways, or sometimes not at all. So, the (somewhat confusing) argument that rising air cools conflicts with natural observation. It fails to resolve the paradox.

Now consider the electrostatic paradigm. Imagine the air attracted to the earth. In that situation, the tendency for warm, near-earth air to rise would depend on how much electrostatic force keeps that air stuck to the earth. The rise is not a given. Especially if that air contains lots of positive charge (**Fig. 3.1**), it could remain tightly stuck to the earth's surface despite its warm temperature. Thus, within the framework of the electrostatic paradigm, hot air need not *necessarily* rise. It might, but it might not.

A side issue in this discussion seems worthy of consideration. **Figure 3.5** shows that thermal gradients may *reverse* in regions above the lower atmosphere (*i.e.*, the troposphere). Higher up, it's evidently hot enough that you might need air conditioning.

Why temperatures in the upper region exhibit these peculiar distribution patterns has never been made clear, to my knowledge. But an answer could come from the very definition of the word "temperature." What does temperature really mean? An example helps pinpoint the issue.

In central Russia, the mid-winter air can turn frigid. Yet infrared images of clouds suspended in that extremely frigid air reveal that the clouds seem hot — hotter even than warm smokestacks on the ground beneath (see *Pollack*, 2013[2], Fig. 10.6). We draw that conclusion from the abundance of infrared energy emanating from those clouds: The clouds seem "hot." But how could a cloud remain indefinitely so much hotter than the frigid air surrounding the cloud? That makes little sense.

This paradox arises from the reflexive assignment of "temperature" to the amount of emitted infrared energy. "Temperature" is certainly a useful everyday term – you can exclaim, "the stove is hot!" In a more formal context, however, the term suffers ambiguity. I addressed this issue extensively in Chapter 10 of my previous book,[2] and try to minimize my own use of the word. More precisely defined is the term "infrared energy." A cloud may emit gobs of infrared energy, but that does not necessarily mean that it's hot.

Fundamentally, infrared energy arises from atomic-level charge movements. The more intense those movements, the higher the emitted infrared energy. Way up above the earth, in the thermosphere's domain (a layer of the earth's atmosphere) of highly charged plasmas, intense charge movements create huge amounts of infrared energy. We reflexively equate those charge movements with "high temperature," but that linkage can be just as misleading as it is in the Russian cloud example. Certain frequencies of charge movement may equate to "temperature," but not others. In **Figure 3.5**, the ordinate (y-axis) might better be labeled "infrared energy" instead of "temperature." That assignment could potentially solve the problem.

For now, suffice it to say that the thermosphere likely contains many moving charges. Protecting yourself from those intense charges may be a capital idea, but you will be relieved to learn — especially if you were inclined to drag it up that long ladder — that an air conditioner may be less useful than a charge deflector.

As for the mountain-slope paradox, the issue may resolve within the context of charge. Because electrostatic forces may easily overpower gravitational forces, charge matters a lot. Positively charged warm air doesn't necessarily rise above cooler air. Whether it does or doesn't could well depend on electrostatic forces.

Paradox 3: Air Density. Do Heavier Gases Lie Beneath Lighter Ones?

Having dealt with pressure and temperature paradoxes, we move next to the density paradox. You might ask: *What* density paradox?

The atmosphere contains large quantities of nitrogen and oxygen; it also contains carbon dioxide, argon, helium, and some trace gases. The densities of those gases vary. In order of increasing density, we have: helium, nitrogen, oxygen, argon, and carbon dioxide.[w2] You might expect that the denser gases would settle to the bottom of the atmosphere. Since carbon dioxide is 25 percent denser than oxygen, the perplexing question arises: Why don't we breathe mainly carbon dioxide? Or, perhaps for a little spice, maybe some of that dense argon as well?

Of course, the winds will mix those gases, and mixing could over-whelm any gravity-based stratification. That view currently prevails, although some models do imply the presence of stratification.[w3] Consider the reverse argument: Even with the wind intermittently stirring the mixture, enough time has surely elapsed to allow the heaviest gases to follow their natural tendency to reside mainly toward the bottom. Wind will stir the mixture, but the default is clear: Denser molecules prefer to reside downward. It's much like the vinegar in your salad's vinaigrette.

You may shake the mixture vigorously, but the lighter oil will eventually float to the top, leaving the heavier vinegar at the bottom. Molecules stratify.

The stratification issue has the potential to be serious in urban areas that continuously generate lots of CO_2. Imagine one of those areas during an extended summer period without noticeable wind. It happens. With little or no mixing, shouldn't the high-density CO_2 linger at the bottom, putting residents at risk for mass asphyxiation (**Fig. 3.6**)?

Fig. 3.6. The specter of the densest atmospheric gas settling to the bottom of the atmosphere.

A potential rescue from this dilemma can be found in the surprising constancy of the oxygen-to-nitrogen ratio. That ratio remains stubbornly constant at altitudes ranging from deep valleys to the tops of mountains and many kilometers upward.[4] So invariant is the oxygen-nitrogen ratio that to allow discernment of even trivial variations, the ratio is commonly expressed as four or five significant figures; see **Figure 3.7**. Even in the winter in Siberia or Antarctica, where few plants can pump oxygen into

ATMOSPHERIC GAS CONCENTRATIONS

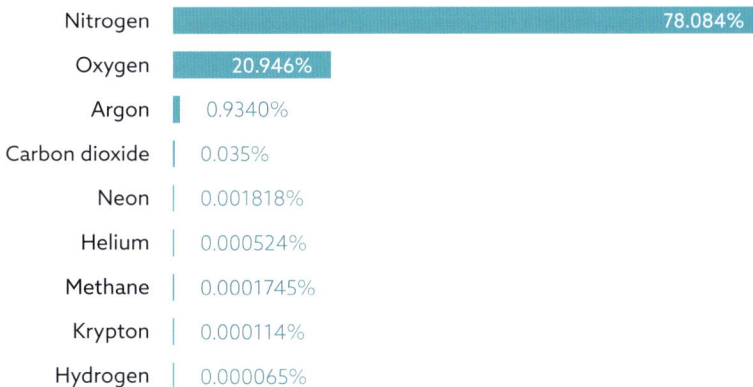

Nitrogen	78.084%
Oxygen	20.946%
Argon	0.9340%
Carbon dioxide	0.035%
Neon	0.001818%
Helium	0.000524%
Methane	0.0001745%
Krypton	0.000114%
Hydrogen	0.000065%

Figure 3.7. Several gases fill the atmosphere. Relative amounts are given with accuracy to four or five significant figures.

the air, the oxygen-nitrogen ratio remains as high as elsewhere. Atmospheric scientists would consider variation by as much as 0.1 percent to be eye-popping.

A possible explanation for that consistency is that oxygen and nitrogen (and perhaps other gases) may be at least weakly linked in some fixed (stoichiometric) ratio. Ample precedent exists. Gases often internalize themselves into polyhedral cage-like enclosures known as gas clathrates.[w4] Those clathrates contain fixed numbers of gas molecules plus ordered water. The most recognized example is methane hydrate, known primarily as a nuisance because it clogs pipelines in deep-sea drilling operations. Other examples exist. Hence, molecules in the atmosphere may tend to pack together in fixed ratios.

I mention this kind of stoichiometric packing only to reinforce the argument that it may be misleading to invoke density as an independent variable. We do it routinely, but danger lurks. If atmospheric molecules pack, but packing is ignored, then we may wind up with paradoxes: For example, we should routinely breathe carbon dioxide because of its high density in the atmosphere.

In the electrostatic interpretation, densities obviously exist; however, they are less relevant than charge-based forces. Hence, some potential for averting the density paradox may lie in considering a more central role of charge forces — as we shall see in subsequent chapters.

In sum, we commonly presume that pressure, temperature, and density govern much of atmospheric dynamics, but in the last three sub-sections I've attempted to show that those variables do not always provide neat and tidy explanations. Rather, they lead to paradoxes. In the coming chapters, we will explore the possible roles of charge-based forces in understanding the diverse workings of our atmosphere. That exploration will include the genesis of weather, of which we currently understand so little.

Section Summary and Perspectives

The earth contains an excess of negative charge while the atmosphere contains an excess of positive charge. Scientists have long known of these charges, yet their implications are mostly ignored.

Some of those implications seem rather obvious — like the positive atmosphere's attraction to the negative earth. Said attraction, I argue, may be the very force that keeps our thin atmosphere in place.

Meanwhile, conventional thinking on atmospheric dynamics leads to paradoxes. Invoking pressure, temperature, and density as primary variables creates enigmas that challenge easy escape. For example: What keeps the atmosphere from blowing away? Why doesn't cooler, denser, air consistently slip down mountainsides? And, with dense carbon dioxide expected to settle to the bottom of the atmosphere, why don't we run on CO_2?

A route toward averting those paradoxes may come from acknowledging the fact that atmospheric positivity attracts earthly negativity. We shall soon see evidence in support of that notion.

Meanwhile, I end the chapter by offering a teaser related to the negatively charged earth. Think of what might happen if a suitably outfitted skydiver were dropped from the stratosphere. Besides experiencing some measure of panic, the diver will experience an earthly draw. Now, suppose that the skydiver carried electrical charge. If that charge were positive, then the attractive force arising from the pull of the negative earth might drag the diver increasingly rapidly downward — with an unhappy ending. If the charge were negative, then the diver might fall more slowly, perhaps very much more slowly during the close approach to the negatively charged earth. At some point the diver might not fall at all.

Imagine!

SECTION II

Earth Dynamics

This section builds on the previous chapters' foundation. We deal with everyday features of earthly systems whose origins are rarely considered, including wind, magnetic field, and rotation.

Why does each day have 24 hours?

CHAPTER 4

What Makes the Wind Blow?

As I draft this chapter from my lovely Seattle hilltop home, I peer out at the whitecaps covering Lake Washington below. Wind drives those pretty white tufts. The image serves as a reminder of the sheer omnipresence of wind on this planet, be it fierce or gentle, or sometimes even absent.

So, let's get to it: What makes the wind blow? Might that flow of wind arise from charge forces, perhaps related to the vertically separated atmospheric charges considered in the preceding chapters?

The question of the wind's origin seems almost too obvious to ask. Accustomed as we are to the everyday phenomenon of wind, we rarely stop to inquire as to its cause. Propelling the wind requires energy, but energy from where? Weather gurus may point to local differences of atmospheric pressure as the wind's source, but how might such pressure differences arise? What energy creates them? And how could those pressure differences sometimes remain so confined as to produce localized wind gusts? Do we feel satisfied that we really understand how wind works?

This chapter offers a simple hypothesis to explain the genesis of wind. I will suggest that the driving energy may come ultimately from the global battery, whose terminals lodge respectively in the atmosphere above (positive) and in the earth below (negative). From that separation we expect mainly vertical forces, but I will argue that differences of atmospheric positivity in laterally adjacent regions can create *horizontal* charge gradients, which may then propel airflow. That flow of air could be the phenomenon we commonly call "wind."

Charges High in the Sky

The previous chapter described how energy from the sun creates a positively charged atmosphere above the negatively charged earth. Those separated charges form a vertically oriented battery, arguably recharged by the sun's energy. So long as the fires of the sun continue to burn, this giant battery should remain well charged (**Fig 4.1**).

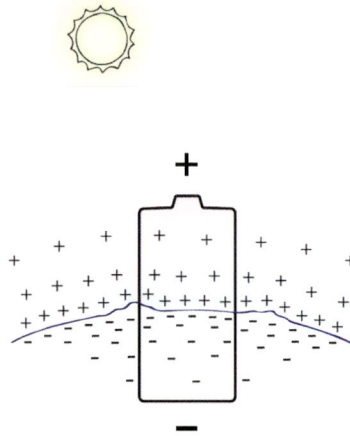

Figure 4.1. Energy from the sun separates charges, leaving the earth negatively charged and the atmosphere positively charged. This separation effectively constitutes a battery.

Any force that develops out of those separated charges would seem *a priori* to lie in the vertical direction. Vertical charge separation implies vertical forces, as opposite charges attempt to recombine. But suppose atmospheric positive charges rose higher above one earthly region than over another nearby; then, concentrated positive charges would lie adjacent to sparse positive charges, creating a horizontal charge gradient. Any such gradient ought to propel horizontal flow. Could such flow be the same as what we call wind?

A scenario of that sort raises several questions: Does atmospheric positivity rise higher in certain regions than others? If so, then why? And might any such horizontal charge gradients suffice for creating the wind?

To begin answering these questions, consider first the process of evaporation, which arguably releases charged hydronium ions into the atmosphere (Chapters 2 and 3). Sunlight drives that process. During nighttime, the absence of sunshine implies relatively little evaporation; but as the sun peeks its head above the horizon, water's charges begin to rise. As the day progresses, evaporation increases, and by early to mid-afternoon it reaches its peak, only to diminish again toward evening. This describes the cycle of evaporation, and, hence, the expected 24-hour cycle of upward-driven charges.

I must mention, however, that it's not *just* the positive hydronium ions that rise during evaporation. Negatively charged droplet clusters may rise as well, quickly dispersing, and ultimately re-condensing to form clouds. To avoid clouding the issue (so to speak), I reserve discussion of those negatively charged clusters for Chapter 7, where we begin discussing weather.

Here we limit our discussion to fine weather, *i.e.*, dry, calm weather, and focus on those rising positive charges. Since like-charges repel one another, repulsion should drive the released charges progressively higher into the atmosphere as the day progresses. It is these high positive charges on which we now focus, the charges that create the upper terminal of the earth's electric field (**Fig. 4.1**).

The term "electric field," for those who may have forgotten from Chapter 2, is a concept used to help define the nature of local electrical forces. Michael Faraday, who originally coined the term, set the stage. He defined the local electric-field strength as the force experienced by a theoretical positive "test" charge placed in the field. Where that test charge feels high force, the field strength is inferred to be high. Low force implies correspondingly low field strength. The field's direction is conventionally drawn from the positive toward the negative.

Knowing the values of electric field strength over space permits us to understand how much force a unit charge will experience at any point in that space — a useful expedient when trying to understand what's happening up there in the atmosphere.

To see how these concepts play out in our situation, envision an imaginary positive test charge lying at some height in the atmosphere

(**Fig. 4.2**). When situated immediately above the earth (*panel a*), that test charge should experience a strong earthward attraction: The negative earth pulls directly on that positive charge, while from above, many positive charges push the positive test charge downward. Those combined forces should impel the charge decidedly earthward. This strong force denotes high field strength near the earth's surface — as routinely measured.

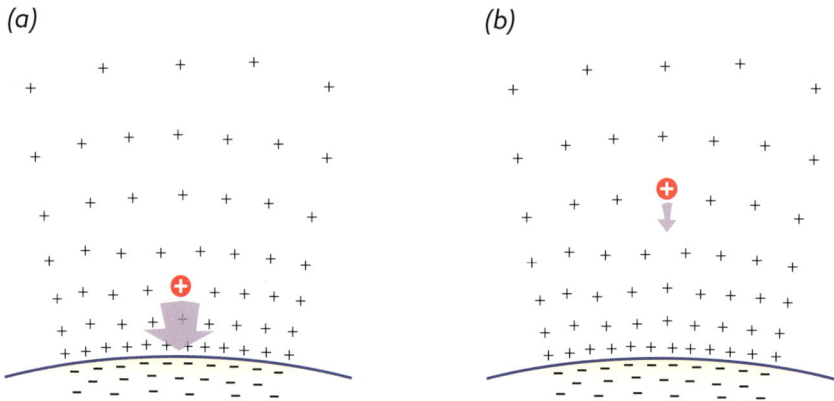

(a)   *(b)*

Figure 4.2. The force (purple arrow) experienced by a positive test charge (red) defines the local electric field strength. The field is highest near the earth's surface (*a*), and diminishes with increasing altitude (*b*).

Now suppose our positive test charge lies farther up in the atmosphere (*panel b*). Three factors diminish its attraction to the negative earth: (i) larger separation from the earthly source of negativity; (ii) intervening positive charges below the test charge, which counterbalance some of the earth's negativity; and (iii) from above, fewer positive charges pushing the test charge downward. So, our friendly test charge positioned high up hardly feels the earth's pull. Diminished pull equates to lower electric-field strength, which, again, is amply confirmed (Chapter 2). Field strength diminishes with increasing altitude.

I hope this exercise adds a little more understanding to the seemingly arcane subject of electric fields. Knowing something about the relation between electric fields and the driving forces they produce permits us to move on toward this chapter's goal: discerning whether *lateral* electric-field gradients could propel wind.

Do Electric Fields Really Vary During the Day?

Now that we understand something of electrical forces and fields, let's return to evaporation and to the growing stack of atmospheric positive charges that evaporation is expected to produce each day. Is this expectation supported by evidence?

Figure 4.3a shows the time course of the atmospheric electric field, measured just above the ocean surface in fair weather.[1] Two independent, century-old measurements (Carnegie, Maud expeditions) show much the same result: (I) the field strength is lowest at nighttime; and (II) it increases progressively during the day, reaching a near plateau in mid-afternoon and finally diminishing toward evening.

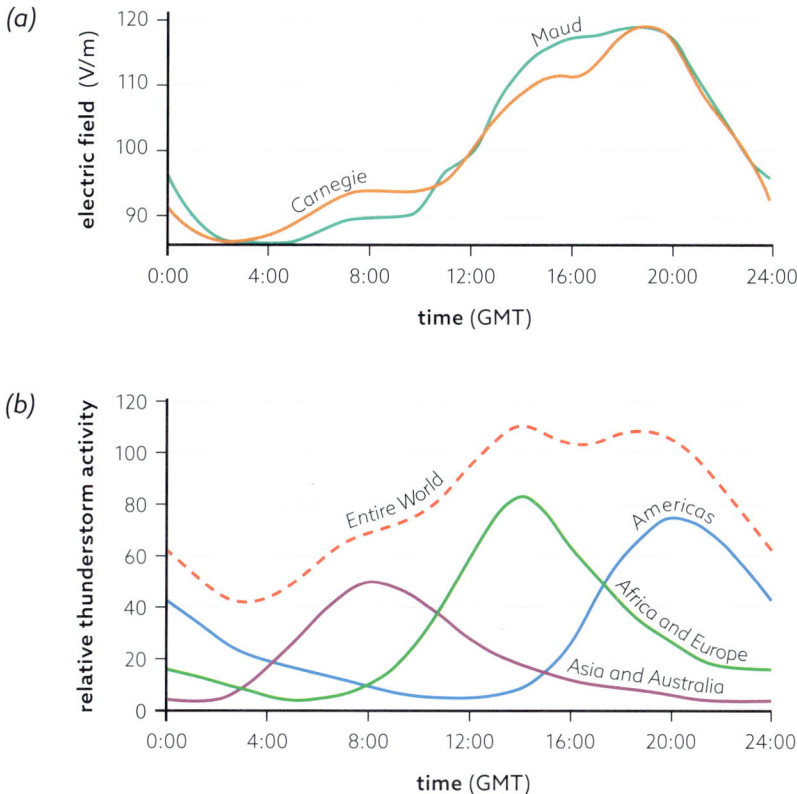

Figure 4.3. *(a)* Diurnal variation of electric field strength above the ocean as measured in two expeditions (Maud, Carnegie). *(b)* Timing of global thunderstorm activity. Redrawn from Roble and Tsur[1]. Note: GMT = Greenwich Mean Time.

This result confirms the anticipated correlation between the growth of electric field and the expected time course of evaporative release of positive charge. The two match nicely. If indeed evaporation drives positive charges upward (Chapter 2), then the scenario makes sense: Evaporative pumping of positivity into the atmosphere drives those positive charges higher and higher as the day progresses, increasing the atmospheric field strength at any altitude.

Yet, a confounding issue arises. A prevailing assumption among atmospheric scientists has been that the field-intensity peaks such as those in **Figure 4.3a** occur synchronously at all locations over the globe. The rationale? Any such field-strength variation was presumed, early on, to have a cosmic origin. If so, then the variation ought to occur at the same time everywhere on Earth.

On the other hand, those early measurements were carried out principally at two loci, whose longitudinal positions did not differ by a whole lot. Subsequent measurements have been few, and I am not aware of any measurements that have systematically tracked the diurnal time course of electric field over the broad face of the earth. Until that happens, it's not possible to determine with any degree of certainty whether the field changes occur synchronously over the earth, or whether, instead, the timing of those changes depends on the position around the earth.

Figure 4.3b weighs in on that issue. It shows the pattern of annual thunderstorm activity in different locales. From the phase differences among the different regions, we can discern that thunderstorms occur most frequently during late afternoons, everywhere on the earth. Common experience concurs. So, if thunderstorm activity correlates with high electric field and high atmospheric charge, as expected, then electric fields should have a mid- to late afternoon maximum everywhere on Earth. That is, the field strength should largely follow the sun's intensity.

If all the above is correct, then we arrive at the following simple paradigm: The more intense the solar energy, the higher the evaporation rate and the higher the stack of positive charges in the atmosphere.

Day-Night Differences in Vertical Charge Stacking

The sun-mediated stacking described above makes sense if you think about how the earth's negative and positive charges get built and, also, how they spread.

First the negatives. Those charges lodge principally within the earth's peripheral geosphere. That includes the earth's surface, its seafloor, and some distance below; see Chapter 2. The peripheral shell might be thought of as a dense matrix of electrical wires running from the earth's surface to points somewhat deeper, interconnected over the entire shell. The highly conductive nature of this network means that any "excess" negative charge building locally should quickly spread to other regions around the shell. As a consequence, charge should be distributed practically uniformly around the earth, at all times. Whether measured in Moscow, Chicago, or Beijing, net charge of the earth's shell should largely stay put at some fixed value.

The atmospheric situation, by contrast, should differ fundamentally. Air is an insulator, not a conductor. The air molecules themselves carry the charge, so these charges can move laterally only as fast as the air molecules bearing those charges can carry them. I'm not suggesting the absence of lateral movement; in fact, the next section explores the atmospheric charges' lateral flow. I'm merely asserting that, unlike the swift flow of charge expected in the earth's highly conducting shell, the lateral flow of charge in the atmosphere ought to occur more slowly.

I now return to the crucial point: the growth of charge in the vertical direction. As positive charges emerge from the water beneath, they mainly push up against existing positive charges. Repulsion drives those positive charges progressively higher into the atmosphere. Thus, a strong propensity should exist for positive charges to stack vertically, reaching higher into the atmosphere during daylight hours.

That dynamic brings me to the central theme of this chapter: Daylight stacking should create major charge-height differences between day and night zones (**Fig. 4.4,** *see next page*). At night, with scant driving energy, few positive charges should build high into the atmosphere. But, as daylight appears, positive charges ought to begin stacking, piling up until mid- to late afternoon, when the sun's influence begins to wane and the height of the stack begins to diminish.

Hence, if you were a positive test charge lodged in a balloon at some height above the earth, you'd experience the electric field depending on time of day (**Fig. 4.4**). First, nighttime (*a*): With few charges above you to push downward, the contribution of high positives should be minimal; they should supply only a modest downward push, if any. Later (*b*), as the sun begins beating down, more positive charges should build, eventually piling up above you. Those accumulating high positive charges will push you more strongly earthward — translating to increasing electric-field strength. So, the electric-field strength up there ought to oscillate with time of day (**Fig. 4.3a**).

(a)

(b)

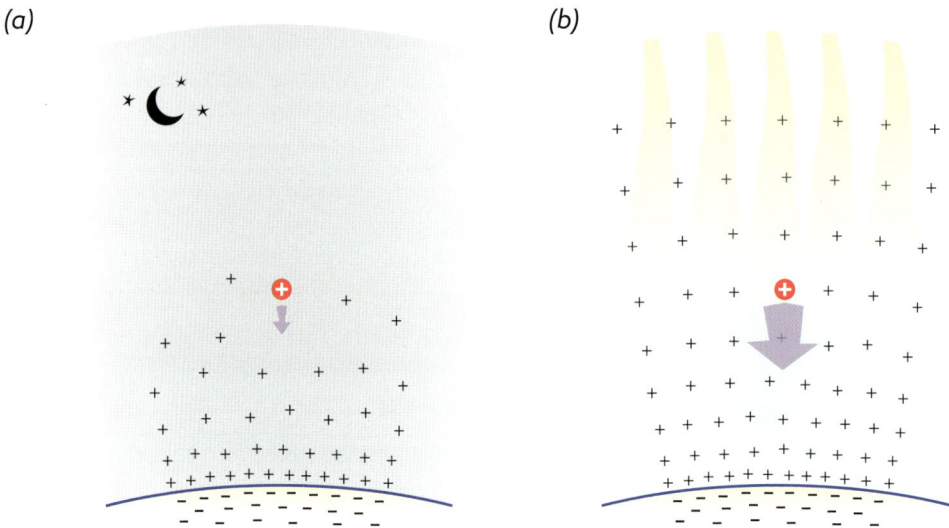

Figure 4.4. During nighttime (*a*) the absence of solar energy implies little or no positive charge buildup in the atmosphere. During daytime (*b*) energy from sunlight builds positive charges, driving them high into the atmosphere. That pileup increases the downward push on a positive test charge, indicating a stronger electric field.

Why should that vertical charge stacking matter? To answer that question, consider the physical boundary between day and night, *i.e.*, between the zone where positive charges stack high into the atmosphere and where those charges have already receded or have yet to begin growing. Could those boundaries imply something worthy of attention?

Lateral Charge Gradients and Wind

Imagine standing at some point on the rotating earth (**Fig. 4.5**). If the prospect of 24 hours' standing seems burdensome, then imagine reclining on your lounge chair, say somewhere in the mountains of Ecuador during July's dry season. It's pleasant. A warm breeze wafts by, and the sweet smell of local vegetation intoxicates your senses. You yawn lazily and pass, relaxed, into contented somnolence. Then, night approaches. The sun vanishes until dawn breaks, and the bounty of radiant energy once again illumines your region of the earth. Whether lounging or standing upright, you and the earth have rotated through almost one full cycle.

Figure 4.5. View of the earth, as seen from above the South Pole, showing the earth's rotation from west to east.

During the course of that cycle, the atmosphere's positive charges above you waxed and waned. They've risen upward into the atmosphere like a newly emerging geyser; then they retreated, backing down to their foundational level. Accomplishing this rise and fall is possible only because those positive charges have freedom to move, driven by forces pushing them upward or permitting them to recede downward.

That's the critical point: No serious obstacle prevents those free atmospheric charges from moving. When a positively charged region lies adjacent to a region with less positive charge, lateral spreading must occur. It may occur more slowly than the corresponding earthly negative charges, but it should still occur. Packed charges will flow naturally to nearby regions with less positive charge, dragging the air with them.

Such flow is not merely conjectural. Electric-field measurements made in balloons have shown substantial horizontal gradients, beyond those associated with weather.[2] Some of those lateral gradients, the authors argue, may in fact drive the currents responsible for creating the northern lights.

Lateral charge flow, I suggest, *may be the very phenomenon we call "wind."* It might not be the *only* source of wind, but if that charge-driven airflow is any different from wind, it's hard to understand how, or why.

With so bold a hypothesis, you might ask about the conventional view of wind generation. Why doesn't it suffice? "If it ain't broken," the common expression goes, "then why fix it?"

Wind generation is said to rest on pressure differences. Wind flows from a region with higher pressure to one with lower pressure. On the face of it, that hypothesis seems sensible. But as for the *source* of those pressure differences, it's hard to find explanations beyond vague references to differential heating (equator *vs.* poles) and the spherical shape of the rotating earth (the Coriolis effect). Missing is the kind of step-by-step progression of understanding that Occam and Newton might have preferred.

I invite motivated readers to check for themselves. To keep matters simple, I would challenge you, the reader, to employ current views on wind genesis to derive satisfying explanations for common phenomena. They could include local wind gusts, as well as the prevailing easterly and westerly winds (whose origin I will attempt to describe below). I expect that such challenges may prove daunting.

Thus, we press on. Could wind arise from lateral charge gradients?

Morning Gradient and the Trade Wind

Let's begin with the sunrise boundary. (We'll consider the sunset boundary in a moment.) **Figure 4.6** illustrates the early-morning scenario: As a freshly illumined atmospheric region begins to become stacked with positive charge, some of that charge will flow to the neighboring, yet-to-be illumined and hence uncharged, region. The charge flow, and the resulting wind, moves from light to dark.

Figure 4.6. Positive charge gradients at the sunrise boundary could create wind.

Consider that wind's direction. Because the earth turns from west to east, east sees the sun first (**Fig. 4.7**). Paris may bathe in morning sunshine while New York remains dark. Hence, that positively charged air should always flow from east to west.

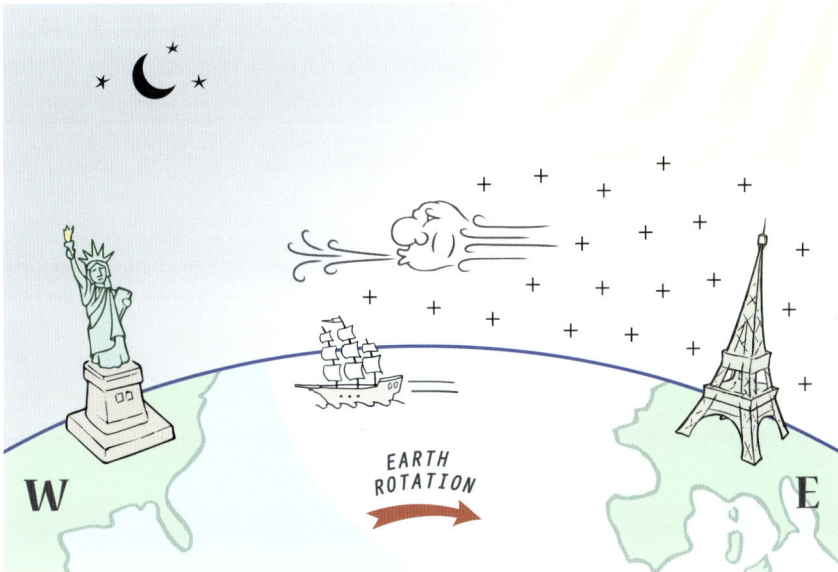

Figure 4.7. In the morning, charge should ordinarily flow from east to west because the east is the first to see light and build positive charge.

While most of us know the locations of Paris and New York, those cities lie somewhat removed from the place where the sun's effect should be most pronounced: the equator. At equatorial latitudes, the effects of sunshine should be most potent because in this region the sun's rays hit the earth almost perpendicularly. Hence, the impact of radiant energy on morning-wind production should be most evident there. Since the equatorial belt contains a lot of water (check the world map), which ought to facilitate positive charge production, the equatorial region should experience a robust morning-charge gradient and abundant east-to-west airflow.

Although that airflow may *originate* near the light-dark boundary, it should persist around the earth. The westward flow will push the air ahead. It will also leave a vacuum in its wake that sucks the air from behind; so, the wind should flow continuously. Further, the gradient's origin moves progressively around the earth, perpetuating that more-or-less continuous flow. So, the flow forms a continuous belt encircling the earth.

Sailors have long known about, and exploited, this morning-driven wind. Known as the "trade wind," this reliable east-to-west wind gave Christopher Columbus the confidence to sail from Spain to the New World. Trade winds could well arise as a natural consequence of the early morning day-night boundary.

Upper Westerly Winds and the Jet Stream

Beyond that early-morning boundary, a complementary boundary should form as day recedes into night. Similar wind-generating principles should apply, except in the opposite direction.

At first blush, that scenario would seem to be problematic: Winds generated in both directions should lead to cancellation. Isn't that a problem for the mechanism under consideration?

As we shall see in a moment, that's not necessarily the case. The reason lies in the altitude difference. As the day presses on, positive charges get driven higher into the atmosphere. With the lateral charge gradient situated higher up, the west-to-east wind should blow at higher altitudes than the trade winds. If so, then no cancellation problem should exist. Let me now explain in a bit more detail why that west-to-east wind should flow higher up.

For the first of two reasons, consider the situation at some earthly locale during mid-afternoon (**Fig. 4.8**). By that time, said region will have enjoyed ample exposure to sunlight. The sunlight should have driven positive charges high into the atmosphere. The most towering buildup ought to occur where the sun shines most intensely, *i.e.*, within the equatorial belt. By late afternoon, that zone should be rich with high atmospheric positivity (*panel i*). North and south of that belt, where the sun's rays hit more obliquely, atmospheric charges will have attained lesser heights. So, as the day progresses, an equator-to-pole height difference of atmospheric positive charge should build.

i.

Sun builds positive charge around equatorial belt.

ii.

Charges disperse to the north and south.

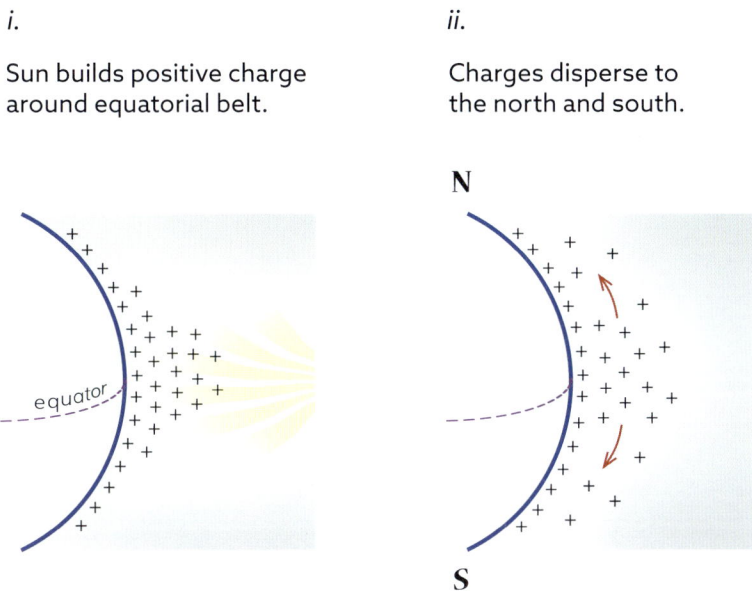

Figure 4.8. Mobile positive charges should always flow toward regions with less positive charge. During late afternoon, that would include from equator toward poles.

With what consequence? Positive charges towering high above the equatorial belt should flow toward the two poles (*panel ii*). That flow should limit equatorial positivity, while enhancing positivity in regions north and south.

Such airflow happens. High, poleward flows are well recognized since the time they were first proposed by George Hadley in 1735[w1]; appropriately, they are called "Hadley Cells." Those poleward wind flows have long been attributed to atmospheric heating and cooling. If, instead, they are driven by atmospheric charges, then the arrival of those positive charges ought to enrich regions north and south of the equator with high positivity. That would augment any local buildup of positive charges — both of which stand by, ready to act.

Ready to act how? Are we prepared for the second critical argument for high atmospheric charge flow?

Please consider the sunset boundary (**Fig. 4.9**). West of that boundary, sunlight still impinges; hence, reasonably high, intense atmospheric positivity should persist. East of the boundary, the sun will have already set, and the high charge will have begun its sharp decline. The resulting horizontal gradient, powerful because of the substantial charge difference, should have a predictable outcome: At the upper atmospheric levels, all the high positive charge in the west should flow toward the more-depleted east, resulting in strong west-to-east flow. Such positive charge

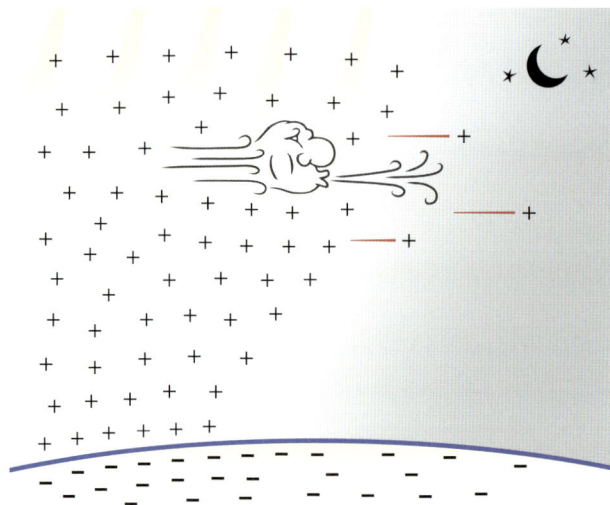

Figure 4.9. Charge flow at the afternoon - evening boundary. High positive charges anticipated to flow from west to east.

flow has been experimentally confirmed. High atmospheric charges do flow from west to east.[3]

That strong, upper-level flow, I suggest, constitutes the prevailing-westerly wind. According to the arguments presented above, that charge-mediated wind should whoosh around the earth continuously, fueled by the shifting sunset boundary. So long as the sun continues to shine and persists in driving evaporation, the powering energy should be at hand. The high west-to-east wind should persist.

Now, returning to the opposing-wind cancellation question raised early in this section, I suggest a simple answer: The prevailing westerly doesn't neutralize the prevailing-easterly (trade wind) for the simple reason that the westerly wind blows higher in the sky. The two winds blow at differing altitudes.

Now to the strength of those high-up westerly winds, which flow throughout the northern and southern hemispheres. They flow strongest at the middle latitudes, off the equator. Why so?

The culprit may be the poleward feed of positive charge discussed earlier (**Fig. 4.8**). Arrival of that high positive charge to mid-latitudes should augment the positive charges locally generated (**Fig. 4.4**). The charge stack becomes taller. That amplification should enhance the lateral charge gradient, strengthening the west-to-east wind flow. So, the westerly wind should flow powerfully at these middle latitudes.

Meanwhile, such high flow appears to be virtually absent in the earth's equatorial regions, and one may ask why. Frankly, I'm not sure. Perhaps this is because the ocean water beneath remains warm. Warm waters ensure ample evaporation even during nighttime. So, positive charges may get driven up into the atmosphere during the day as well as at night. Absent a strong afternoon-night boundary difference at the equator, the lateral charge gradient ought to remain weak; hence, little upper-level wind should flow. For that reason, the upper-level westerlies may be confined to latitudes removed from the equator.

You probably know the consequences of that westerly: Flying from New York to Paris takes one hour less than flying from Paris to New York. Belonging solely to the upper atmosphere, where planes fly, those prevailing westerly winds flow without end. Most people know them by another name: the jet stream.

Charges do seem involved in that stream. At least at northern earth latitudes, electrical currents are known to flow along latitudinal paths, from the dayside to the (roughly) midnight zone. This charge flow was first identified by the distinguished Norwegian scientist, Kristian Birkeland (symbols of whose discoveries appear on Norwegian bank notes).[w2] Birkeland went on to argue that those electrical currents were responsible for producing the auroras, visible at extreme northern and southern latitudes (also known as the aurora borealis or northern lights in the northern latitudes and the aurora australis or southern lights in the southern latitudes). Near the polar caps, the electrical potentials routinely amount to some 40 kV, and can grow to as high as 200 kV.[w3] Thus, electric fields *parallel* to the earth's surface certainly exist, and modern measurements have confirmed them.[4] On the other hand, I've yet to hear of anyone linking those parallel electric fields to jet-stream winds, as suggested here.

So, two principal winds encircle the earth (**Fig. 4.10**) — an easterly trade wind running mainly near the equator, and a westerly jet stream running strongest away from the equator at mid-latitudes. According to the proposed paradigm, the easterly blows gently because early morning charge gradients have not yet risen high enough to be substantial; hence, this remains a largely near-surface wind. The westerly, by contrast, blows stronger and higher. This follows mainly from the late-afternoon charge-gradient, arising from the vast quantities of high positive charge that have had the better part of the day to build. Hence, a strong westerly wind pervades the high altitudes, while a weaker easterly pervades the lower skies, especially near the equator.

A good video of these prevailing winds, including the upper-level poleward flows, may be seen on long-exposure satellite images obtained from NOAA.[w4] The video nicely shows the east-to-west flow confined largely to the near-equatorial region, while the west-to-east flow pervades both the northern and southern hemispheres.

Occasional Winds

The curious reader may wonder about winds that blow only occasionally rather than continuously. Puffs of wind can come practically out of nowhere and they can be extremely localized. Perhaps you've wondered why they blow. The default explanation has centered on pressure

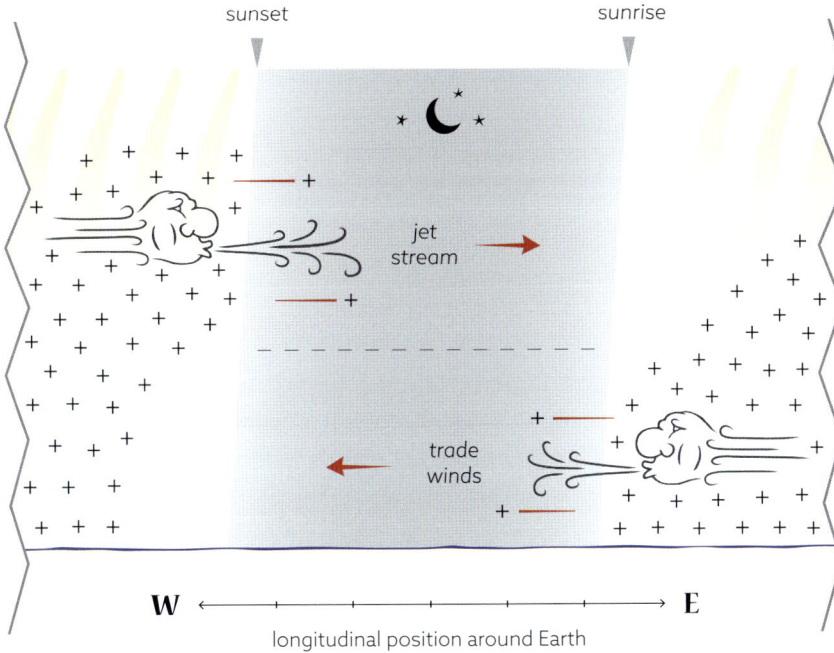

Figure 4.10. At the sunset boundary (*left*), lingering late-afternoon high atmospheric charge flows toward low-charge darkness; this should generate high-level west-to-east wind flow. The morning-generated trade wind (*right*) blows oppositely, from east to west.

gradients; but, how could atmospheric pressure gradients arise so suddenly and remain so localized? I'm unaware of any existing explanation.

Interpretations based on charge gradients seem easier to envision because atmospheric charges exist practically everywhere, and their distribution need not be uniform. According to the arguments above, *any* local charge gradient should generate wind.

Common examples include the wind gusts often blowing beneath a dark overhead cloud. The dark cloud blocks the sun from impinging on the earth beneath, while the regions surrounding the cloud do not block at all. This sharp illumination difference can create lateral differences of atmospheric positivity, initiating the wind that you may experience under such conditions.

Another instance, in my own experience in the northern hemisphere, comes from ambling from a sunny area into a long-shaded area, such as one just north of a tall building. Entry into that shaded area, I've noticed, often brings a sudden breeze. The same, according to my Austrian friend, under a large isolated tree situated in an otherwise sunny terrain: always a dependable breeze. Any such persisting boundaries between light and dark may act similarly to the day-night boundaries, creating lateral charge gradients and hence localized wind.

A third example may be experienced during summer at the beach. Perhaps you've noticed: A pleasant breeze often blows from the water toward the beach during the day, reversing course as day passes into night. A possible explanation for this characteristic pattern may lie in local differences of rising positive charge. During the day, sun-induced charges should rise from both sea and land, but more intensely from the sea with its practically endless supply of free positive charges. The resulting lateral charge gradient should drive the wind from sea toward land.

When evening arrives, the situation reverses. Those positive charges lying above the water should retreat rapidly as they recombine with the sea's abundant, free-floating negatives. Meanwhile, recombination over land should occur less readily, since most of the earthly negative charges, firmly buried well beneath the earth's surface *per se*, are less readily accessible for recombination. With higher positive charges remaining above land compared to above the sea, the atmospheric charge gradient reverses, and the wind source shifts from offshore to onshore.

The several examples above offer representative explanations for the genesis of localized winds. All involve localized charge gradients. I'll touch more on those charge gradients in later chapters when we deal with weather.

This chapter has focused mainly on the origin of persistent winds, with a brief detour into localized winds. While I have proposed explanations that I believe are straightforward, I invite the reader once again to compare textbook explanations to see how they match up. Those persistent winds may well annoy you by lengthening your westbound flight time, but I hope they may soon cheer you by providing fresh potential for tackling other poorly explained but taken-for-granted phenomena — including such contenders as the origin of the earth's spin, and the origin of its magnetic field. We deal with those phenomena next.

Summary

Anything that moves requires a driving force, and that includes the wind.

This chapter argues that the wind we experience may arise from atmospheric charge gradients. Such gradients seem practically inevitable if solar energy drives earthly positive charges high into the atmosphere. That rise evidently happens principally during the day. As nightfall approaches and sunlight fades, those high positive charges should begin receding. Hence, the boundaries between day and night should manifest lateral charge gradients, and evidence indicates that they do.

Those lateral gradients are replete with power. The laws of electrostatics tell us that like-charges, repelling one another, tend to escape to regions with less charge. Hence, horizontal charge gradients across the day-night boundaries should bring lateral charge flow. Those flows could represent "wind."

Two day-night boundaries, morning and evening, should generate different types of wind flow. The morning boundary, with limited earlier sunlight exposure and hence relatively few positive charges accumulated in the atmosphere, should generate relatively weak winds. Those east-to-west winds would seem to correspond to the well-known trade winds. The evening boundary should contain more abundant, higher level, charges because of the entire day's sun-driven buildup. The steep gradient between late-afternoon daylight and dark nighttime should create a strong, persistent west-to-east wind. That wind seems to correspond to the prevailing westerly, or, the so-called jet stream.

The persistently strong west-to-east wind could have consequences beyond just quickening eastbound airplane trips. Flowing charge creates magnetic fields. We consider in the next chapter whether the charges flowing around the earth could contribute to, or, even create, the earth's magnetic field. And, in the chapter following that one, we consider an even more provocative question: whether the strong westerly wind flowing persistently past the earth's mountainous surface creates enough drag to help drive the earth's rotation.

CHAPTER 5

Is the Earth a Giant Magnet?

For anyone hopelessly lost in the forest, nothing beats a compass. The compass helps you find your way because its magnetized needle senses the direction of the earth's magnetic field. It tells you which way is north.

Clearly, that magnetic field belongs to the earth. It maps so tightly to features of the earth's surface that we presume its origin must lie within the earth itself. The logic seems inescapable; yet nobody's plumbed deeply enough to check that presumption. Does the earth's core *really* contain some kind of embedded magnet?

Well, perhaps. But certainly not a solid magnet like the ones we normally think of. The temperature of the earth's core is estimated to be 7,500 °K (kelvin) — well above the critical temperature at which materials lose their magnetism. Known as the Curie temperature, that critical temperature differs for different materials. For iron and nickel, the main constituents of the earth's core, the Curie temperatures are 1,043 °K and 627 °K, respectively. Since the earth's estimated core temperature is substantially higher than either one, there's no way that a solid iron-nickel core could bear responsibility for the earth's magnetism.

Appreciating this fact, geologists have largely dismissed the solid-magnet theory. Instead, most suppose that the earth's magnetic field comes from charge-bearing fluids flowing within the earth's outer core. Moving charges create magnetic fields, so this explanation would seem to make sense. But precisely how any such flowing charges might create the earth's magnetic field with its well-known characteristics remains poorly understood.

This chapter considers a possible alternative origin — that the magnetic field originates *outside* the earth rather than inside. Why not? If hefty atmospheric charges flow from west to east around the earth (Chapter 4), then any such steady-current ring will create a magnetic field around it. Could that magnetic field possibly be what your compass senses?

Which Way is North?

Compass needles always point north — at least, that's how we think about it. The needle is understood to be "north-seeking" because it points roughly in the northerly direction.

However, a couple of issues confound this seemingly simple under-standing. First, the compass needle doesn't point to the actual north pole. "True north" lies at the north terminus of the earth's rotational axis, which deviates slightly from the earth's magnetic axis (**Fig. 5.1**). Their separation, reckoned on the earth's northern surface, amounts to about 500 kilometers[w1] (~300 miles). In other words, the earth's magnetic axis and the earth's rotational axis don't line up; so, the respective poles are spaced slightly apart, leading to occasional confusion.

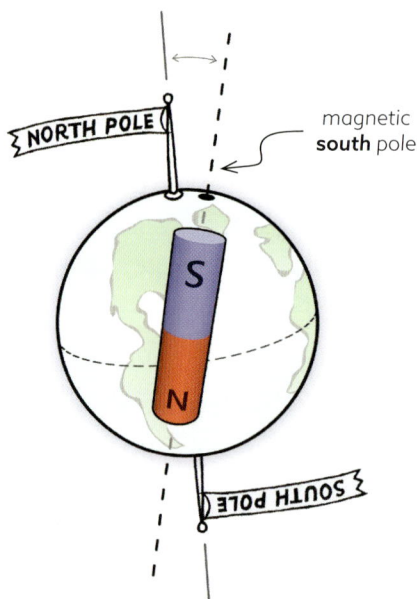

Figure 5.1. The earth's magnetic and rotational poles differ somewhat in position. And, it's the magnetic south pole, not north pole, that lies near the rotational north pole.

A second problem with our intuitive picture arises from the magnetized needle inside the compass. While it does point roughly north, in fact, the needle points to the *south* pole of the earth's magnetic field, not its north pole. It's the magnetic south pole that's located up north (**Fig. 5.1**). This terminology can utterly confuse.

Nevertheless, the diagram in **Figure 5.1** shows the common textbook picture, with the magnetism originating inside the earth. In such a scheme, the magnetic field lines would exit one pole of the earth and enter the other, in a continuous fashion. A diagrammatic representation is shown in **Figure 5.2**.

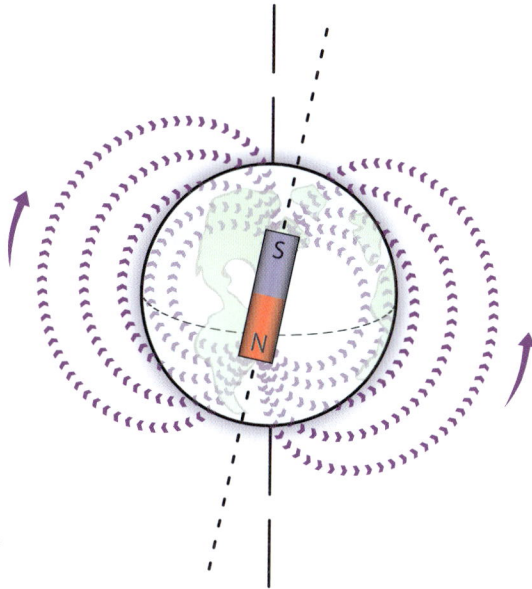

Figure 5.2. Magnetic field lines outside the earth (purple), running from magnetic north to magnetic south.

So now that we're on the same page regarding roughly what the earth's magnetic field looks like, I raise the chapter's central question: Could that field arise, not from some phenomenon inside the earth, but from something outside? Here, I'm referring to the massive flow of charge (current) carried by the prevailing westerly wind (Chapter 4).

You might first want to know what's wrong with the prevailing inside-the-earth hypothesis, and I will address that issue later in this chapter. So, please bear with me.

I wish to deal first with whether the newly proposed hypothesis is even remotely plausible. I believe it might be: After all, the west-to-east positive charge flow constitutes an electrical current, and currents create magnetic fields around them. Physicists have recognized that linkage for centuries. Confounding that simple idea, however, is the presence of the opposite, east-to-west trade wind. But that flow is much weaker (Chapter 4). So, for the moment we focus mainly on the prevailing west-to-east winds and the charges deduced to flow within those winds. The first test of reasonableness is to check whether the direction of charge flow matches the direction of the magnetic field that we read on a compass.

To graphically illustrate this question, we invoke a rule of thumb (no pun intended) that students of physics and electrical engineering call the "right-hand rule." Along a wire, steady current flow generates a circular magnetic field (**Fig. 5.3**). If your right-hand's thumb points in the direction of the current flow, then your grasped fingers point in the direction of the magnetic field lines.

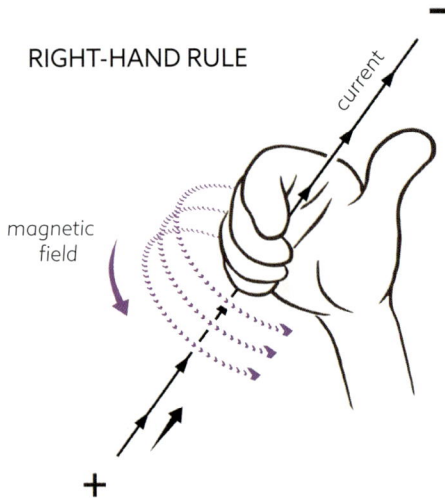

Figure 5.3. Right-hand rule indicates direction of magnetic field around a flowing current.

Now, with that understanding, consider the flow of atmospheric current. The west-to-east charge flow considered in the previous chapter represents a steady current, whose flow must create some kind of enveloping magnetic field. Does that field point in the same direction as the actual one measured around the earth?

Though perhaps I'm delivering the punchline too soon, the answer is yes. The strong upper-atmospheric charge flow around the earth should produce the magnetic field shape shown schematically in **Figure 5.4**. It's a little harder to illustrate than in **Figure 5.3** because the current moves in a circle instead of a straight line, but, as long as you align the thumb of your right hand in the direction of the current, your fingers tell you that closest to the earth's surface, the direction of the field should run northward, *i.e.*, from physical south to physical north — just as observed (**Fig. 5.2**). Hence, we're on track with this hypothesis so far. The field direction is proper.

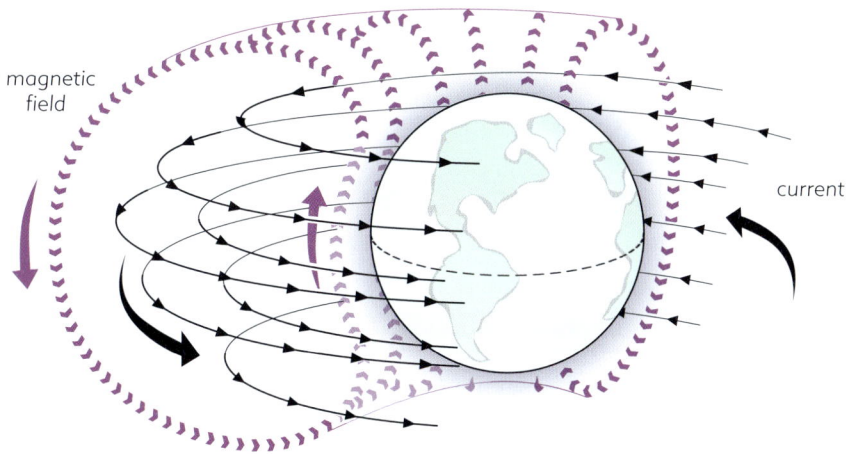

Figure 5.4. Persisting west-to-east flow of positive charge (thin black lines) constitutes a current, which produces a magnetic field (purple arrowheads). Arrowheads merely denote direction, not strength or distribution of magnetic field.

It's worth noting that Jupiter also has a jet stream. And for that planet, a known linkage exists between the jet-stream current and the external magnetic field.[1] So, the linkage proposed here is not without precedent.

Is the Current Strong Enough to Do the Job?

Apart from direction, a second issue is the magnitude of the proposed field. How strong a magnetic field could the atmospheric current generate? Is it sufficient to do the job?

Unlike the illustration of **Figure 5.3**, which implies a modest current flowing along a single line, the jet-stream currents flow over a broad region of the atmosphere. That breadth implies a substantial flow. Consider, for example, only the fraction of atmospheric current responsible for creating the northern lights. That's estimated to be approximately 100,000 amperes [w2] — a whopping blast of charge flow. Presumably, all jet-stream currents integrated over their full breadth must substantially exceed that already colossal value; hence, impressive amounts of current flow.

On the other hand, the complexity of that current distribution makes the magnetic field's spatial distribution difficult to assess. Some rough estimate is nevertheless possible: An infinitely long wire carrying that same 100,000+ amperes would generate a field in excess of 2×10^{-5} Tesla at a radial distance from the wire of 1 km (0.6 miles). That's similar to the value of the magnetic field (between 2.5×10^{-5} and 6.5×10^{-5} Tesla) on the earth's surface. Given the many uncertainties, the computed value does not seem to lie seriously outside the realm of reality.

A third point, beyond strength and direction, concerns the disparity of the earth's magnetic field axis and its rotational axis. Why shouldn't they coincide? By definition, the rotational poles lie along the axis of the earth's rotation; if the earth were a spinning basketball, the poles would correspond to points where you put your fingers to hold the ball. The magnetic poles, according to the proposed scheme, fundamentally differ in origin. They correlate with the distribution of winds (Chapter 4), which could well depend, among other features, on the distribution of waters over the earth's surface, all of which are not so easily predictable. Nevertheless, the fact that the earth's magnetic poles and physical poles don't coincide should not be surprising.

Nor should it be a surprise to learn that the magnetic poles shift over time (**Fig. 5.5**), presently about 60 km (37 miles) per year.[w3] Measurable shifts may occur even throughout the day.[w4] Such shifts could easily arise from variations in sunlight or tides, either one of which would create variations in atmospheric charge flow, and hence pole locations. (More

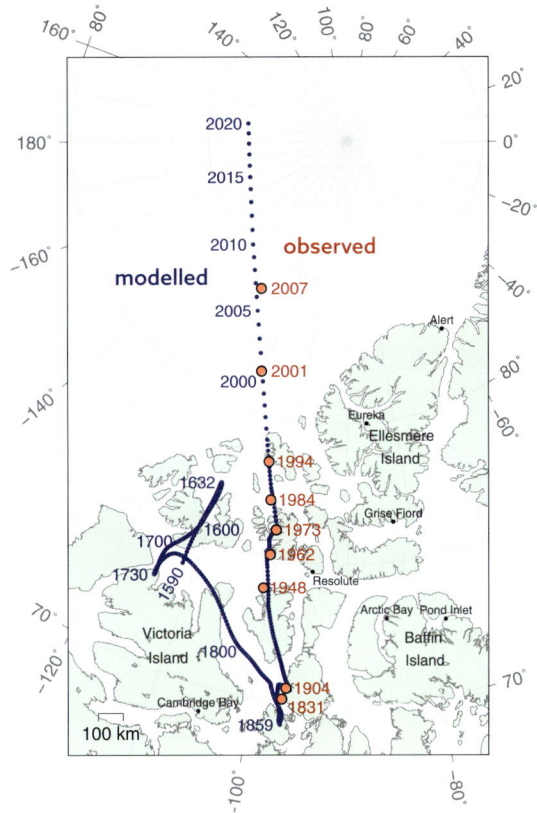

Figure 5.5. Magnetic north pole shifts over time. Observed pole positions taken from Newitt et al.[2]

dramatic shifts in pole locations, including apparent flipping of poles, will be considered in a subsequent chapter.) The observed pole shifts seem compatible with an external origin of the earth's magnetic field, while seemingly less compatible with an origin fixed inside the earth.

If atmospheric currents, indeed, create the earth's magnetic field, then some variation in field strength might be expected between the earth's light and dark sides. Why so? Recall that those currents are proposed to originate at the late afternoon's light-dark boundary (Chapter 4). By the time they progress to the earth's dark side, modest dissipation would seem inevitable. Such dissipation would reduce the local magnetic field strength on the dark side, relative to the light side. Indeed, measurements of magnetic field strength taken at fixed earthly loci support that expectation: They reveal a 24-hour periodicity.[w5]

While only modest, that periodicity seems potentially significant. Cumbersome to explain with a pure, inside-the-earth origin, geoscientists have found it necessary to consider the addition of outside-the-earth mechanisms to help account for that periodicity. Contributions from external currents arising from sunlight-separated ionospheric charges have been invoked.[w5] In other words, contributions similar to those suggested here have seemed *necessary* to account for standard experimental observations. Hence, the proposed outside-the-earth origin of the earth's magnetic field may be less extreme than considered.

Finally, a point of correlation. It's well known that the earth's magnetic field is strongest around the poles and weakest around the equator. Correspondingly weakest around the equator is the prevailing westerly wind (Chapter 4). Thus, the place of lowest atmospheric current flow correlates well with the place of lowest observed magnetic field — a correspondence that reassures. Moreover, a further weakening of that equatorial magnetic field should come from the trade winds, which blow opposite to the prevailing westerly in that same equatorial zone. Although relatively modest in strength, that opposing charge flow should nevertheless weaken the equatorial magnetic field further. So, reasonable spatial correlation exists between the weakest anticipated net atmospheric charge flow and the weakest magnetic field.

Further to the point, it's at the two poles that the magnetic field is strongest. There, massive charges from the solar wind converge on the earth (see Chapter 12). Those charges are intense enough to create visible auroras in those polar regions. Once again, therefore, we note a spatial correspondence between atmospheric current flow and magnetic field strength.

At least qualitatively, then, the proposed origin of the earth's magnetic field seems to fit the general characteristics anticipated from the wind-based origin. And the numbers offer a crude but reasonable match.

The Van Allen Belts

To those accustomed to imagining a magnet inside the earth, it must surely seem radical to envision a magnet arising from the flow of current outside the earth. But let me assure you that the concept is not so radical: A similar principle has been adduced to explain the presence of large bodies of charged particles known as the Van Allen belts.

Discovered in 1958 and named after the space scientist James Van Allen, two giant belts envelop the earth — the so-called Inner and Outer Van Allen belts (**Fig. 5.6**). The inner belt extends from 0.2 to 2 earth radii above the equator, while the outer one lies more distant; it's centered approximately four to five earth radii above the equator and spans a vertical distance of three to ten earth radii. The belts are huge — hard to miss.

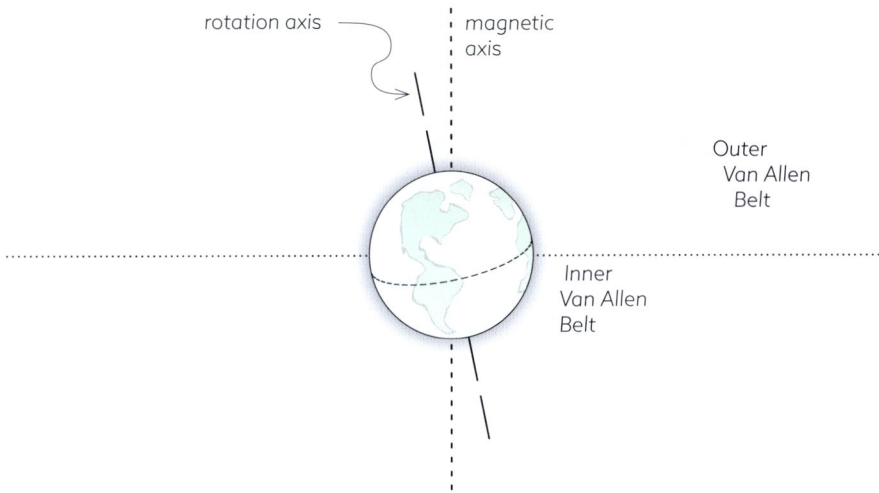

Figure 5.6. Inner and outer Van Allen Belts surround the earth.

Interestingly, the positioning of those belts and the earth's magnetic field correspond both vertically (along radii) and horizontally. In the vertical direction, the earth's magnetic field tapers off to zero at about 65,000 kilometers above the earth. That's about ten earth radii, and hence in the same range as the outer Van Allen belt. The belts also correspond horizontally. Because of their rather precise axial alignment above the equator (**Fig. 5.6**), the belts have been linked by scientists to the similarly aligned magnetic field of the earth. Given those spatial correspondences, the prospect of some linkage between the two phenomena is not beyond reason.

Why Might Such Linkage Matter?

The Van Allen belts are thought to consist of electrons and protons, "trapped" in place by the earth's magnetic field. What's meant by the term "trapped"? And how does trapping relate to our proposal that flowing charges might *create* the earth's magnetic field?

Magnetic fields don't ordinarily trap isolated, charged particles. If those charged particles are flowing, however, then some "trapping" is possible. Charge flow (current) sets up a circular magnetic field around the charges' line of travel (**Fig. 5.3**). If that magnetic field repels a surrounding magnetic field (like the repulsion between two magnetic south poles) then those flowing charged particles can be effectively trapped and prevented from entering the surrounding field. That kind of thinking may be a little challenging to digest, but it is commonly used to explain the Van Allen Belts' existence. Central to that line of thinking is the flow of charge.

Hence, implicit in the conventional understanding of the Van Allen belts' existence is the necessity for charge flow *outside* the earth. To create those belts, scientists argue, charges must flow somewhere above the earth. Those very flows are what we are proposing to underlie the origin of earth's magnetic field.

In sum, the presence of atmospheric charge flows is indicated not only through the discussion presented in the prior chapters, but also by scientists dealing with Van Allen belts. Both sources point to the presence of charge flows high in the atmosphere. If indeed those charge flows exist as argued, then we are obliged to deal with the consequences of their presence. That would include the magnetic fields that must surround those flows. On that basis, I hope the proposed framework will seem a little less radical than otherwise. You might even view the proposed thinking as humdrum conventional.

Magnetic Fields on Other Planets?

Do other of our solar system's planets contain magnetic fields? And if so, does their origin tell us something more about the likely origin of the earth's magnetic field?

Although their strengths vary, all the solar system's planets boast magnetic fields (**Fig. 5.7**). Mercury, Venus, and Mars have weak fields, while the other, more distant planets have fields much stronger than that of the earth. If the earth's magnetic field derives from flows in a molten iron-alloy core, as widely thought, then we might guess that the fields on other planets would have similar origin — no guarantee, but reasonable likelihood. However, those planets have differing elemental

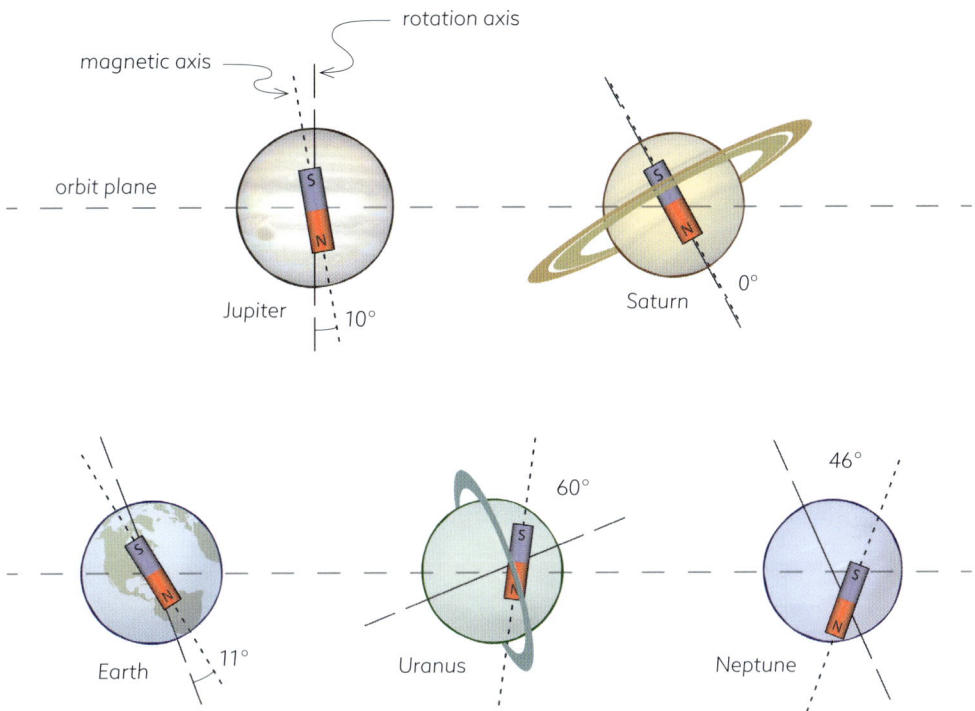

Figure 5.7. Magnetic fields exist on many planets. Bar-magnet representations shown.[w6] Note the off-center location of magnets on some planets. Planet sizes not drawn to scale.

makeups. Some, like Earth, are made of rock and metal, whereas the gas giants consist primarily of hydrogen and helium, and the ice giants comprise other volatiles. That such diversely constituted planets would all somehow develop internal magnets would be quite a coincidence.

On the other hand, if the magnetic fields of those diverse planets come from external currents, then a common origin may be more plausible. Those fields could all come from a generic source: ion-containing winds, as just argued for the earth.

Further to this point, a wind origin predicts reasonable correlation between wind velocity and magnetic field strength. That appears to hold true, at least roughly. The outer planets, beyond Mars, contain the fiercest winds[w7]; and those same planets contain the strongest magnetic fields.[w8] Of course, correlation proves nothing; but it does lend credence to the notion that all of these magnetic fields, including that of the earth, could arise from a common mechanism: charges flowing around the respective planets.

Clearly, much more needs to be done to test this chapter's central theme. With a distinct hypothesis specified, experiments could be sharply focused on testing the merits and demerits of the hypothesis. I don't claim that what's proposed here is the ultimate answer, but only that an externally generated magnetic field may be a theme worth exploring. If it turns out to be well supported, then the concept could carry implications beyond what we may commonly think.

Summary

The deduced west-to-east flow of charged air in the upper atmosphere constitutes an electric current. Currents are known to produce magnetic fields around them. Plausibly, much or all the earth's magnetic field could arise from that robust west-to-east current flow. Known data provide at least a rough fit.

If so, then postulating the existence of magnets lying within the bowels of the earth may be unnecessary. The magnetic field would arise elsewhere.

An atmospheric current-based origin of the earth's magnetic field would carry implication: It would mean that like its electric field, the earth's magnetic field might ultimately arise from the energy of the sun. Solar energy arguably drives the evaporative events that create the west-to-east winds, which in turn create the magnetic field. Hence, a simple paradigm: one kind of energy transitions into another.

We now come to our final question about the earth's most prominent physical features, one that seems almost too obvious to ask. What makes the earth go 'round? What energy bears responsibility for turning the Earth?

CHAPTER 6

What Makes the World Go Round?

Day fades into night, and night into day. Periods of darkness alternate with periods of light. This repeating pattern, as we all know, arises from the earth's daily spin as it slowly circles the sun. But how much do we really *know* about the earth's rotation? Do we know its cause? And, is the spin so reliable that we can count on it continuing into eternity?

Most spinning objects slow down and eventually grind to a halt. A twist of the fingers may set a top in motion, but that energy gets sapped, not only by pivot friction, but also by the air dragging on the top's moving surface. Such sapping of energy slows the top's spin (**Fig. 6.1**). The Ferris wheel is another example. Those grand structures may seem to rotate endlessly but pulling the motor's plug immediately removes the driving force and the wheel soon stands idle. For continuous rotation, a steady input of energy would seem necessary.

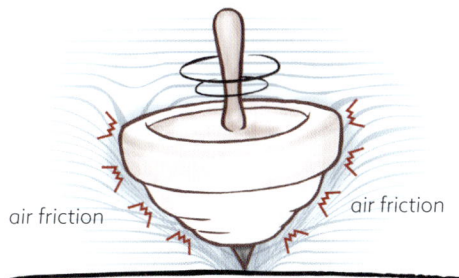

Figure 6.1. A rotating top rubs on the surrounding air, creating friction and slowing the spin.

air friction   air friction

Why does the earth *appear* to behave differently? Why does it keep rotating, seemingly forever, and without obvious energy input? Low friction surely facilitates; in the supposed vacuum of space, there's not much air resistance to slow things down. Inertia helps as well, for the earth's colossal mass ought to keep it spinning for a very long time.

But, eventually, everything runs down and the earth shouldn't be any different — unless we were to identify an energy source that continuously drives the earth's rotation, akin to the motor on the Ferris wheel.

The seemingly endless rotation of the earth constitutes the theme of this chapter. We will consider the role of the dominant wind. Could that wind potentially help spin the earth in the same way that a swipe of your hand helps spin a globe? If that wind gets driven by solar energy (Chapter 4), then the energy for spinning the earth would come from the sun and so long as our sun continues to shine, the earth should continue to rotate.

I do appreciate that the idea of a wind-driven turning of the earth may seem radical. Perhaps it is. But let's begin by looking at the prevailing view — continued spinning without continuous energy input. How realistic is that stance?

What Drives the Earth's Rotation? The Conventional View

Physicists suggest that the rotation of the earth may have arisen as a natural consequence of its formation. Early in its history, the earth is thought to have acquired the angular momentum needed to set it spinning on its own axis. If so, then the obvious question arises: Why does it *keep* spinning? Billions of years have elapsed since the earth formed. That's not thousands, or even millions, but *billions* of years — comprising trillions of rotations.

Most processes wind down because of inevitable energy losses. In the case of the earth's spin, those losses could stem from friction of various types: air layers shearing over one another; tidal waters flowing past one another; *etc.* Those losses continually diminish the system's energy. If no identifiable source replenishes that energy, then the rotational speed should diminish progressively, with the system eventually running out of steam. The earth should stop rotating.

Obviously, it hasn't. However, the earth's rotational speed has indeed diminished: Over the last century it has slowed by about 1.7 thousandths of a second per day — or something less than one percent since the time of the Middle Ages. That kind of rundown could fit the conventional hypothesis of initially acquired angular momentum, at least qualitatively. Imagine, however, that the same slowdown rate had prevailed over the longer haul, say, since the time of Earth's birth. To accommodate such continuing slowdown, the original rotational period must have been a lot shorter than we experience today. Early on, simple arithmetic shows, one full day would have persisted for less than a second. That's one earthly rotation in the blink of an eye. Hold your hat!

Let's take that estimate a bit further. A simple calculation shows that a one-second rotation would create so much centrifugal force that the earth might fly apart. Consider the numbers. The critical angular velocity of the earth above which it would break up can be obtained by setting the centrifugal force equal to the gravitational force holding everything together. The centrifugal force is given by $m\omega^2 r$, where m is the earth's mass, ω is its angular velocity, and r is the earth's radius. The gravitational force is given by mg, where g is the gravitational constant. Thus, $m\omega^2 r = mg$. Solving for the angular velocity gives, $\omega = (g/r)^{0.5}$, equivalent to the period, $P = 2\pi (r/g)^{0.5}$.

For the earth's surface, this formulation gives a critical period, $P_c = 5{,}070$ seconds. Shorter periods would cause the earth to fly apart. This number is obviously much greater than the early-on one-second estimate; so, if the same slowdown rate were carried into the distant past, the earth should potentially fail to exist. It would have torn apart.

While such numbers may provide speculative amusement, a potentially more substantive argument against the acquired angular-momentum hypothesis stems from measured fluctuations in the earth's spin rate

(**Fig. 6.2**). If we posit that the earth got an initial kick, and keeps spinning because of its massive inertia, then we expect a progressive, steady slowdown. How much slowdown is uncertain, but we should certainly not expect to see periods of acceleration.

Nevertheless, the measured rotational speed does fluctuate. As **Figure 6.2** shows, rotational speed increases and decreases on a seasonal, monthly, even daily basis.[1, w1] If massive inertia supposedly keeps the earth rotating, with perhaps modest slowdown, then what could explain periods of speedup?

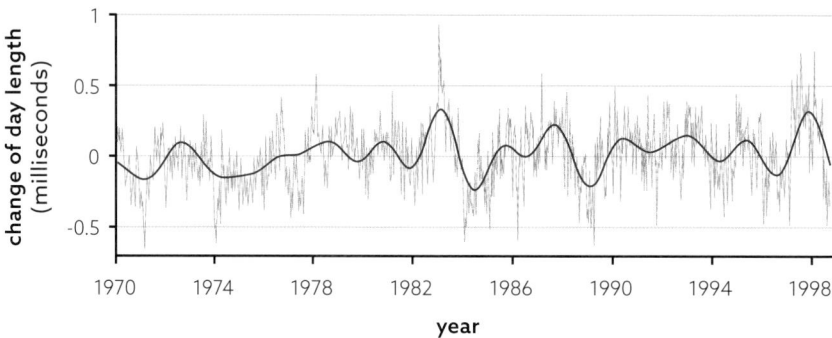

Figure 6.2. Length of day (inverse of earth-rotation speed) as a function of time.[2] Accelerations and decelerations are evident.

Well, I don't wish to advance this argument too far, for a possible rescue exists within the bounds of convention. Physicists think in terms of conservation of angular momentum. Envision the twirling ice skater, speeding up by pulling her arms inward. If something in the earthly system were to move similarly inward, toward the earth, then earthly rotation could speed up. That could happen within the atmosphere; or, within the earth itself, as, for example, from large earthquakes that thrust mass downward. Any of the above phenomena could cause the accelerations, just like the twirling skater. Physicists debate these options, some of which could plausibly hold validity. How they could explain the observed time course, with accelerations extending over many months (**Fig. 6.2**) seems less clear.

A Wind-Based Hypothesis for Earthly Rotation

A distinctly different hypothesis posits that the earth keeps spinning because of a continuous feed of energy. I'm suggesting that the energy in question comes from the sun. I'll explain how in a moment.

If the sun's energy drives the earth's spin, then the observed fluctuations in the earth's rotational speed could be explainable in terms of fluctuations of incident solar energy. Solar energy does indeed fluctuate: The reported time scales range from days to years.[3] The size of those energy fluctuations, roughly 0.1% of mean solar energy, substantially exceeds the variations of mean earth-rotational speed shown in **Figure 6.2**. Such difference seems explainable: The earth has immense inertia, which acts to preserve its rotational speed. Therefore, any variation in the energetic drive should exceed any earthly speed variation. On that basis at least, the proposed linkage between solar energy and earthly rotation seems potentially viable.

But how could the sun drive the earth's spin?

Here's a train of logic that follows from the material offered in previous chapters: The sun's energy drives the separation of charge between earth and atmosphere. Atmospheric charge ultimately drives the wind, which in turn flows past the earth's surface. The robust, prevailing westerly wind flows broadly over the earth (Chapter 4); as it flows past high mountain peaks and even past the lower atmospheric layers that stick to the earth, it exerts a shear force, which arguably helps maintain the earth's spin. The wind keeps the earth turning.

To imagine how that could happen, envision one of those old-fashioned globes suspended between the two ends of a curved bracket (**Fig. 6.3**). With well-lubricated pivots, even a feather could induce the globe to spin.

Figure 6.3. With limited friction, even a small force can drive rotation.

Initiating that rotation, therefore, requires only a modest force, with still less force needed for maintenance.

Similar reasoning could well apply to the wind-based shear force on the earth. The wind's driving force might seem modest at first, because, relative to the earth's surface, the prevailing westerly wind velocity amounts to only 100 to 200 km/h (62 to 124 mph). While that's hurricane force, it's perhaps modest enough to raise the question of adequacy. However, that velocity is not the relevant one: It's merely the differential speed between wind and earth, as between feather and globe. Think instead of the wind speed reckoned relative to some point in space. That amounts to *more than twice the speed of a jet plane* (Chapter 10). That hefty wind is what's proposed to drive the earth — which can't keep up, so it falls behind by 100 to 200 km/h.

To create perspective on the wind proposal, imagine a fleet of supersonic jet planes, similarly directed. Envision a huge fleet of them hitting the earth's mountain peaks simultaneously. Conceivably, the impact could kick start some trivial earthly rotational movement. Now think of additional fleets, endlessly battering those mountains with sacrificial hits. With ceaseless hits and limited resistance from air friction, those impacts could arguably drive, or at least contribute to, the earth's persisting rotation (**Fig. 6.4**).

Figure 6.4. Could the prevailing wind drive the earth's rotation?

I'm suggesting here that simple drag could help turn the earth. The wind, shearing past lower layers of air, and particularly past geological protuberances such as mountains, would drag the earth to move along with it. Some of that driving wind would inevitably slip by; so, the earth would not keep pace with the wind. That would explain why the earth's surface velocity remains some 100 to 200 km/h lower than the apparent wind velocity.

You might have noted a couple of hitches in this argument: First, the prevailing westerly wind flows higher than most mountains. You'd think the wind might miss those critical peaks. However, wind doesn't abruptly diminish from full blast to zero in contiguous regions; it tapers gradually. Areas of highest wind velocity drag adjacent air molecules, which drag others, *etc.*; so, the downward taper is gradual. Therefore, although generally high in the atmosphere, the prevailing westerly should nevertheless exert considerable influence on the earth's protuberances.

In fact, studies have shown that the obstacles created by mountain protuberances reduce the local wind velocity by up to 80%.[4] And, surprisingly, the Mongolian Plateau has an impact that is approximately four times greater than that of the larger and taller Tibetan Plateau and the Himalayas to the south.[5] Thus, evidence exists that the jet stream does exert drag on earthly protuberances.

A second potential hitch: the trade winds. Those winds blow in the opposite direction, from east to west. That should oppose the prevailing westerlies. If the westerlies drive the earth's spin, then shouldn't those trade winds act to inhibit that spin? A strong impact might be anticipated *a priori* because the trade winds lie close to the earth's surface, possibly influencing the earth more profoundly than the westerly winds higher up. Doesn't this compromise the hypothesis?

Not necessarily. The trade winds are largely confined to the relatively narrow band of near-equatorial latitudes. They don't cover the entirety of the earth's surface. And further, the region they do cover is mostly water, whose flatness should limit the wind's grip. So, the trade winds' impact should be fairly limited.

More importantly, the trade-wind velocity, typically 15 to 20 km/h (9 to 12 mph), is merely a tenth of the velocity of the higher-altitude prevailing westerlies.[w2] Since drag force depends on the square of the velocity, a velocity difference of ten times computes to a force difference of 100 times. Hence, the trade-wind force should be inconsequential — a gentle breeze *vs.* a hammering blast. The powerful westerlies should dominate, tending to spin the earth from west to east.

If the proposed paradigm bears some measure of truth, then any long-term wind-speed variations ought to induce comparable variations in the earth's rotational speed. Stratospheric winds are known to vary with a biennial period. And guess what? The periods of higher wind velocity do correlate with periods of higher earth-spin velocity.[6] Furthermore, a periodic enhancement of sub-tropical jet-stream winds, called the Madden-Julian cycle, also correlates with enhancement of the earth's spin velocity.[w3] While these correlations have seemed perplexing within the context of the contemporary view, such temporal correlations should follow directly if high-altitude winds spin the earth.

In essence, what I propose is this: *The energy of the sun could drive, or at least help drive, the earth's rotation.* Fundamental to this hypothesis is the fact that the sun's energy separates charge in the water beneath. Negative charges mostly cling to interfaces beneath the earth's surface, while positive charges rise evaporatively high into the atmosphere. The rise is not uniform over the earth's surface: Day-night boundaries create lateral atmospheric charge gradients (Chapter 4). Those gradients propel the prevailing westerly wind. That wind creates a horizontally directed force, which drags the earth endlessly round and round.

All of this gets driven by the sun. The sun's energy may fluctuate from time to time, and so therefore, may the earth's spin velocity. Nevertheless, our planet's vast inertia ensures that rotation remains relatively steady: We can confidently set our watches and predict the timing of sunrise and sunset. If the sun were to cease shining, then the earth could well grind to a halt.

Similar Behavior Throughout Our Universe?

If the positive charges needed to drive the wind come from evaporated water, then this raises the question about other celestial bodies. What about planets without water? What makes them rotate?

In fact, water is hardly rare; ongoing research shows that water keeps turning up on one planet after another, including planets outside our solar system. Even in deep space, water and ice are not uncommon according to radio frequency analysis and spectrographic data.[w4] On the other hand, the required charge gradients need not come from atmospheric water. Anything the sun or a star can volatilize into a charged vapor could help spin a planet. Hydrogen, methane, and ammonia exist on the various planets of our solar system, any of which could theoretically suffice for such a role.

Conversely, the absence of *any* atmospheric charge gradient should imply the absence of any rotation. The moon stands as an example. We see the same lunar face all the time because the moon does not rotate about its own axis. Nobody would expect a body so desolate to sustain very much "atmospheric" charge, and indeed, estimates suggest an atmosphere some million times as tenuous as that of the earth.[7] Almost none. So, the absence of lunar rotation could fit nicely with the charge-gradient-driven mechanism.

The planet Mercury offers another data point. That planet rotates extremely slowly. At the same time, the scant data currently available show little or no wind on that planet.

By contrast, vigorous winds and correspondingly high rotation speeds are documented on all other planets from Venus all the way out to Neptune. Neptune's ferocious winds, for example, can exceed 2,000 kilometers per hour[w5] (~1250 mph). Correspondingly, its rotational velocity is substantial: With a diameter approximately four times that of the earth, and a rotational period about two-thirds of an earth day, surface velocity (reckoned on its equator) computes to about six times that of Earth. With such high rotational velocity and high wind velocity, Neptune's behavior fits nicely within the paradigm.

On the other hand, not everything is certain. At the extreme distance of Neptune from the center of our solar system, the sun's energy must be limited in supply; the question arises whether solar energy could suffice for driving Neptune's wind. Energetic responsibility could perhaps fall to some other agent such as the solar wind, a prominent feature that extends beyond even Pluto (Chapter 12). A further uncertainty: In the gas giants, ranging from Jupiter all the way to Neptune, charged gas rather than hydronium ions may densely fill the atmosphere and hence bear responsibility for driving the wind. Thus, any precise correlation between planetary wind speed and rotational velocity may be limited by unknown factors. Nevertheless, the correlation so far seems reasonable.

That doesn't necessarily mean the proffered hypothesis is adequate. Adequacy surely rests on whether additional supportive evidence can be obtained, both for the various planets and especially within the confines of planet Earth.

On the other hand, the underlying energy framework would seem to make sense. The steady, continuous infusion of energy seems to me a more likely paradigm than any that poses an initial kick start with no further energy supply. After all, the rotation of earth has likely persisted for as long as billions of years. Given the anticipated losses, the challenge of keeping the earth spinning indefinitely ought to require a driving energy of some kind — just as it does for the Ferris wheel.

For the required energy source, I nominate the sun.

Section Summary

Water exists practically everywhere on our planet. It lies in oceans, lakes, rivers, trees, and even beneath seemingly dry land. That aqueous medium absorbs the energy of the sun, arguably creating a rechargeable battery whose negative terminal lies within the earth's waters, while its positive terminal fills the atmosphere above.

That battery, with a voltage of up to half a million volts, can hardly be insignificant. It's got plenty of juice. More surprising would be if that battery bore no consequence at all; and yet, with the exception of the modest atmospheric currents that flow as a consequence, most scientists presume that that huge electrical potential plays little role in earthly matters.

In the last several chapters I have attempted to show how features of this solar battery could underlie various physical features of earth dynamics: its prevailing winds; its magnetic field; and perhaps even its spin. Conceivably, these phenomena could all stem from differences of atmospheric charge between the earth's bright and dark sides, driven by the sun's energy.

While this paradigm may seem radical — perhaps too much to swallow in a single gulp — I have earnestly tried to identify shortcomings and have modified the relevant chapters in accordance with the flaws I could find. Perhaps some of the phenomena in question can be explained by alternative mechanisms. If so, I'm not yet able to identify them.

The next section continues along this same course. We will consider additional consequences that might follow from the sun's energy and the earth's charge, including the earth's weather and its revolution around the sun. I hope you'll brace for more provocation.

SECTION III

Weather

To carry an umbrella, or not to carry an umbrella, that is the question. If anything is unpredictable, surely it's the weather. Forecasters may warn us of an imminent deluge, but sometimes with only 50% probability.

A possible reason for this uncertainty is that the determinants of weather remain poorly understood. A prominent atmospheric scientist recently whispered to me quite an embarrassing secret. Years of research notwithstanding, experts still don't understand at least two important weather-related phenomena: how water evaporates; and how clouds form. It's a bit like trying to predict how a car will run without comprehending the roles of the gas pedal and the brake.

Colleagues who have read early drafts of this book have suggested that I start this section by presenting the widely accepted weather paradigm, and only then begin to explain where I may harbor doubts. That strategy makes good sense. However, finding a coherent view of weather that builds from fundamental principles (a route that Occam might prefer) has been a challenge. I've certainly tried.

Instead, let me commend you to do what I could not manage to do: Immerse yourself in any relevant website or textbook. Learn about weather. See if you can track the buildup of a meaningful weather paradigm, one that not only explains ordinary weather events from fundamental principles, but also answers simple questions you might have — questions like why clouds float, what determines if those floating clouds will produce rain, why gentle-looking clouds can turn ominous, *etc*. If the textbook picture satisfyingly answers your questions, then, reading through this section of chapters may be superfluous.

This section attempts to build a fresh understanding of weather. It combines simple physical principles, logic, and experimental findings to build a coherent paradigm. As you might have come to expect, I will argue for a weather paradigm with a dominant role played by electrical charge. Of course, I will consider widely held views along the way, by either integrating them into my development or providing critiques where they don't offer rigorous understanding. My goal is to develop an accessible understanding of why it might rain, and why a gentle rainfall can sometimes evolve into terrifying typhoons.

That weather goes hand-in-hand with climate goes without saying. Excessive heat, ravaging wildfires, and spreading desertification are seen in some regions, while increasingly devastating hurricanes and extreme flooding are seen in others. Why so? While this section does not deal extensively with the issue of climate and not at all with climate change, any serious consideration of that change demands an understanding of weather at its most fundamental level and that is what the following chapters are designed to achieve. What determines the weather?

CHAPTER 7

Cloud Origins: Packing Moisture into the Atmosphere

Winds blow, the earth rotates, and clouds float. While that's obvious, questions immediately arise. Among my favorites: *Why* do those clouds float? What keeps them suspended up there in the sky?

Maybe you envision the cloud as a billowy, cotton-like substance with a natural proclivity to float, like a dandelion fluff? But that image can mislead. By offering a flimsy measure of understanding, it may provisionally satisfy, but it fails to deal with the underlying physics: Namely, what force keeps the cloud afloat?

A more relevant comparison of a cloud than to dandelion fluff is perhaps to some volume of water. Envision a torrent of rainfall deposited 5 cm (2 inches) high onto city streets with clogged drains. How much would that citywide slice of rainwater weigh? Can you guess? Now think of that same weight of water lodged in the giant cloud from which it came. What force could keep that weighty mass suspended up there, high in the sky?

Clouds consist mainly of submillimeter-sized droplets. Droplets are denser than air. As such, they should descend steadily toward the earth from the pull of gravity. Sometimes they do descend, in the form of precipitation; rain does fall. But during other times, those droplets remain lodged within the parent cloud, suspended indefinitely in the sky. How could they defy the force of gravity?

My atmospheric science colleagues try to assure me that cloud droplets never really remain suspended; they always descend steadily — but their small size retards the descent rate to the point of imperceptibility. So, clouds don't really float, according to the prevailing explanation; they merely *appear* to float.

Wrapping my head around that concept has proved daunting. The small-size-slow-descent argument might suffice for a droplet in isolation. But the droplets in question don't lie in isolation — they cluster together to create clouds. Cluster size may equate to a football stadium, or even a small town. What prevents so heavy an aggregate from plummeting onto your head?

Figure 7.1. Puffy white cloud in an otherwise cloudless sky — a familiar sight, wanting for explanation.

And, if that's not puzzling enough, then there's the question of why clouds form at all. What drives atmospheric moisture to "condense" into the structures that we know as clouds? Perhaps that is happening spontaneously when the moisture content is sufficiently high. Yet, if that's possible, then how does one explain the emergence of a single, isolated cloud in an otherwise clear blue sky (**Fig. 7.1**)? Does moisture hang out only in spots?

This chapter begins to address those enigmatic issues by describing several surprising discoveries from our laboratory. The discoveries in question will be critical for building a paradigmatic understanding of weather.

The immediate goal is to answer the following questions:

- How does moisture evaporate?

- What form does atmospheric moisture take?

- How does that moisture "condense" into clouds?

- And, what keeps those clouds suspended in the air, against the pull of gravity?

As you might have guessed, I'll be suggesting electrical charge as a central protagonist in this drama.

That suggestion is hardly radical, for cloud electricity was long-ago identified as an integral feature of atmospheric science.[1] That view took a back seat, however, as the atmospheric science field shifted its emphasis from cloud physics to computerized weather prediction. Electrical charges pretty much vanished from the scientific scene. Nevertheless, the charges themselves have *not* vanished. They are present everywhere in the atmosphere and if we are to seek full understanding of weather, then their consequent role cannot be ignored.

That is why I will be looking toward electrical charge as a potentially central player in the drama of weather. The charges are ever-present. We need only find out what they might or might not do.

Because the full picture of cloud formation comprises multiple issues, I have broken the chapter into sections. Pausing after each section is a strategy worth considering.

THE RAINBOW'S MESSAGE

Rainbows' Treasure: Pots of Golden Insights

To begin our search for understanding, we consider something both ephemeral and mysterious: the rainbow. Who doesn't fancy a rainbow (**Fig. 7.2**)?

Surely this spectral phenomenon has captured your imagination. Perhaps the prospect of finding a glistening pot of gold lying at the rainbow's terminus had once struck that sense of wonder in your heart. What I wish to convey here is something more mundane, but centrally relevant to the material of this chapter. It deals with the scattering of light.

Figure 7.2. Rainbow, as seen from a Seattle hilltop.

For a rainbow to appear, the sun will ordinarily lie behind you, with some kind of moisture situated in the air ahead. When that moisture properly disperses the incident sunlight, you may see a rainbow. This description might seem simple enough, but an important factor makes it a lot less trivial than it appears. That is, the ability to scatter light depends on size.

Consider plain old air. Too small are the molecules of that gas, even when clustered, to scatter very much light. That's why light from the sun fails to bounce around before reaching our eyes — it passes directly to us. So, the sun appears undistorted. By contrast, a cloud of dust particles suspended in that beam of sunlight is something we *can* detect. We detect those particles because they are large enough to scatter appreciable amounts of light; and, it's that deflected light, eventually reaching our eyes, that tells us that something must be present.

A particle starts to become visible roughly when its size just exceeds the wavelength of the illuminating light. Visible light has a wavelength of approximately half a micrometer, so micrometer-sized particles will be marginally detectable. Particles ten times larger can be easily perceived. And, the human eye readily discerns a 100-micrometer grain of sand, about the diameter of a human hair, as it deflects light. Size matters. From being barely detectable to being easily identified depends on how much light the object scatters, and that, in turn, depends on its size.

How does this technical detail relate to rainbows?

Lay observers may think that rainbows are the result of sunlight getting scattered by individual molecules of airborne water. But the arguments presented just above imply that this can't be true. A single water molecule is miniscule. Like an air molecule, the water molecule spans approximately one ten-millionth of a millimeter, roughly 10,000 times smaller than the wavelength of visible light. Thus, the lone, suspended water molecule cannot scatter very much visible light. Whatever scatters the light to create the rainbow must be of supramolecular size — measurable in micrometers — and, because rainbows are generally intense and easy to see, probably closer to hundreds of micrometers. For those unfamiliar with metric terminology, we're talking about a size roughly equal to that of a tiny water droplet.

Hence, the very existence of the rainbow implies that *the atmosphere must have the capacity to contain sizable droplets, each one comprising vast numbers of water molecules.* This understanding is nothing new; I should rightly point out that it had been known since the time of René Descartes, the 17th century philosopher. Please note: A one-micrometer droplet contains *billions* of water molecules, not just a few. Creating rainbows requires the presence of those bountiful droplets.

I have found it convenient[2] to refer to those droplets under the more generic rubric: "vesicles." This term avoids any confusion that might arise in distinguishing droplets from bubbles. The two may look alike, but they differ in an important way: droplets contain liquid; bubbles contain vapor. Nonetheless, one species is easily mistaken for the other.[2] The term "vesicle" averts confusion by encompassing both.

The main point here is neither about this proposed terminology, nor about those soul-stirring rainbows, but more about the underlying technical issue: light scattering depends on size. Large particles, including "vesicles," scatter lots of light. Smaller particles fail to do so, and therefore remain undetected.

So, when it comes to visibility, size does matter. But why should you care about vesicle size and light scattering?

Vesicles and Humidity

By considering that at least some atmospheric moisture could come in the form of vesicles, some everyday phenomena come to clearer understanding. One such phenomenon is the haziness that accompanies intense summer humidity. If you've ever spent much time in New York or Kyoto during humid summers, then you'll know what I mean. You've probably experienced the visual fuzziness that commonly accompanies high summer humidity; the loss of contrast keeps you from seeing distant objects as clearly as you might like.

You're probably thinking of summer pollution. Inevitably, that's got to be part of the haziness story. Nevertheless, a major part is humidity itself: Almost a century ago, when experimental weather stations could be situated remotely from any local pollution, atmospheric scientists concerned themselves with the role of humidity in light scattering. Those two phenomena turned out to be well correlated.[3] Later, with the advent of more modern instruments, scientists could better separate any residual pollution-based scattering from scattering caused by humidity, and the earlier conclusions were amply confirmed. Increases of humidity alone brought sizeable increases of light scattering. [4]

This evidence tells us that *whatever brings humidity to the atmosphere may well scatter light and thereby create visual fuzziness.* If summer humidity arises from water-containing vesicles, then those vesicles will also bring

Why Is the Sky Blue?

Blue sky, red sunset. Why so? How do these features relate to the scattering of light?

Visible light consists of a series of wavelengths ranging from violet / blue at the short end of the spectrum, all the way to orange and red at the long end. Prisms render those wavelengths visible (see prism figure below). To simplify the foregoing discussion, let's just consider the two near extremes of wavelength — blue at the short end and red at the long end.

A critical feature is that light may get scattered along its path of travel. It may deflect multiple times, heading into different directions each time, some of it never reaching our eyes. This scattering process is formally described by the "Rayleigh formulation." This formulation was named after Lord Raleigh, an English mathematician and recipient of the 1904 Nobel Prize in physics for his discovery of the noble gas, argon, and other properties of gases. He is credited with equations describing light scattering that describe why the sky is blue.

The Rayleigh formulation specifies the magnitude of light scattering as being inversely proportional to the fourth power of wavelength. That means blue (shorter wavelength) will get scattered much more than red (longer wavelength). Blue light likes to disperse.

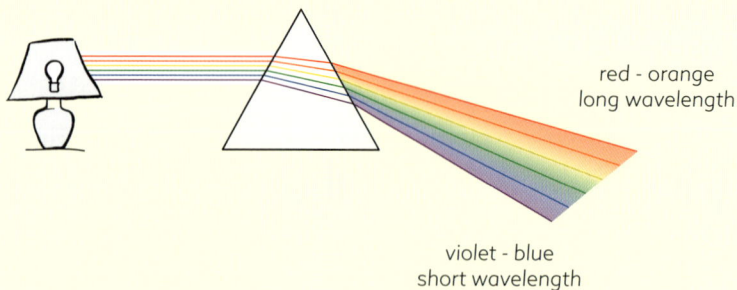

red - orange
long wavelength

violet - blue
short wavelength

summer haze. Or, put another way, the presence of summer haze is a sure sign of sizeable atmospheric light scatterers, which could easily be water-containing vesicles.

Such vesicles are not merely theoretical abstractions. In a moment I will describe direct evidence for vesicles evaporating into the atmosphere.

So, as the figure below shows, red light may pass from source to observer relatively unscathed; it hardly scatters (*panel a*). Blue, on the other hand, gets appreciably scattered along the way, reaching the observer with diminished intensity (*panel b*). When the source contains both red and blue, the observer, therefore, detects more red than blue, registering a violet interpretation (*panel c*). Over longer distances between source and observer, the augmented blue scattering will leave mainly the red color reaching the observer (*panel d*). In terms of perceived color, then, distance matters (*panel c* vs. *d*).

What does this understanding imply?

continues next page

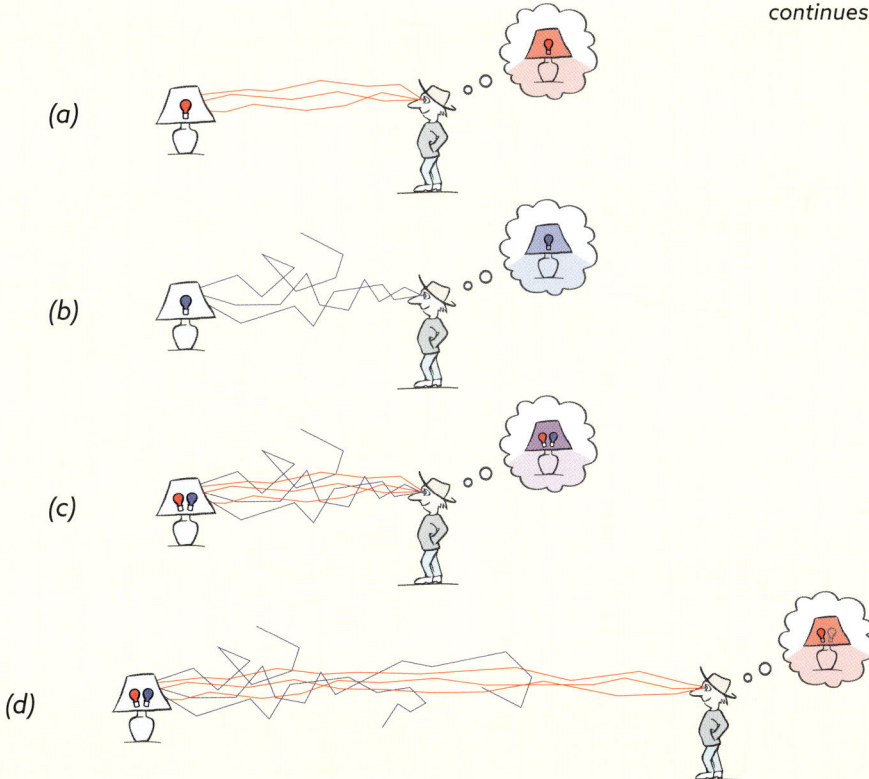

(a)

(b)

(c)

(d)

(a)

atmosphere

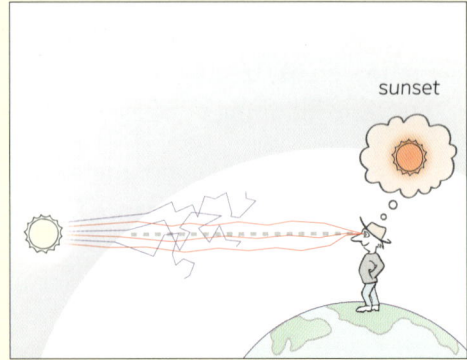

(b)

sunset

(c)

blue sky

(d)

When you look directly to the sun above, the solar reds come through to you directly with relatively little atmospheric scatter, while the blues do scatter (*panel a, above*). So, you perceive the sun as its "true" color minus some blue; it comes across as yellowish white.

At sunset, that loss of blue light grows more significant because of geometric factors: The sun is low on the horizon, so its light passes obliquely through the atmosphere. Because of that obliqueness, the path to reach your eyes is relatively longer. With that longer path, more blue light gets scattered, leaving the residual with red dominance. So, at twilight, the sun looks reddish (*panel b*).

Why then is the sky blue? When looking upward at a narrow region of the daytime sky, angled away from the direction of the sun, the only light you receive is the sunlight that's been scattered and redirected toward your eyes. That would include the amply scattered blue, but not much red. Since you're gazing in some skyward direction, you interpret that blue light as coming from said region of the sky. So, the sky "looks" blue (*panel c*).

Finally, there are the sunset clouds, which share the setting sun's reddish color (*panel d*). They do so because the path length from sun, to cloud, to observer is pretty much the same as the path directly from the sun to observer. The blues get similarly scattered away, leaving red as the remaining light reaching your eyes. So, sunset clouds look red.

EVAPORATIVE SURPRISES

Caught in the Act of Evaporation

We referred earlier to water-containing vesicles, which scatter light. Are those vesicles real? And if so, can you imagine how they might get into the atmosphere?

In our laboratory, we found that such vesicles rise directly from evaporating water. When I first saw the evidence, described in the sections below, I almost fell out of my chair. My understanding was that evaporation occurred one *molecule* at a time. That's what textbooks tell us. And, in fact, in Chapters 2 and 4 of this book, I argued that single, positively charged molecules — hydronium ions — do rise directly from the water. There's no reason to challenge that conclusion. But who would have thought… rising vesicles as well?

It began with a curious undergraduate student working in my laboratory and independent enough to tackle a problem without direction (interference) from me. When Denny Luan began his experiments, which focused on warm water, he presumed that evaporation occurred one molecule at a time. His results quickly confuted that notion. Remember those old cartoons showing wispy vertical lines above warmed food or warmed water to denote rising moisture? That's what Denny photographed — but in horizontal cross-sections.

Those cross-sections showed mosaic-like patterns, as in **Figure 7.3**. The white boundary lines contained all the evaporating water, while the dark regions were empty. The images looked indeed like mosaics — patterns built of multiple, independent, ring-like pieces, stuck together to vaguely resemble a pretzel.

We found later that each ring was built from numerous tiny vesicles of water, closely packed to give an almost-continuous appearance. The rings scatter light, conferring visibility on the rising vapor. The dark empty spaces lying within the curved boundaries contain nothing to scatter light; hence, they appear dark.

While those patterns surely seem unexpected, my guess is that you've already seen them from a different perspective, perhaps while sipping coffee at Starbucks. I'm referring to the familiar cloud rising from a cup of hot coffee, an example of which appears later in the chapter (**Fig 7.8**). The cross-sectional images shown in **Figure 7.3** cut through those clouds, revealing detail that neither you nor I might have anticipated.

That detail will be critical for understanding the nature of evaporation.

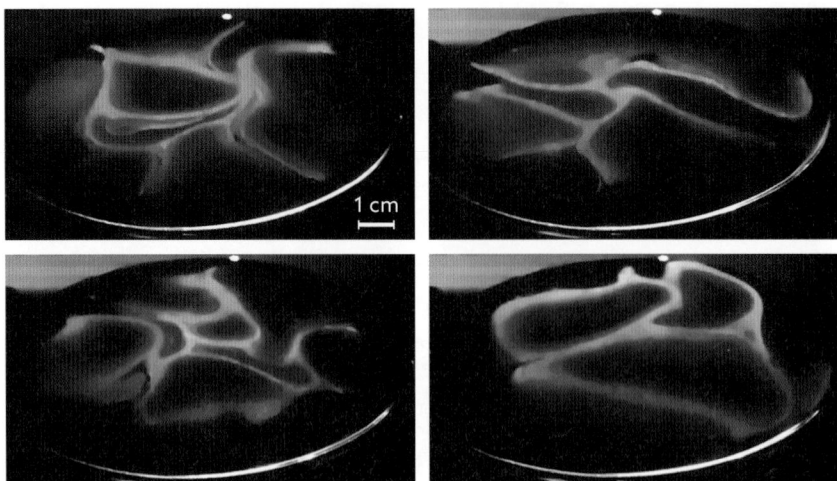

Figure 7.3. Horizontal cross-sections of vapor rising from warmed water. The white rings represent high concentrations of vesicles, which form the visible vapor.

Genies Rising From the Water?

It took no great feat of intellect to appreciate that those mosaic patterns could not have materialized out of nowhere. They likely came from similar shapes in the warm water beneath, which would then spawn the rising vapor. Once we knew what to look for, we could see those mosaic patterns in water merely by looking. You may have even seen them yourself. Any warm water sitting still for a period should contain those mosaics. Ordinary cameras easily capture them (**Fig. 7.4**).

When we proceeded to study those water mosaics, we could confirm that the vapor mosaics rose *directly* from the liquid mosaics.[5] We could detect those vapor mosaics by focusing a video camera on a plane immediately above the warm water surface, as closely as possible. Evidently,

Figure 7.4. Mosaic pattern in a pot (left) and a cup (right) of warm water.

those mosaics did not merely form in the air; they sourced from similar patterns in the water beneath.

Clearer images of those water mosaics could be obtained by illuminating the chamber with very low-angle light. The light hit the bowl from all directions, almost parallel to the water surface, but aimed slightly downward. With some measures taken to combat lens fogging, we found those images to be striking, almost qualifying as a piece of art (**Fig. 7.5**).

The character of those images told us something important about the water surface: It was not flat. Between the dark boundaries, dome-like surface mounds bulged upward like fingers pushing onto a surface membrane from beneath. You could actually see those mounds with the unaided eye. (I'll speak to the origin of those mounds later.) Because the illumination was low angle, the bulges blocked incident light from reaching the intervening mosaic boundaries; hence, the hills lit up while valleys remained dark, producing such eye-catching images as the one in **Figure 7.5**.

Figure 7.5. Mosaic pattern seen in warm water using low-angle illumination. Borders are dark because mounds (white regions) prevent the low-angle light from reaching them.

By verifying the presence of mosaic patterns in the water itself (**Figs. 7.4, 7.5**), we could confirm that the characteristic mosaic patterns in the vapor must indeed have arisen from similar patterns in the water. The vapor pattern was no longer mysterious; we could see where it came from.

How the water mosaics, themselves, come into being is an issue I'll deal with shortly. Please hang in there.

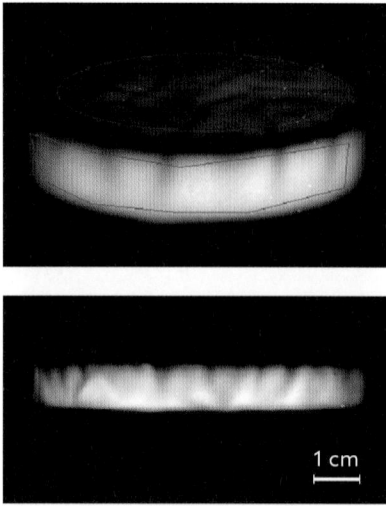

Figure 7.6. Oblique and side views of warm water obtained with an infrared camera, showing the downward projecting lines. Darker regions imply lower temperature, which in turn implies more order.

Rising Mosaic Tubules?

Critical information about the nature of those water mosaics came from images of the water obtained with an infrared camera. When taken from directly above, those images showed water mosaics just as clearly as the striking one in **Figure 7.5** — nothing new, except reassuring confirmation and more pretty pictures. But oblique and side views showed something more telling: The mosaic shapes *extended vertically downward*, into the water's depths (**Fig. 7.6**). Now that was a wow! observation.

So, rather than planar, the water mosaics were evidently tubular — like snorkels extending down beneath the water surface. They might curve a bit, but essentially, they descend toward the bottom. That implied that the mosaics rising into the vapor were likewise not planar, but more probably, tubular (**Fig. 7.7**).

The tubular character of the rising vapor was confirmed by videos taken at some fixed plane above the water surface. Frame-by-frame analysis showed the vapor mosaic pattern hardly changing over an interval of time as the vapor rose past that fixed plane; only minor shifts occurred. Hence, the vapor structure had to be tubular, not planar. Then, following each such pass, the characteristic pattern would vanish, only to get replaced by a new mosaic pattern, sometimes with a different shape altogether. Evaporative events seemed to occur discretely, one after another, propelled by increasing repulsion between successive tubules.

While readily observed in those fixed-plane videos, the vapor-mosaic patterns generally can't be resolved by the unaided eye. When viewed from the side, they look merely like rising clouds of vapor. With a dark background, you can easily see those clouds as they belch discretely, one after another, from your hot cup of coffee (**Fig. 7.8**). Only when one examines cross-sections of that rising vapor do those mosaics reveal themselves.

Here, I must interject a few words about the conventional interpretation of those mosaic patterns. I refer not to the mosaics seen in the vapor (which, to my knowledge had never been observed prior to our own studies), but to the mosaics situated in the water beneath. Those mosaics have been well studied and are commonly termed "Bénard cells." Viewed from above, by infrared (*i.e.*, "thermal") imaging, those mosaic cells have quite naturally been interpreted in a thermal context, *i.e.*, in terms of temperature-induced convection: Thus,

Figure 7.7. Vapor rises from tubular mosaics forming in the water. We argue later in this chapter that the vapor bears negative charge. Cross sections through the rising vapor yield images similar to those illustrated in Figure 7.3.

Figure 7.8. Vapor rises in a series of puffs.

warmer, less-dense regions of water are thought to rise, cool, and subsequently descend, setting up perpetual up-down circulations that create the mosaic patterns. That's the commonly held view.

Vertical convection currents certainly flow in warm water — we could easily detect them microscopically, as have others. Hence, vertical flows constitute a natural explanation for creating the tubular mosaics. However, we found that convection cannot be a general explanation. We confirmed this by refrigerating one of those experimental flasks to slow down the vertical flows. We saw that the mosaics could persist even when no vertical flow at all could be detected.[2] (Fig. 15.8)

So, rather than being convective in origin, mosaics appear to arise differently. The evidence presented above and below point to an origin based on real superstructures lodged in the water, detectable because their physical properties differ from the rest of the water. You might liken those structures to floating ice cubes — surely made of water, yet distinct from the water in which they float.

During evaporation, those tubular structures evidently rise directly from the water, like missiles lifting from underground silos, one after another (**Fig. 7.7**). They rise to create the vapor, whose superstructure remains largely unchanged from its aqueous origin.[5]

I recognize that the prevailing view of evaporation differs. Chemical physicists suppose that evaporation occurs when a molecule of water acquires enough energy to escape from the liquid. That would occur statistically, molecule by molecule. The arguments above don't preclude any such mechanism; in fact, a mechanism of roughly that nature may account for the evaporative release of positive charge (see Chapters 2 and 4), molecule by molecule. The arguments here merely describe an *additional* contribution to evaporation that fits with the visual evidence of rising mosaic tubules. This mechanism can evaporate lots of water in significant blobs.

I appreciate that this latter mechanism may seem unfamiliar, even bizarre; but it's the simplest interpretation of what we see experimentally. Soon, we will examine how those mosaic tubules may be built, why they ascend from the water to the vapor, and what fate befalls them as they ascend.

How Much Do Clouds Weigh?

The weight of a cloud is a question rarely asked. Clouds float so effortlessly that we forget that, at least in large part, they consist of vesicles (droplets) of water, each of which has weight. By adding the weights of constituent droplets, you can compute the weight of the cloud. That's easily done.[w1] A typical cumulus cloud weighs in at just over two million kilograms. That's well over four million pounds.

Those immense numbers don't provide much "feel," so some atmospheric scientists have begun converting the weights to elephant-equivalents. A modest, garden-variety fluffy white cumulus cloud might weigh in at about 100 elephants,[w2] while a tall, ominous-looking cumulonimbus cloud might tip the scales at roughly a million elephants.

If a million elephants threaten to come crashing down from the sky at any moment, then perhaps a nuclear holocaust should be the least of our worries. At a minimum, this animal-equivalent terminology may give new meaning to the common expression, "It's raining cats and dogs."

Or (stay tuned), has our thinking about "cloud weight" gone astray?

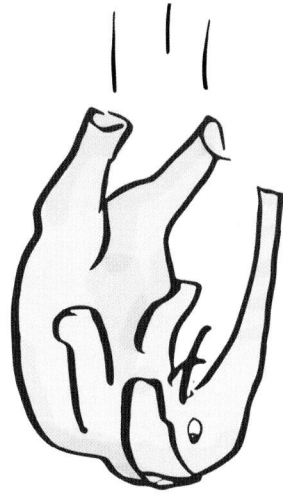

RISING TUBULES ARE BUILT OF VESICLES

Tubular Building Materials

Exactly what are those rising tubules made of? They evidently contain water, for that is their mother material. But water doesn't ordinarily form tubules, as far as we know. On the other hand, it can form vesicles.

We would eventually learn (see below) that the walls of those mosaic tubules comprise numerous small water vesicles (**Fig 7.7**). We would further deduce that those vesicles were negatively charged, and held together by positively charged hydronium ions, those ions acting like mortar and holding together the composite tubular structure.

The first hint of this kind of studded structure came from video microscopic images of warm water, obtained with an infrared camera. The images showed that just outside the tubes' walls were incessantly flowing, vesicle-like blobs of micrometer size. That made sense: Since the tubes, or segments thereof, evaporated regularly into the atmosphere, they evidently needed continual replacement, which necessitated fresh materials. Those blobs seemed to fill that bill. Whether they constituted droplets or bubbles remained unclear (and was never pursued for lack of time). Nevertheless, it seemed that vesicular blobs composed those tubular walls.

The second hint that tubules were made of vesicles came from direct observation. The vesicles, or vesicle clusters, could sometimes be detected without a microscope. We could often recognize them gathering to form the water-based mosaics (**Fig. 7.9**). Sometimes they were faint (*panel a*), other times more obvious (*panel b*). The scene definitely reminded us of rocks being mortared together to make a stone wall — except that this wall was evidently built of vesicles.

(a)   *(b)*

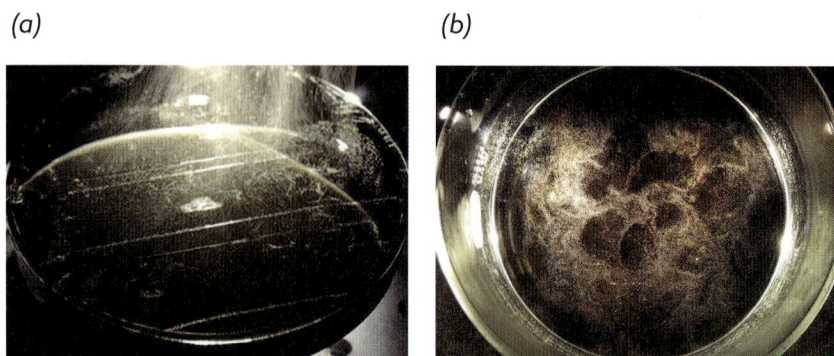

Figure 7.9. Incipient mosaic patterns. *(a)* Warm water poured into a container. Right part of the image shows the mosaic rings forming from vesicles. Parallel lines are experimental artifacts left by removed tape; please ignore. *(b)* Tap water, 60°C, run slowly into a clear bowl placed on a black cloth to enhance contrast. Vesicles accumulate near boundaries, leaving each cell relatively empty.

It was not only in the water that we could visualize the vesicles coming together, but with suitable lighting we could sometimes see them persisting in the vapor as well (**Fig. 7.10**). Clearly, whether inside the water, or evaporated to outside the water, tubules were made of vesicles.

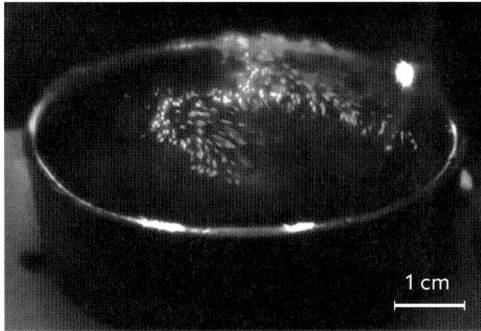

Figure 7.10. Vapor above pan of warm water. Illumination was set to reveal the vapor's vesicular nature.

Having visualized vesicles both in the water and in the vapor, we began looking more closely at the transition region just above the water's surface. There, we could often detect what looked like sheets of rising vesicles.[5] While the optical system's limits didn't allow us to confirm the anticipated tubular nature of the material rising through the water-air interface, we could at least confirm the passage itself, supporting earlier inferences that water mosaics rise to create vapor mosaics.

Those rising vapor patterns do not persist indefinitely. As you've probably noticed at your local coffee hangout, the patterns dissipate as they rise higher into the air, eventually becoming practically invisible. Invisibility implies that the scattering structures must have diminished in size, otherwise the vapor would continue to remain detectable. Presumably, the vesicles that had clung to one another to form the extensive, visibly detectable structure, had dispersed. Whatever glue holding the vesicles together seemed to have loosened its grip as the vapor rose, allowing the mosaic superstructure to disintegrate into scattered vesicles.

Now, project this vesicle-dispersion phenomenon more broadly. Since evaporation is persistently occurring from warm water over the surface of the planet, scattered vesicles like the ones rising from your coffee cup should populate the atmosphere. When those vesicles are in low concentration, we call that "low humidity"; higher concentrations indicate elevated humidity. Under the right conditions, those free atmospheric vesicles may ultimately coalesce to form clouds (see below) — not unlike the clouds above your coffee (**Fig. 7.8**). As we shall see later, coalescence of vesicles will prove to be key to cloud formation.

Oddly enough, this evaporative process rather resembles the precipitation process in reverse. When it rains, droplets (vesicles) fall from clouds to the earth; during evaporation, vesicles do the opposite — they rise from the earth. Eventually, those rising vesicles coalesce to re-create the rain-generating clouds. In so doing, the earth's water gets recycled. The water molecules you are drinking might be the very same molecules imbibed by Genghis Khan.

In sum, the raw material for constructing those tubules appears to be the vesicle. Vesicles create tubules in the same way as stones create walls. The lingering question, to which I alluded earlier: Why, in the first place, should those vesicles draw together and stick? What's the mortar?

Vesicle Charge?

Stickiness often involves opposite charges, and we need therefore to know whether vesicles might be charged. You might expect otherwise, given the water molecule's electrical neutrality. But perhaps not: Second thoughts may come from the previous chapters' emphasis on net charge, and if you're thinking along those lines, then I believe you might be on track. Vesicles appear to have a net negative charge.[2] I'll explain why in a moment — this is important.

Let me first set the stage by focusing on an obvious feature of the vesicle: its tendency toward roundness. Amply confirmed in larger, more easily visualized vesicles, the consistently near-spherical shape implies an internal pressure pushing out against a restrictive membrane, like that of an air-filled balloon (**Fig. 7.11,** see next page). Might any such membrane be a shell of EZ water? (For a detailed description of EZ water, please see Chapter 2.)

The EZ-shell concept first came to mind while peering into a champagne glass. There, I could observe something seemingly inexplicable: endless streams of bubbles originating from a single isolated point on the glass surface — a crack, scratch, bump, or some other asperity. How could bubbles stream endlessly from a single point and nowhere else?

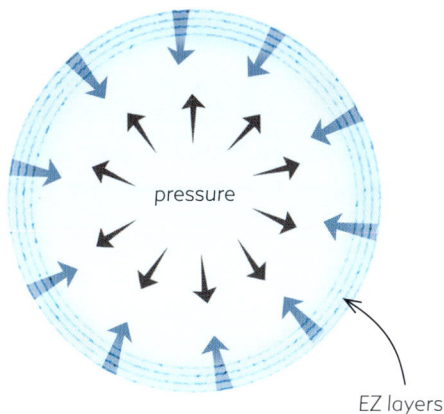

Figure 7.11. Droplet shape. The force of the distending pressure balances the inward force of membrane tension, promoting a spherical shape.

We came to recognize that those asperities might be just the kinds of structures that nucleate EZ growth. Bubble formation, it seemed, might be linked in some way to EZs. That set us on a track of investigation.

We could soon confirm that the enveloping shell of the vesicle consisted of parallel layers of EZ. We nailed this down by verifying the absorption of light at the wavelength of 270 nm, the signature absorption characteristic of the EZ.[2 (pp 225-231)] We could also see those shells by direct imaging.[2 (Figs. 3.6, 3.8)] Like layers of an onion, the enveloping EZ layers appeared able to sustain the pressure developed within the core of the vesicle, presumably leading to the vesicle's roundness (**Fig. 7.11**).

Having dealt with the shell, we next explored the possible source of internal pressure. Within droplets and bubbles alike, we found that the pressure arises from positive (proton) charges, distributed throughout the core of the vesicle, repelling and pushing outward against the EZ membrane. We knew those charges were present: We could detect them spewing from just-broken bubbles.[2 (pp 240-241)]

If those interpretations are correct, then both components of the vesicle bear charge: The core contains a net positive charge from protons (hydronium ions), while the enveloping sheath bears the characteristic negative charge of EZs. The two species presumably arise together, from the splitting of water molecules that create the EZ.

In theory, those charges should balance. However, the positive charges lie free. Hence, positive-positive repulsion could easily drive some of those positive charges to escape before the growing vesicle closes,[2] implying that the closed vesicle could well contain *net negative charge* (**Fig. 7.12**).

The possible loss of positive charge as vesicles form may seem trivial, but it's important for much that follows. To better understand the rationale, consider the way vesicles likely form (**Fig. 7.12**). EZ layers begin building next to solid asperities (*panel i*). As those layers grow thicker and broader (*panel ii*), complementary protons get released into the water (see Chapter 2, **Fig. 2.4**). Some of those positive protons (or hydronium ions) will attract the dangling negative EZ nearby, inducing modest curvature (*panel iii*). This process continues (*panel iv*), producing increasing growth and curvature. With additional proton buildup and still more curvature, the hydronium ions eventually become enclosed within the sphere (*panel v*).

But there's a hitch. Prior to EZ-shell closure, the mutual repulsion of all those positive hydronium ions will have driven many of them outside the still-open vesicle (*panels ii – v*). With such natural loss of positivity, the soon-closed vesicle (*panel vi*) should retain a net negative charge.

While direct tests of this residual-charge hypothesis would certainly be worthwhile, the chain of logic implies an inevitable loss of positive charge during vesicle buildup, resulting in a negative-charge residual. If individual vesicles bear net negative charge, then, logically, the mosaic's tubular walls should likewise bear a net negative charge.

Now the inevitable question: How could those negatively charged vesicles stick together to form the tubules? Middle-school science teaches us that negatively charged entities repel; vesicles should fly apart from one another. So, what could conceivably bond those like-charged vesicles together to form tubules? Could responsibility lie in those escaped hydronium ions, positively charged and dwelling happily nearby? We will consider that option shortly.

LIKE-LIKES-LIKE: A MOLECULAR GLUE

Can Like-Charges Attract?

Now, onto the issue of how those vesicles can attract to build the tubular superstructure. Given that those vesicles all bear the same net charge and that their exposed shells bear that same negative charge, how could

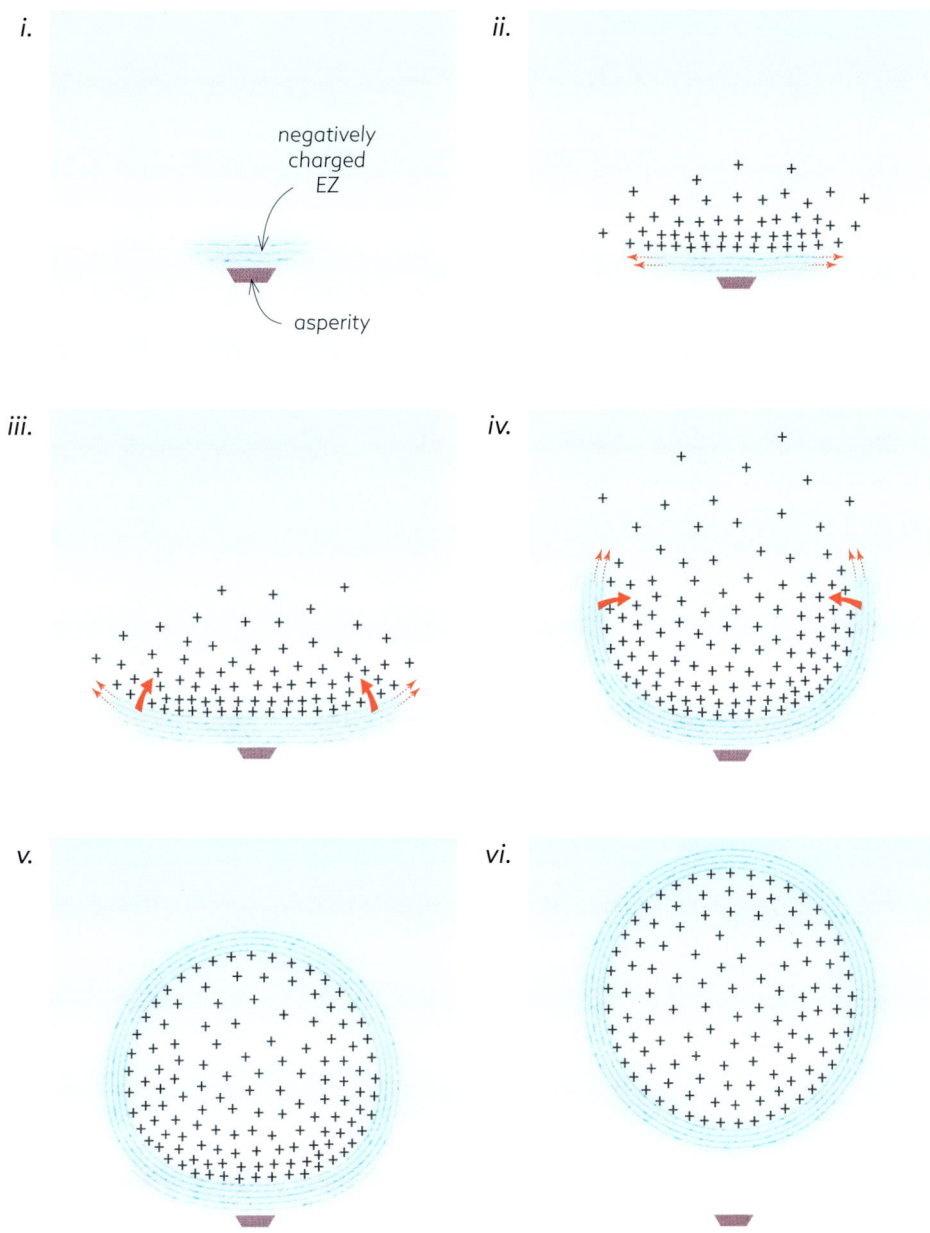

Figure 7.12. Vesicle formation. EZs form next to a nucleator, such as a glass asperity (*i*). As the negative EZ forms, complementary positive charges build next to it (*ii*). Those positive charges attract the negative EZ, inducing curvature (*iii*). Meanwhile, hydronium ions repel one another, tending to disperse. Hence, some should exit the still-open vesicle (*iv*). Finally, the vesicle seals as open ends of the EZ connect (*v*). Positive-charge loss leaves the closed vesicle (*vi*) with net negative charge.

Spouting Rain? A Diversion

Ahh, the pleasing hum of a Tibetan singing bowl for meditation and relaxation. That purr soothes the spirit. But for a display that can throw the viewer into paroxysms of awe, it's hard to beat its lesser-known cousin, the Chinese spouting bowl.

What makes the spouting bowl so worthy of our attention? When you fill it with water and then rub the rim with wet hands, the bowl starts vibrating. Vibration produces a pleasing low-frequency humming sound much like the Tibetan bowl. But something additional happens. If you continue to rub, the water begins spouting upward, like rain but upside down. Droplets may rise to heights up to a meter.

Spouting bowl.

Observers casually interpret the rising droplets as coming from some kind of "resonance" effect. The sound makes the water resonate. But that off-the-cuff explanation leaves open the nature of the physical force underlying the droplet rise. What force propels those droplets upward? Also less than obvious: What creates those droplets to begin with?

Intrigued, we purchased two such bowls to study in our laboratory. In the short duration of the study, we couldn't discern all the secrets of those spouting droplets, but we did uncover one clue. The inside of the bowl was not smooth; it boasted four sizeable protuberances. We found that those protuberances spawned all the rising droplets. None originated elsewhere.

This observation left us wondering whether those protuberances might be acting like the champagne-glass asperities, sites of EZ formation producing endless streams of bubbles. If so, then those forming droplets would likely have negative charge. The negatively charged vesicles would then rise into the air by repulsion, coming not only from the freshly minted vesicles beneath, but also from the negative earth — perhaps like the clouds of negatively charged vesicles rising from your cup of hot coffee. That feature could conceivably explain the upside-down rain.

they possibly attract? Like charges *attracting* one another? Lest you surmise that the author must be on some kind of drug, let me assure you that like charges cannot attract. However, they *can* draw toward one another if unlike charges lie between them. The effect is analogous to two warring parties. They will remain at some distance, unless drawn together by an alluring attractor (**Fig. 7.13**).

Figure 7.13. Entities that would ordinarily remain apart can come together under the right circumstances.

I first came to this principle when I stumbled across the work of Norio Ise, emeritus professor from Kyoto University. Ise studied colloids — micron-sized particles suspended in water. He found that those like-charged colloidal particles could be drawn together by interposed opposite charges. By demonstrating this principle both theoretically and experimentally, Ise earned Japan's highest scientific prize: dinner with the Emperor. (I'm told that the food was good.)

Ise confided to me that this paradoxical attraction had been long known. Half a century ago, the Nobel physicist Richard Feynman coined the expression "like-likes-like." Like-charged particles evidently "like" each other because they tend to come together. Feynman surmised the reason: unlike charges lying between them. Thus, Feynman[6] opined: "like-likes-like through an intermediate of unlikes."

Mildly skeptical that like-charged particles could ever draw toward one another, I asked my students to test the phenomenon in our own laboratory. Employing two like-charged 0.5 mm gel spheres, we could easily confirm their apparent attraction: Even at initial surface-to-surface separations as large as half a millimeter, the two spheres gradually drew together.[7] Using pH-sensitive dyes, we could also confirm the presence of unlike charges lying between those spheres, presumably arising as EZs grew around each particle, releasing positive charges as they did.

Those experimental findings affirmed the reality of Feynman's like-likes-like phenomenon. The attraction was not a fluke at all, but a genuine feature of nature, whose application could potentially extend from the cosmos down to the atom. You will hear more about this phenomenon later in this book. I view it as central for all of nature, as I believe did Feynman.

The like-likes-like phenomenon allows us to understand how the water-borne vesicles we are considering could coalesce. If individual vesicles have similar negative charge polarity, as well as negatively charged shells, they should surely repel; but opposite charges lying between them could drive their coalescence. Those opposite charges may reside in the protons that escaped during vesicle formation (**Fig. 7.13**); those proximate protons may then act as "glue" for binding together the negatively charged vesicles. Many such bonded vesicles could then build to create the tubular mosaic superstructures.

So, yes, negatively charged vesicles *can* draw toward one another, the interposed positive charges constituting the drawing agent. Ordinarily, positive charges should lie virtually everywhere surrounding each vesicle; however, the regions between vesicles will enjoy contributions coming from the two flanking vesicles, not just one; hence the concentration will be highest between those vesicles, so the vesicles will inevitably draw toward one another, creating the superstructure.

Only dabs of positive-charge glue should suffice for spot welding, just where the vesicles touch. While additional positive charges may be attracted to those negative vesicles over time, the modest minimal requirement means that the tubular superstructure could well retain a net negative charge, at least at the outset. Such presumptive negativity will be important for what follows.

Before moving on, I can't resist one speculation. Remember those lovely rainbows? Rainbows can sometimes occur *within* clouds, especially high cirrus clouds. Perhaps you've seen them on occasion. Those light displays are commonly attributed to the presence of ice crystals, which confer some kind of regularity within the cloud.[w3] But such attribution raises the question of why ice, denser than air, should not plummet to the earth instead of remaining suspended within clouds — and hence whether ice could really be the agent responsible for creating those cloud rainbows.

A possible alternative mechanism may lie in the like-likes-like phenomenon. Individual vesicles repel one another, while the positive charges lying between them attract. If attractive and repulsive forces balance, then that balance could produce a stable, ordered, vesicular array, in much the same way as it produces an ordered colloid-particle array.[8,9] Ordered particles are known to create rainbow-like iridescence.[10] Hence, cloud-rainbow formation may have a simple theoretical basis, and hence a presence that might not be quite so magical as we might have hoped.

Returning to the main thread, we move on from the question of vesicular bonding to the final critical question: what inspires the tubular superstructure in the water to rise into the atmosphere and evaporate as vapor? The rising of vapor will be the final link in the chain of events.

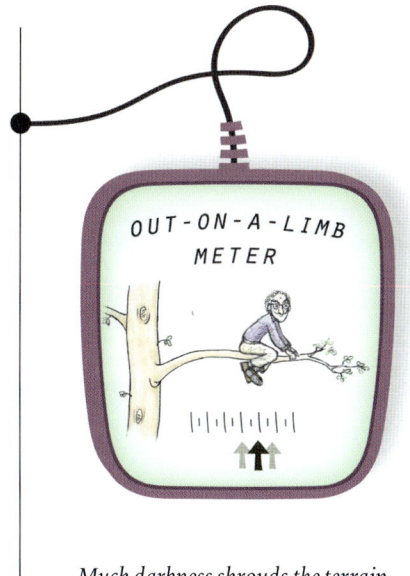

Much darkness shrouds the terrain of understanding. Where attempts to understand remain particularly speculative, I will post "out-on-a-limb" alerts, as above.

TOWARD A PRINCIPLE OF CLOUD FORMATION

What Drives the Tubular Mosaic Upward?

I suggest that evaporative events may be driven by charge repulsion.

Consider the tubular mosaic structure assembling in the water. As more and more negatively charged vesicles build onto that growing superstructure (with small dabs of positively charged glue holding those vesicles together), the net negative charge of the ensemble should increase. That negative charge cannot build indefinitely: Once the charge magnitude reaches some critical threshold, internal repulsion should tear the structure apart at its weakest point. An upper section of the superstructure should then break free, creating the mass for the unitary evaporative event — the cloud above the coffee (**Figs. 7.7** and **7.8**).

Why should that mass rise from water to air?

Low density would seem to be the knee-jerk answer — but if vesicles are built principally of liquid water, then their density must surely *exceed* that of the air above. Even if each vesicle's core were to transition from liquid to vapor, still, the solid EZ shell would contribute enough compact mass to keep the overall vesicle density higher than that of air. Gravitational forces ought to preclude any rise.

Another option is charge. Recall (Chapter 1) that charge-based forces overwhelm gravitational forces, by many orders of magnitude. While gravitation may pull downward, an upward-directed charge force could easily overwhelm the weak gravitational force, and thereby drive the vesicles upward. We saw a vivid example of that upward-directed charge force in **Figure 1.5**.

The existence of such a levitational force would seem practically inevitable. If the breakaway superstructure, built principally of vesicles, has net negative charge, then the residual mosaic remaining in the water after breakaway should likewise have net negative charge. The two entities ought to repel. Charge repulsion should help drive the breakaway structure upward.

Augmenting that upward push are two additional upward-directed forces: repulsion from the negative earth beneath and, the pulling force from the atmosphere's positive charges above. All three of those electrostatic forces should sum up to propel the rise of the tubule, and hence the tubular aggregate that makes up the mosaic. Charge forces should all push upward.

If you're still with me, you might be thinking: Something's wrong here. If the water beneath keeps losing negative charge in the manner outlined above, then wouldn't the residual water become increasingly positive? Could such a scenario be realistic?

Well, probably not — but please remember (Chapters 2 and 4), along with the belching of those negatively charged aggregates, comes the continuous release of the other evaporative component, the positively charged hydronium ions. Companion release of positive and negative charge doesn't necessarily ensure precise maintenance of water's neutrality, but it does provide a means by which any progressively increasing net charge remaining in the water is not an inevitable outcome. It speaks to the issue of reasonableness.

It also emphasizes that evaporation comprises not one but two components: negatively charged vesicle aggregates and positively charged hydronium ions (**Fig. 7.14**). Both species get driven upward by like charges immediately beneath. Surely, some significance must be attached to this dualistic feature — opposite charges rising from the earth into the atmosphere. What might that be?

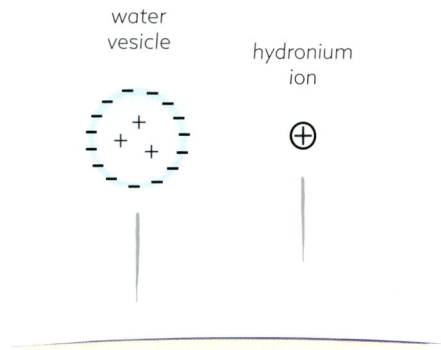

Figure 7.14. Evaporative components: negatively charged vesicles and positive hydronium ions.

Clouds obviously contain gobs of moisture. Whether those clouds lodge in the atmosphere or emerge from your cup of coffee, obtaining that moisture implies the presence of moisture-containing vesicles. But vesicles repel. Bringing those repelling vesicles together requires positive charges. Thus, the natural question: Might clouds build in situations *when moisture-containing vesicles are plentiful, and when atmospheric positive charges exist in high enough concentration to draw those vesicles together?* Such a requirement would explain why clouds form sometimes, but not at other times. It would also explain why evaporation involves two components, not just one.

But not so fast. After their lift from the earth, those vesicles must fall back down as rain. How could that happen? If you're wondering about that paradox, then you're perhaps one step ahead. The answer, I believe, is not complicated, but it does require a critical step of discussion, which comes in the next chapter.

Why Your Laundry Won't Dry on Humid Days

Atmospheric humidity retards evaporation. Your wet clothing doesn't dry so readily in humid weather. How does water know to avoid evaporating when the air is seriously humid?

Humid air ought to be saturated with negatively charged vesicles. You can tell because distant objects seem fuzzy. To scatter the light coming to you from those distant objects, many vesicles should be present in the atmosphere — creating something akin to ground fog but less extreme.

Under such conditions of abundant humidity, the negative charges in the lower atmospheric vesicles can be expected to repel any negatively charged tubules lodged in the water beneath. Repulsion should inhibit their rise. So, your wet underwear takes longer to lose moisture (*a*). By contrast, dry atmosphere poses no such barrier. In fact, the air's positive charge helps draw those negatively charged mosaic tubes upward, enhancing evaporation and drying your laundry in reasonable time (*b*).

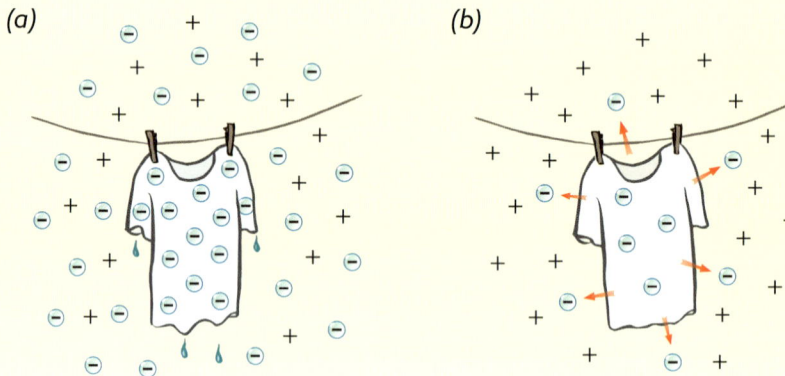

Tying Loose Ends

While the previous section provided a natural conclusion to this chapter, please indulge my proclivity to resolve lingering issues. Two loose ends need to be tied. First, why wouldn't those rising positive charges immediately combine with rising negative charges, annihilating all separated charge? And second, if both positive and negative charges rise simultaneously, then how could the atmosphere ever acquire its net positive charge? Let me address the second question first: How does the atmosphere become positively charged?

To answer that question, consider first the rising tubules. Initially, those tubules lodge inside the water. There, the tubules' negatively charged surfaces ought to be attractive to the many positively charged hydronium ions quartered in that water. Some of those oppositely charged ions will stick to the tubules. Enhancing the propensity to stick should be the mutual repulsion among the free positive charges: Repulsion should drive those charges away from one another, and hence toward the oppositely charged tubular walls. Some may adhere, or at least hover nearby.

But any such hovering should be short lived. Once those tubules rise from solution into the air, the repulsive push of those ions toward tubules will have vanished. Without that repulsion, hovering positive charges ought to readily disperse into the surrounding air. It's like taking a deep breath of positive charge in the water and letting it out in the air. Such release constitutes one possible source of atmospheric positivity — a natural consequence of the tubular rise.

A second and probably more potent source of atmospheric positivity is the direct escape of hydronium ions from the water into the air. Recall the finger-like surface bulges that created those visually dark valleys (**Fig. 7.5**). To create those bulges, something must push upward. A likely agent, once again, is repulsion, *i.e.*, the repulsion among the free hydronium ions lying within the intratubular water. Repulsion creates a pressure that not only produces the upward bulges, but may also drive hydronium ions to cross the bulging interfaces into the air above. In other words, the evaporative process may literally *push* the positive hydronium ions out of the water.

Both of those processes should build atmospheric positivity.

Now to the second question: Why do those rising positive charges not just cancel the negative charges rising at the same time? Opposite charges should ordinarily annihilate one another. Some annihilation could well take place. However, the two evaporative phenomena in question are separated in both space and time.

First consider time. From those persisting surface bulges, we can infer that positive charges must be pushing continuously, with some hydronium ions powering through the interfaces and rising more-or-less continuously. Steady rise is anticipated. The tubules, by contrast, rise discretely, one cluster at a time. At the high temperatures typical of most of our experiments, clusters of tubules may rise in rapid succession, but at lower temperatures characteristic of the natural environment they occur more sporadically.[5] Hence, most of those continuously rising positive charges should escape the clutches of the sporadically rising negative charges. The timings are disparate.

Regarding space, the two evaporative phenomena are also distinct. From a water surface containing many visible mosaics, only a modest fraction of them will rise synchronously (**Fig. 7.7**); the rest remain stationary.[5] So, while the positive charges may rise from all over the water surface, the negatively charged vesicles rise only in localized regions. Once again, the two species should remain largely separated. With separation in both space and time, negative-positive recombination should remain limited.

With the onset of sunlight to drive these processes, then, we anticipate two distinct evaporative events. Vesicles rise, carrying abundant moisture and positive charges rise, carried by hydronium ions. Both negative and positive components should be needed for cloud formation, so you can perhaps envision how the dynamics of those two processes might play out in the game of weather — the subject of the next chapter.

Summary

To begin our consideration of weather, we examined rainbows. We learned that rainbows arise from the light dispersed by droplets (vesicles). Each droplet bends the light in much the same way as does a prism. Collectively, the many droplets reinforce the effect, creating the enchanting display of colors that we know as a rainbow.

From those alluring light displays, we learned something simple: Size matters. Large vesicles deflect, or scatter, lots of light. Smaller ones scatter less. For objects as small as air molecules, scattering is so modest that we can't see those objects at all.

This understanding proves useful, not only in comprehending why the sky is blue and sunsets red, but also in deducing the nature of what we see. When distant images appear fuzzy, we know that sizable scatterers must lie between those objects and our eyes. The scatterers deflect light, rendering distant objects blurry.

Evidence implies that those scatterers may include water-containing vesicles. In high enough concentration, such as during summertime mugginess, those vesicles may scatter enough light to fuzz long-distance vision — a common summertime experience. We can understand why ample humidity (vesicles) tends to blur vision.

Humidity arises from evaporated water. According to experimental evidence from our laboratory, one component of evaporation occurs as a series of discrete events. Each event involves a set of vertically oriented tubules rising from the water like a series of missiles rising from a silo. Each tubule comprises numerous tiny vesicular water droplets, the completed tubular assembly resembling a beaded pipe. Parallel assemblies of those beaded tubules rise into the atmosphere. They soon break apart, dispersing constituent vesicles into the air and creating atmospheric humidity (while also scattering light).

It is those dispersed vesicles that arguably condense into clouds, returning to the earth in the form of rain — only to evaporate again to create new airborne vesicles and fresh humidity. In that way, evaporated water recycles.

Vesicle condensation into clouds may at first seem paradoxical. Vesicles of the same charge should repel, remaining as far apart as possible; they should resist any temptation to come together and condense into clouds. Yet they apparently do condense. Clouds do form. The paradox may resolve with the like-likes-like mechanism — positive charges lying between the negatively charged vesicles. Those charges exert a pulling force that draws the vesicles toward one another. That condensing mechanism, as we shall soon see, could well underlie cloud formation.

But clouds don't always form. They form only in some situations, not others. The next chapter deals with the conditions required for cloud formation. We will explore what must happen to turn fair weather into foul. Also, what transforms those puffy, white cumulus clouds into entities dark and ominous? And, why do those dark clouds sometimes manage to hold their water, while other times relieving themselves with prodigious outpourings of rainfall?

In short, we attempt to unravel the mysteries of cloud behavior.

CHAPTER 8

Floating Gray Elephants: Clouds and Rainfall

As a long-term resident of Seattle, I begrudgingly anticipate the arrival of winter's gloomy skies and relentless rain. Summer means bright, sunny days; but winter's foul weather follows almost as predictably as night follows day.

Why so much rain during Seattle winters?

I posed the rain question some years ago at a scientific conference strategically housed at a charming resort in the Caribbean. As my wife and I made our way between buildings, the darkened skies above suddenly unleashed an unexpected torrent of rainfall. We dashed for cover. As we did, I asked her loudly, over the din of pounding rainfall and her chronically impaired hearing, *"Why* does it rain?"

That semi-shouted question was meant to be purely academic: What turns on the rain? It was a simple scientific query. Nevertheless, a hotel attendant scurrying immediately behind us, overheard my question. Misinterpreting my "why" question as a disgruntled lament, she replied plaintively: "I'm *awfully* sorry." When rain threatens to sink Caribbean vacations, locals harbor boatloads of guilt. Rain should never spoil vacations.

We enjoyed chuckling over the incident, but the question remains: Why does it rain? Literally, why does water fall from the sky sometimes, but not others? Thick, dark clouds may loom ominously above, but in terms of unleashing their contents, we never know: Sometimes they do, but sometimes they don't.

Even in this advanced technological age, it seems surprising that we cannot consistently calculate the weather. Forecasters predict weather based mainly on historical trends and variations of temperature and pressure: The sun's energy influences temperature, while the columns of air pressing on the earth's surface theoretically create pressure. A complex mix of these two variables supposedly determines the earth's weather patterns.[w1] Yet, we still can't be sure if tomorrow will require a raincoat.

The standard pressure-temperature weather paradigm omits the critical variable that this book emphasizes: electrical charge. Thunderbolts may jar us from our complacency, but we usually relegate charge-based phenomena to mere sideshows in the drama of weather. Few see electricity as playing a pivotal role in atmospheric phenomena.

Here in this chapter, we consider the possibility that electrical charge plays a dominant role in all weather patterns. We have already considered the role of charge in evaporative processes. Next, we focus on the moisture-filled vesicles that rise up during evaporation (Chapter 7). We continue to ask whether ambient positive charges could condense those negatively charged vesicles into clouds.

If you're thinking that the previous chapter's like-likes-like mechanism may be relevant, then we're in synch. I plan to explore the possibility that that versatile mechanism may lie at the center of all major aspects of cloud formation.

CLOUD GROWTH

Cloudy Concepts

First, what's the prevailing view of cloud formation?

Common understanding implies that clouds form when moist, rising air cools. Cool air can't hold as much moisture as warmer air; hence, the water-vapor molecules in that rising air clump into droplets, which then aggregate into clouds.[w2]

Why can't we simply accept this understanding and stop there?

We can't stop because the concept is incomplete. Omitted from this paradigm is the issue of exactly *how* those vapor molecules might clump into droplets and *how* those droplets might then aggregate to form clouds. What underlies those actions? And, if those questions are not enough to prompt ample head scratching, then here's another: Why don't those coalescing water droplets form ordinary liquid water, which would immediately plummet to the earth? Why thinned-out, billowy, floating clouds instead of a dense liquid?

In considering how clouds form, I don't challenge the prevailing view that moisture rises; plenty of supportive evidence appeared in the previous chapter. The lingering question focuses more on *how* that rising moisture (mainly packed into dispersed vesicles) gets condensed into feathery clouds. What forces of nature could prompt such behavior?

To answer that question, we begin with the simplest weather condition: fair and dry. Fair weather generally connotes clear, blue skies. "Clear" makes sense: If vesicles are in short supply, then the raw material ordinarily needed for building those suspended entities must be largely absent. Without water-containing vesicles, serious clouds cannot form. The sky remains clear blue.

On the other hand, highly localized clouds can form with even a modicum of isolated, fair-weather humidity. A good example is the sky over isolated lakes (**Fig. 8.1**). Strong midday sun can promote massive local evaporation. Evaporative species rising just above the lake would include abundant negatively charged vesicles along with positively charged hydronium ions (Chapter 7). Since the positive charges could theoretically draw the negatively charged vesicles into a condensed mass, it's not difficult to imagine the formation of a cloud of "condensed" vesicles

Figure 8.1. An isolated cloud formed above localized water.

just above the lake, especially on a windless day in an otherwise dry region. Indeed, by midafternoon, localized clouds will often form above small, isolated lakes.

Figure 8.2. Cumulus clouds often show cauliflower-like caps.

Like that lone cloud, the ordinary fair-weather cloud could be nothing more than a collection of negatively charged vesicles, all drawn together by free positive charges. The cloud could remain of modest size; or, with enough positive and negative constituents, it could grow larger. We call those fair-weather structures cumulus clouds (see Box). Best viewed looking down from an airplane, the cauliflower-like tops of such mature cumulus clouds can seem gentle enough to qualify as abodes for angels (**Fig. 8.2**).

Regarding those angelic cumulus clouds, certain features may at first seem paradoxical. For one, their size doesn't always match the size of the body of water beneath. Sometimes it does, other times not. Above vast ocean expanses for example, one often sees flat, stratus clouds of practically endless extent. That fits with expectation: extensive evaporative water sources beneath producing a huge cloud mass above. Other times, small, punctuated, cumulus clouds may appear instead. With a virtually endless supply of water beneath, one may rightly ask why small, separated clouds should *ever* appear instead of the one vast cloud that we'd anticipate.

Put another way, why the empty spaces between clouds?

The Like-Likes-Like Principle, Again

To address the question of cumulus cloud formation, please think back to Chapter 7's introduction of the like-likes-like principle and its application to colloidal suspensions (**Fig. 7.14**). When initially mixed with water, colloidal particles commonly lie randomly dispersed throughout the liquid. However, that pattern changes progressively. As each particle grows a

Common Cloud Types

Clouds can be classified into types, each with distinctive shape and altitude range. You've seen most of them, but the figure conveniently illustrates some of the most common varieties.

"Cumulus," by the way, comes from the Latin for "heap." Cumulus clouds appear "heaped up" on top, like mounds of cauliflower. "Nimbus," on the other hand, refers to rain. Cumulonimbus clouds may dump huge quantities of rain.

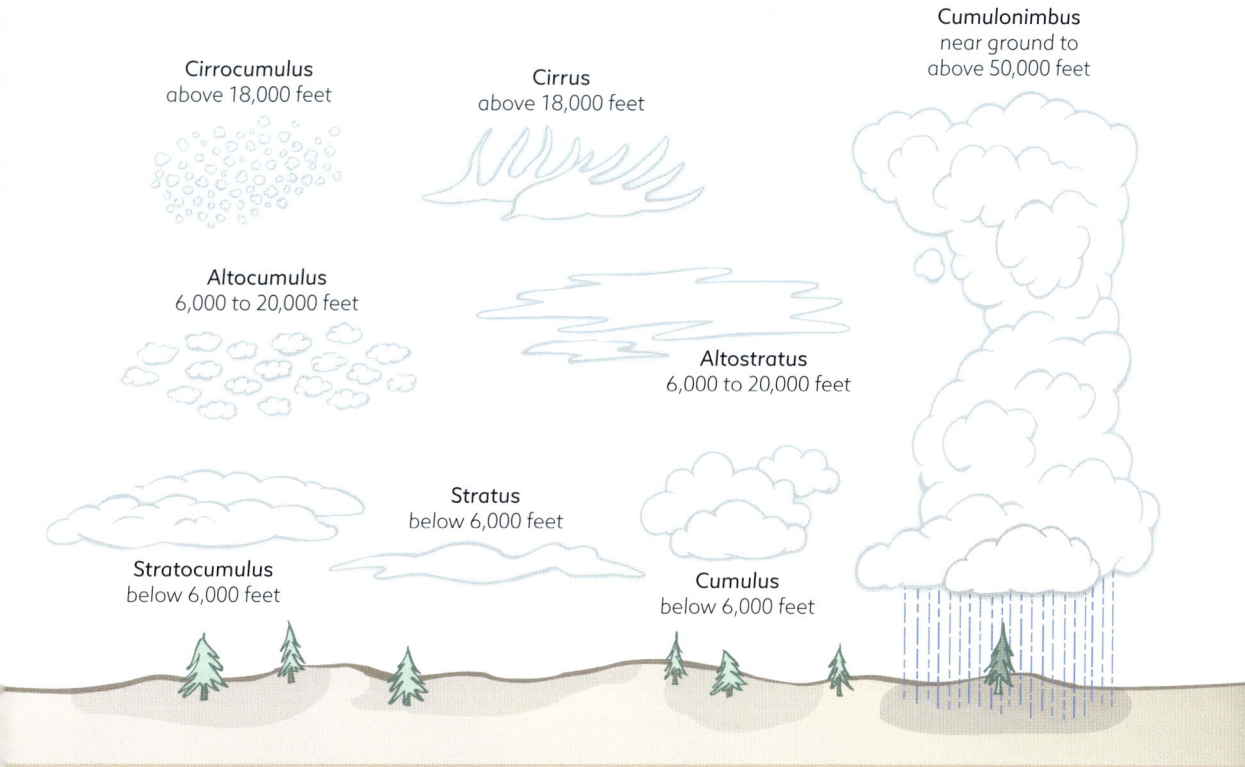

Cumulonimbus
near ground to
above 50,000 feet

Cirrocumulus
above 18,000 feet

Cirrus
above 18,000 feet

Altocumulus
6,000 to 20,000 feet

Altostratus
6,000 to 20,000 feet

Stratus
below 6,000 feet

Stratocumulus
below 6,000 feet

Cumulus
below 6,000 feet

negatively charged EZ shell and releases protons in the process, those released protons begin attracting the negatively charged EZs through the like-likes-like interaction.[1] The attracted particles slowly gather into ordered, multi-particle clusters, leaving voids between those clusters. Those voids contain no particles at all. They remain clear.[2]

A similar situation could well exist in the atmosphere. There, negatively charged EZs envelop atmospheric vesicles (Chapter 7), while positive charges loom in the form of hydronium ions (Chapters 2 and 4).

Those oppositely charged species (the vesicles being of comparable size as the colloids) contain the raw materials required for the like-likes-like attraction. Just as happens in the colloidal suspensions, *that attraction may draw the scattered vesicles together to "condense" into clouds*, leaving clear voids between those clouds. Thus, punctuated clouds can easily appear, even over vast bodies of water. In fact, we'd expect them routinely.

Given such a like-likes-like drawing mechanism, it would be natural to find voids not only between clouds, but sometimes even within clouds. Strong like-likes-like condensation in some interior regions would leave voids in contiguous or nearby regions. While we can't notice such internal voids from outside, droplet-free "holes" are indeed confirmed to exist inside of clouds,[3] just as they do between clouds. In terms of the like-likes-like mechanism, that feature makes sense.

With this developing understanding, we can return to the critical question raised earlier in this chapter: Why do clouds exist in the form of dispersed droplets? Why are they not merely liquid condensates of water, condensates that might crash down on us like overturned pails of water?

Perhaps that rather fundamental question has never crossed your mind.

The reason for the clouds' obvious character, I argue, is that the like-likes-like mechanism inevitably produces those dispersed entities. In the framework of that mechanism, positive charges initially draw negatively charged entities toward one another. However, the draw ordinarily terminates. Once the attraction pulls entities close enough that repulsion between those like-charged entities begins to equal the attractive draw, movement halts. Forces lie in balance.

That's why colloidal particles, such as those in milk, remain suspended at some distance from one another, and not aggregated into one massive blob. The like-likes-like force balance keeps those particles apart. The same should happen in the cloud: Cessation of like-likes-like draw ought to keep cloud vesicles spaced *apart* from one another, and not condensed into the liquid state. That principle confers the clouds' feathery texture.

The issue of texture is critical. Any mechanism, be it the like-likes-like or some alternative, must explain the cloud's billowy texture. If like-likes-like turns out to be inadequate, then we are obliged to identify some alternative mechanism that not only draws vesicular droplets together but ensures that they are kept at some distance from one another. Whether any such alternative exists, I'm not certain. But, so obvious a feature as the clouds' feathery nature cannot be ignored.

So, while the individual droplets that compose the cloud may contain liquid water, clouds, as a whole, do not. They contain suspensions of liquid droplets, likely held in place by a balance of opposite charges. Apart from those droplet interiors, liquid water is nowhere to be found in clouds. We need not worry about inattentive angels kicking over buckets of water onto our heads.

Cloud Nuclei?

Is it only vesicles that can draw together in this way? The arguments advanced in the previous section would seem to imply that virtually any particles reasonably endowed with negative charge and suspended in the air could substitute. So-called "cloud nuclei" should have the capacity to build clouds by similar means as vesicles.

Good examples are clouds of smoke. In January 2020, the biggest bush fires in Australian history wreaked havoc on much of the country, but not in the city of Melbourne. According to my Melbourne friend, that city was spared. Yet, for several days, Melbourne received rain that contained ash, soot, and dirt — so-called brown rain.[w3, w4] Apparently, remote fires lifted soot directly into the air above; the soot particles presumably nucleated cloud growth; and, the clouds eventually blew toward Melbourne, discharging that soot-filled rain. The calamity was serious enough even to impact the 2020 Australian Open.[w5]

Thus, particles of soot, dust, or even sea salt from ocean waves, could well operate in much the same way as vesicles. They could easily nucleate cloud growth. And so could those silver iodide crystals employed for rainmaking. Perhaps that's why those substances are commonly referred to as cloud "nuclei." To operate in the manner suggested here, they would need to contain net negative charge, and as I show later (please see Chapter 14), that's typical: By virtue of triboelectricity, practically all airborne substances bear negative charge. So, yes, no obvious reason exists to challenge the view that cloud nuclei can help build clouds.

On the other hand, one aspect of that understanding may be erroneous. The frequent appearance of those nuclei inside clouds has led to the widely held view that the nuclei might be *necessary* for cloud formation.

No evidence I've seen convinces me that those extraneous particles are indispensable. In fact, some clouds contain no particulate material at all.[w6] So, cloud nuclei may be nothing more than welcome guests in vesicle-occupied clouds, but not necessarily permanent residents. Primary ownership may be assumed by vesicles alone.

CLOUD CHARACTER

Energetics of Cloud Formation

If you're with me so far, then you might be thinking one step further. Yes, the like-likes-like mechanism may explain the forces needed for condensation. But, setting up those forces requires energy: Charges need to be separated. Where might the needed energy come from?

Almost certainly this energy comes from the sun (Chapters 4 and 7). Solar energy drives the two relevant species out of the water: EZ-enveloped vesicles and positively charged hydronium ions. Creating those two species requires separating charges, which, in turn, requires energy. Hence, it's the sun's energy that ultimately bears responsibility for creating the raw materials needed for cloud formation.

On the other hand, it's not just the sun that supplies energy; the earth itself also radiates. So, clouds may form even during nighttime. But that latter energy supply may be largely of the recycled variety — sourced from solar energy captured by the earth during the day and re-radiated at night.

Ultimately, then, it's our friendly sun that once again does the job, fracturing those unsuspecting water molecules into their component parts and thereby separating charges. Lodged respectively in vesicles and hydronium ions, those separated charges then represent potential energy. That energy eventually gets released as the separated charges approach one another for cloud buildup. Thus, *clouds don't simply accumulate; they build from forces derived from solar energy.*

To summarize the proposed mechanism: A cloud comprises a large aggregate of negatively charged water-filled vesicles, drawn toward one another by atmospheric hydronium ions (**Fig. 8.3**). All of this requires energy. Without the sun's energy, the creation of those two oppositely charged entities could not occur, and we might have skies without clouds.

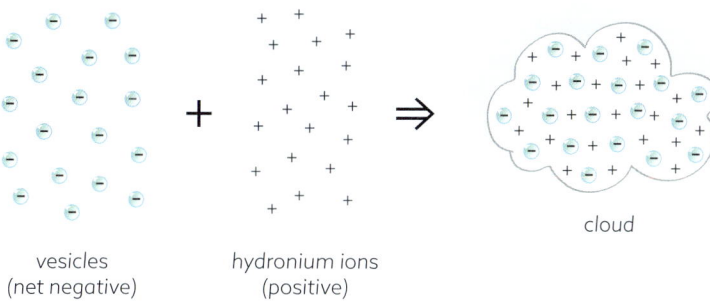

Figure 8.3. Clouds consist of negatively charged vesicles drawn together by positive charges lodged between them.

Cloud Growth

Given our developing understanding of cloud formation, we now advance to specifics. To create a cloud, we need vesicles and hydronium ions — but what fraction of each? What ratio? And, how might that mix allow small, embryonic clouds to grow bigger?

In terms of numbers, only relatively few intervening positive charges should suffice for drawing the negatively charged vesicles into stable proximity (Chapter 7). To appreciate why, think of the contrary: numerous intervening positive charges. If those positives were present in high enough concentration to attract nearby vesicles with force exceeding the repulsive force between those vesicles, then the vesicles would fuse. Cloud components would coalesce into one large mass of liquid water, pouring down from the cloud like the proverbial overturned bucket.

A more modest number of positive charges, on the other hand, could create enough attractive force to balance the inter-vesicle repulsive force, bringing those vesicles into reasonable proximity of one another, but short of fusion. Limited charge should suffice.

The upshot: Because the positive-charge requirement is modest, the addition of that positive charge should not compromise the clouds' overall negative charge. Formed clouds should retain net negative charge. Clouds

Making Serious Rain?

Impossible! Yet ample evidence confirms otherwise. Envision the Sahara turning green, with fruits, vegetables, and trees covering what had once been barren desert.

It all began with an invention of the legendary Wilhelm Reich, a protégé of Sigmund Freud. Reich was controversial — so much so that the Nazis saw fit to burn his books. And when he eventually moved to the US, this time it was the American authorities who burned his books.

Reich's contributions included the "cloudbuster." This rain-making apparatus (figure) consists of a parallel array of long metallic tubes. Their rear ends connect

electrically to a source of water, while their front ends point to a region of the sky. In said atmospheric region, clouds eventually gather, followed by rain. Or so claimed Reich, along with some of his ardent followers.

Among such followers has been the Algerian entrepreneur, Madjid Abdelaziz. His goal: Use this technology to create an east-west green belt running across the Sahara desert. Poverty-struck migrants heading north from central Africa to Europe would then stop to marvel at the unexpected richness. With such abundance greeting them en route, why travel further to an uncertain reception in Europe? Migration peters out.

Abdelaziz has succeeded beyond expectation.[w7] Bountiful greens now stretch from east to west across much of the Algerian desert, with plans to extend that belt across the entirety of north Africa.

Creating rain in this way would surely seem implausible, if not impossible; yet the evidence indicates otherwise.

should therefore repel the earth, much like their constituent vesicles. They should stay aloft, as explained just below.

Why then should clouds grow? As the sun continues to bear down on earthly water, freshly evaporated vesicles should rise into the neighborhood of existing clouds. If hydronium ions are present as well, then the existing clouds can gorge on the newly arriving vesicles, thereby increasing their size. Indeed, videos reveal the presence of particles hovering near clouds,[w8] presumably vesicles or vesicle clusters not yet taken in by the cloud body but ready to be consumed.

Any such growth would depend on circumstances. Conditions with an abundance of vesicles and positive hydronium ions could easily lead to clouds of practically infinite extent — especially as the like-likes-like mechanism bonds individual clouds to one another. Clouds could then grow practically without limit, as do stratus clouds. Or, if materials for construction remain in shorter supply, then clouds could persist as smaller units. This would happen if even modest cloud growth depletes excess vesicles and/or hydronium ions from the cloud's vicinity. Such depletion would then ensure that clouds remain of limited size, even over vast stretches of ocean. Cloud size would ultimately depend on the local supply of construction materials.

In general, then, the proposed cloud-building paradigm is rather simple: the larger the material supply, the larger the cloud.

What Keeps Those Clouds Afloat?

If ever you are lucky enough to peer over a vast expanse of flat prairie land or a sizeable lake, in midafternoon, and see the cumulus clouds looming in the distant sky, you may notice something striking: The clouds often lie at the same altitude (**Fig. 8.4**). More precisely, their flat bottoms often line up. It's almost as though all clouds rest on some great glass coffee table.

This odd behavior raises many questions. Why should the bottoms of those clouds remain so uniformly level across vast reaches of sky? What determines their altitude? And even more fundamentally, why should a cloud even float at all? Remember Chapter 7's consideration of cloud weight, measured in elephant equivalents? We accept the fact that those weighty clouds float in the air, but we mostly ignore the critical question: What keeps them suspended?

Figure 8.4. Multiple cumulus clouds with flat bottoms lined up.

The answer may once again lie in electrostatic repulsion. Vesicles bear substantial net negative charge; and, since clouds are built largely of vesicles, with limited amounts of positive charge to bring condensation, those clouds should retain net negativity. The earth also bears a negative charge (Chapter 2). Hence, the two bodies ought to repel. This repulsion, I suggest, may be the source of cloud's apparent weightlessness. In other words, *clouds should levitate because they are repelled from the earth.*

Could there really be enough repulsive force to keep those clouds suspended above the earth?

The immense power of electrostatic repulsion was demonstrated in the Kelvin Water Dropper. In that apparatus (Chapter 1), charged droplets fall from above, ordinarily settling into a collecting vessel beneath. Once the vessel accumulates enough charge, subsequent descending droplets may slow down, stop, and even reverse course and turn upward (**Figure 1.5**). This demonstration teaches us a lesson worth remembering: *Electrostatic repulsive forces can evidently levitate water droplets.*

That same levitational principle could in theory apply to the vesicular droplets, or, indeed, to the collection of vesicular droplets that we know of as a cloud. Charge forces could propel clouds upward, countering the pull of gravity. If this can occur in the Kelvin apparatus, then why not also in nature?

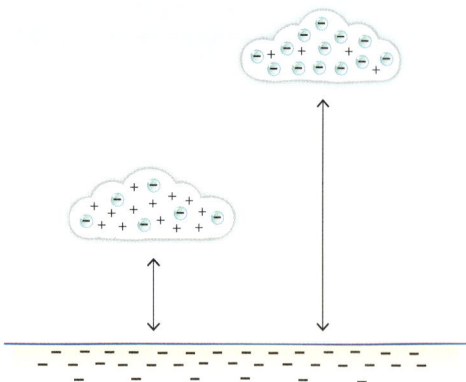

Figure 8.5. Clouds bearing more negative charge (fewer positives) should float higher in the sky.

But how far upward? Simple physics suggests that charged vesicles should rise from the earth until the downward gravitational pull balances the upward electrostatic push. The gravitational pull assuredly drops off with distance above the earth; but the upward push falls off faster, because the electric field drops off so much more dramatically with increasing altitude (Chapter 2). Clouds should rise to the point at which that upward push declines just enough to match gravitation's pull. Barring any extraneous forces, clouds ought to remain at a height set by that balance (**Fig. 8.5**).

These considerations allow us to understand why nearby clouds often sit at the same elevation. Forming under the same atmospheric conditions, nearby clouds should bear similar net charge, experience similar repulsive force from the earth, and therefore reside at a similar height.

The same reasoning can help us understand the converse: why different-looking clouds may sometimes *fail* to situate themselves at the same height. Such clouds may even stack above one another (**Fig. 8.6**). Often arising from different weather systems and blown together from different directions, the respective clouds may contain different concentrations of negatively charged vesicles and/or intervening positivity. Their net charges may therefore differ, and the clouds should naturally settle at different heights.

Figure 8.6. Clouds sometimes stack at varied altitudes, arguably depending on their charge content.

This understanding may help resolve an enduring paradox. Meteorologists' views have long attributed cloud height to prevailing levels of atmospheric temperature and pressure. But if cloud height depends on features of the local atmosphere, then why would clouds stack at multiple levels? The charge-centered paradigm potentially resolves that enigma, as differing cloud heights may correspond to different levels of net charge within each cloud. Heights depend on internals, not the external.

Then, there's the clouds' ephemeral nature. Perhaps you've had the occasion to stare dreamily at the sky, watching a wisp of cloud getting swallowed up by a larger "mother" cloud. That merger could happen if ambient positive charge were available in enough quantity for the like-likes-like mechanism to forge inter-cloud linkage, thereby uniting wisp and cloud. Conversely, if the local positivity were to diminish, say because of diminished evaporation, then not only would the wisps fail to unite, but also the mother cloud might itself dissipate into unassociated vesicles.

Indeed, you may have seen dissipating clouds and wondered, as I had, where the contents went. The contents may be there still, but too sparsely spaced to scatter enough light to make you aware of their presence. Simple dispersal may create apparent disappearance.

To recapitulate, the question as to why clouds float may have a simple answer. Clouds are built largely of vesicles. Vesicles may float because their inherent negativity repels the earth's negativity. Still, you may wonder whether any such charge-based force could be powerful enough to lift large assemblies of vesicles. But please recall the ungodly strength of those electrostatic forces (Chapter 1): A moderate number of charges can lift stacks of jumbo jets. If so, then charges could easily have strength enough to keep even large, weighty clouds suspended in the air.

Thus, we offer a possible answer to the earlier-raised question (Chapter 7, second box): What keeps those weighty elephants suspended in the sky? Those fat elephants may be so bloated with negative charge as to maintain their distance from the similarly charged earth. In that way, they avoid plummeting unceremoniously onto your head.

Why Are Clouds Puffy on Top?

From cradle to grave, cloud building requires energy from the sun. Solar energy, I argue here, may also help explain the clouds' characteristic shapes, often puffy on top and flat on the bottom. Perhaps you've noticed (**Fig. 8.2**)?

A possible explanation for that upper puffiness: Solar energy beats down directly upon the cloud tops. That impinging energy should help grow the negatively charged EZs lodged at the top. The increased negative charge, in turn, should attract nearby atmospheric positive charges, whose addition ought then to attract more proximate vesicles from the atmosphere, *etc.*, leading to the buildup of those characteristic cauliflower-like mounds. With continuous input of energy from above, the tops ought to become increasingly puffy — simply because both the energy and the needed components are freely available there.

The bottoms, by contrast, receive less energy. They receive only the scant solar energy that manages to penetrate through the clouds, along with some weaker energy radiating from the earth beneath. The consequently lower EZ buildup at the bottom should bring reduced tendency for cloud growth there, and hence little or no bottom puffiness. Hence, clouds ought to remain flatter on the bottom.

Clouds As Art?

Occasionally, clouds may present themselves in geometric patterns. The gaps between clouds repeat regularly in either one or two dimensions (see figure). Perhaps you've noticed those intriguing arrays.

Such comb-like and dot-like cloud patterns seem too regular to happen by chance. Contemporary meteorological models relegate their appearance to alternating rainfall and consequent cooling-induced air movements. Properly timed and spaced, those events could plausibly create the observed patterns.

An alternative explanation comes from the similarly repeating patterns seen in colloidal suspensions. Colloidal particles cluster by the like-likes-like mechanism, leaving open voids between those clusters.[4] Those voids may resemble the voids seen between those patterned clouds.

Consider, for example, the regularly spaced white cloud puffs in *panel a*. Each cloud may be thought of as a vesicle cluster, bearing net negative charge. Such negatively charged clusters would then act as "particles," tending to leave intervening voids as they approached one another by the like-likes-like mechanism.[4] The occasional presence of regularly repeating dot-like cloud patterns should therefore not be unexpected.

What I'm suggesting is that the cloud constituents may act in the same way as the colloidal constituents. If the like-like-like principle applies, then voids should be seen in both. Those voids could be two dimensional *(panel a)* or one dimensional *(panels b, c)*.

Reinforcing the cloud-colloid correspondence is the similarity of their optical features. As mentioned in the previous chapter, the high, thin cirrus clouds that sometimes populate the upper reaches of the troposphere often display halos and iridescence. So do colloidal suspensions.[5] Colloidal particles can organize themselves so regularly that they diffract light, much like the cirrus clouds.

Thus, clouds and colloids may well be governed by similar principles. I emphasize this line of thinking because envisioning the cloud as a colloidal suspension of vesicles may be helpful in exploring why clouds behave the way they do.

But I digress. In focusing on technicalities, I don't mean to distract you from the fascination offered by those regularly repeating cloud patterns. Please do search them out next time you fly, either from below or especially from above. I hope they mesmerize you as much as they mesmerize me.

(a)   *(b)*   *(c)*

With those flat bottoms, and uniform charge repulsion from the earth below, we can perhaps understand cumulus clouds' oft-unvarying bottom altitudes.

A test of that flatness-charge hypothesis would check whether flat-bottomed clouds appear mainly over flat surfaces, either prairie land or calm water; there, earthly surface charges should be evenly distributed, creating uniform repulsion. Over rugged terrain such as mountains, by contrast, the bottoms of clouds ought to be well-featured rather than flat. I can at least loosely confirm this distinction from peering out of my living room window: Cumulus clouds floating for some time directly above Lake Washington often have flat bottoms, while those hovering over the hilly terrain beyond do not. A formal study of this charge-based expectation would certainly be welcome.

As products of incident radiant energy, cumulus clouds should form in the presence of sunshine. Observations appear to fit. Over land, those fair-weather clouds generally begin forming in midmorning, grow toward midafternoon, and commonly disappear after sunset.[w9] On occasion, they may persist through the night, perhaps the result of unusually strong radiant energy emanating from the earth, the moon, or perhaps even from sources of cosmic energy, all of which persist through the night. But the general timing of appearance of those cumulus clouds seems to fit the sunshine-based expectation.

Having considered both a cloud's energy requirement and its shape, I should next proceed to the promised discussion of rainfall. I will. But I first wish to interject a few words about a feature usually associated with foul vs. fair weather: atmospheric pressure. Surely, you've heard the mildly ominous warning, "low pressure system on its way" — meaning that rain is coming, or perhaps something more intense.

Why should foul weather be associated with low pressure? And, why should a low-pressure system "come" from somewhere? Who dispatches it?

Foul Weather and Atmospheric Pressure

To characterize the degree of fairness/foulness of weather, forecasters frequently employ the term "pressure." Atmospheric (or barometric) pressure supposedly relates to the weight of the air pressing on the earth. High pressure characterizes fair weather; low pressure, foul weather.

Convenient descriptors perhaps — but why should atmospheric pressure relate to weather?

According to prevailing views (but see Chapter 3), atmospheric pressure arises from the column of air resting on the earth's surface (**Fig. 8.7**). Air has mass. A column of air has a lot of mass, so the column presses measurably on the earth. Higher pressure implies a taller and/or denser column, while lower pressure implies a shorter or less dense column.

high pressure   low pressure

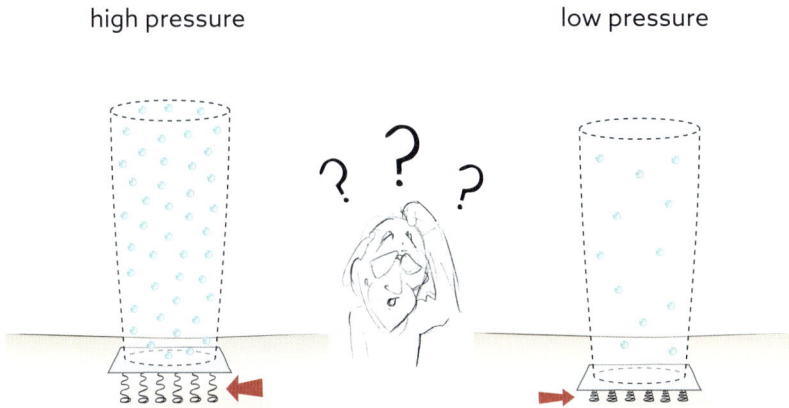

Figure 8.7. A paradox. More moisture in foul weather should create denser air. Denser air, following conventional understanding, should create higher pressure on the earth (red arrows). Yet the pressure is consistently lower in foul weather.

But this explanation brings another paradox. In foul weather, that column should contain lots of water-filled vesicles lodged among the air molecules. Hence, the column ought to be denser. (Arguments have been adduced to support less dense air, but predicated on the assumption that all humidity is in the form of single water molecules, not vesicles.[w10]) Yet, despite this expectation, the pressure in foul weather is not higher, but lower. How can this make sense?

Suppose atmospheric pressure arises from atmospheric charge forces (Chapter 3). In this paradigm, the atmosphere's positive charges attract to the negative earth, so the atmosphere presses on the earth. That compression constitutes "atmospheric pressure."

In this paradigm, the ample quantity of atmospheric positive charges characteristic of sunlight and fair weather should lead to strong air-earth attraction, hence high pressure (**Fig. 8.8**). In wet air, abundant negatively charged vesicles should capture some of those positive charges, condensing them into clouds. With fewer *free* positive charges remaining in the air, the air's attraction to earth diminishes, so the pressure should diminish. Hence, lower pressure ought to characterize inclement weather. Perhaps this reasoning can help resolve the paradox.

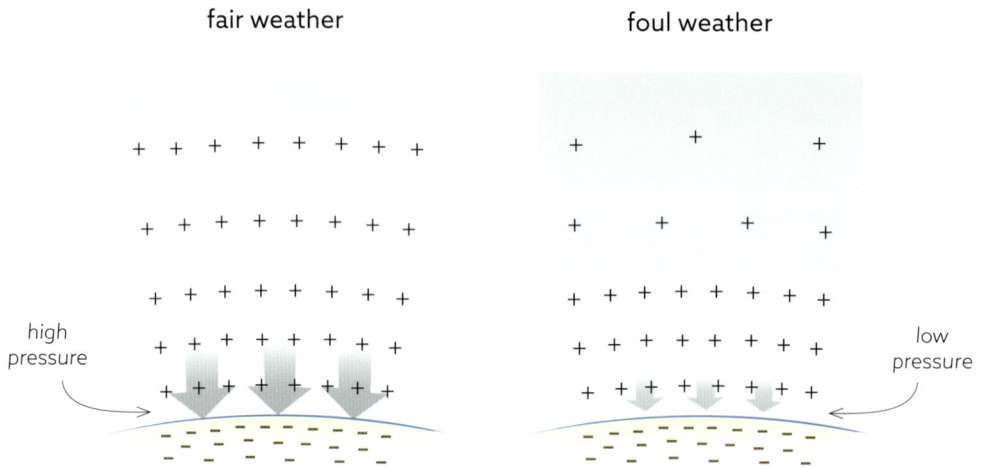

Figure 8.8. Foul weather's low pressure may result from the relatively lower number of high-altitude positive charges attracted to the negative earth.

This proffered framework anticipates only modest pressure differences, even in weather extremes. That's because most of the pressure should arise from the near-earth atmospheric positive charges, which, by virtue of their proximity, attract to the earth most strongly. Those charges ought to remain moderately stable in number, irrespective of weather. By contrast, the higher-altitude charges that do undergo substantial weather-related variations play less major roles in generating pressure because of their greater distance from the earth's surface. Hence, pressure differences between fair and foul weather should remain relatively modest, as routinely observed.

Pressure nevertheless remains a handy predictor of weather, a useful tool that I have no reason to challenge. But, rather than the *cause* of weather, we might, perhaps, better describe atmospheric pressure as the *consequence* of weather. Thus, when ominous clouds form following a pleasant morning, it's not that the low pressure has created the clouds, but rather that the atmospheric conditions leading to cloud formation have created the low pressure.

THE PRECIPITATION DELUGE

A Role of Induction?

To understand why it rains according to the paradigm under development, let us first clarify what transpires when it *fails* to rain — *i.e.,* when the cloud's water-containing vesicles remain lodged inside the cloud. The cloud should retain plenty of negative charge and ought therefore to be strongly repelled by the earth's negativity. Your umbrella can stay home.

Should that repulsion disappear, rain ought to fall — or so you'd think. If the cloud's residence high in the atmosphere results from its ample net negative charge, then adding enough positive charge to neutralize the cloud ought to disable that lift. The contents should logically fall as rain.

A thorny issue remains, however. Acquiring enough of that positive charge to make the rain fall can be a challenge. Negatively charged clouds ought surely to attract nearby positive atmospheric charges, making the clouds less negative. That helps. With diminished peripheral negativity, though, the cloud should experience more difficulty attracting very much additional positive charge. Attaining *net* positive charge, or even neutrality, should thus pose a challenge, and so, within the constraints of this paradigm, rain should not necessarily fall to the earth — ever.

Enter "induction," a feature that could resolve this conundrum (**Fig. 8.9**).

Recall our earlier discussion of induction (Chapter 1). An isolated charge coming into proximity of a surface "induces" opposite charge on that surface. A negative charge will induce surface positivity, while a positive charge induces surface negativity. Common sense underlies this conventional understanding.

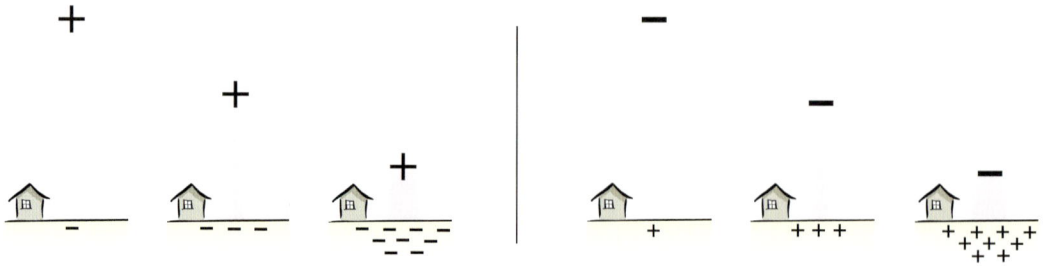

Figure 8.9. Charge induction. Isolated charge in the atmosphere induces equal and opposite charge in the earth below. The closer the charge, the stronger the effect.

Albeit simple, the paradigm comes with consequence. As a result of those induced opposite charges, attraction will follow. Isolated charges will *always* draw toward nearby surfaces. Perhaps you can envision what's coming next, in terms of rainfall.

Please don't think of induction as creating *net* charge. It merely separates charge. An isolated positive charge positioned near a body draws that body's negative charges closer to its proximate surface while at the same time driving positive charges away from that surface. Charges separate. With the opposite charges closer than the like charges, the net result is attraction. It can't be avoided: Induction always creates attraction.

Induction is a widely recognized phenomenon, which can be found in any physics textbook. It's both intuitive and natural, only not yet seriously applied to cloud phenomena as far as I can find. Could it perchance play a critical role?

Will it Rain?

The main point about induction is that it can create an *attraction* between cloud and earth. Like the isolated charge, the charged cloud induces an opposite charge on the surface of the earth below. This effect is well confirmed: Even though the earth is negatively charged, localized areas of earthly positivity can dominate beneath clouds — you can find many illustrations on the web.[w11]

The Balloon Trick

A dramatic example of induction is the charged balloon demonstration (see figure). Charge a balloon by rubbing it on your shirt. Then move it progressively closer to a slow-running faucet. When the balloon gets close enough, the stream will deflect sharply toward the balloon *(photo, left)*. The stream deflects because the balloon charge induces opposite charge on the near side of the falling water-stream, which then draws closer to the balloon *(diagram, right)*. Amaze your friends — the bend can be quite dramatic.

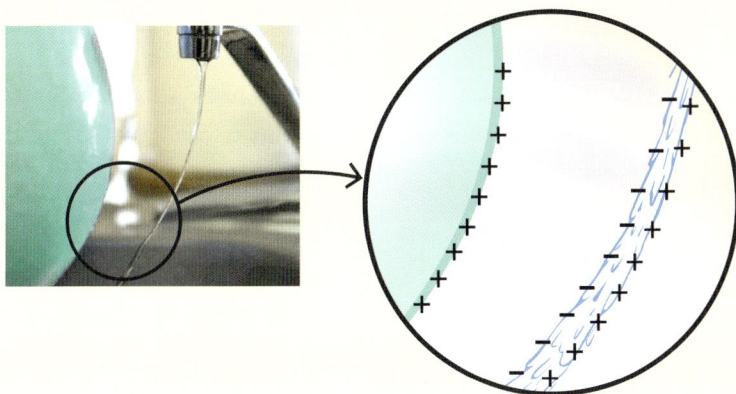

Any such charge separation can have immediate consequences. Negative cloud charge inducing positive earth charge beneath builds electrical potential difference, similar to a charged capacitor. If sufficiently large, that potential difference can give way to electrical discharge. In that way, induction could easily bring cloud-to-ground lightning.

But the main point involves less drama. It's the attraction. The positive charge beneath should draw negatively charged cloud contents downward. With enough attraction, the cloud's contents should draw toward the earth — as rainfall.

Any such dynamic, however, is not a given. Induction can play a determining role only when clouds and earth lie in close enough proximity; otherwise, the inductive effect may remain negligibly weak. The distance consideration takes on significance because the attractive force varies inversely with the *square* of the separation. So, distance counts a lot. Inductive attraction should start to matter only after the clouds have descended close enough to the earth.

Once induction does take hold, it should become self-perpetuating. The inductive attraction should pull the cloud closer to the earth, which in turn ought to increase the induced earth-surface charge, pulling the cloud even closer, and so on. A progressively increasing earthward pull is practically guaranteed. Once the cloud falls into the grip of induction, vesicular droplets should be irreversibly drawn downward. Those falling droplets constitute the familiar phenomenon known as rain.

According to this scenario, then, rain does not simply fall. Rather, the earth *pulls* the raindrops. We commonly presume that rain falls freely, but the induction mechanism implies otherwise — a distinct earthward pull. This active pull, in turn, implies something entirely unexpected: Droplet-descent velocity should exceed the theoretically predicted free-fall velocity. This expectation has been experimentally confirmed. High-speed video measurements reveal that raindrops can fall by up to *ten times* the calculated free-fall velocity.[6]

So, rain does not merely fall. It is evidently drawn downward from the clouds. No wonder pounding rain can be so loud!

The question as to why it rains can now be answered, at least hypothetically. According to the paradigm under consideration, the propensity for a cloud to deposit rain depends initially on its capture of positive charge. That's step one. Once enough loss of negativity permits the clouds' descent to a critical altitude, induction takes over. Inductive attraction then pulls the cloud's droplets toward the earth, creating rainfall.

That mechanism can be tested, at least theoretically. Always, rainfall ought to begin with a single low-hanging droplet. Once the inductive pull on that droplet exceeds the cohesive force holding it within the cloud, the droplet should be plucked from the cloud's grip. It ought to begin falling. The two determining forces should be calculable: The cohesive force retaining the droplet can be estimated from the like-likes-like principle. The downward pulling force can be calculated from the elevation of the cloud's bottom and the induced positive charge measured on the ground beneath. When that latter pulling force just exceeds the cohesive force, rain should begin, always. Of course, in any such test, one needs to be mindful of any underlying assumptions. Nevertheless, at least in theory the proposition should be testable.

This explanation leaves unanswered the question of startup: How might the cloud obtain the necessary measure of positive charge to enable its descent into the inductive realm? Is this a random act of God? Or might those positive charges arrive from elsewhere?

The Source of Positive Charge

If the positive charge arrives from elsewhere (local sources are considered in subsequent chapters), then a possible source may be the wind. In Chapter 4, we argued that wind arises from positive charge gradients: Highly positive regions of the atmosphere flow toward regions with lower positivity. Such flow enriches the charge-poor regions with additional positive charge, especially at high altitudes where those charge gradients can grow to substantial levels. So, wind should commonly bring positive charge.

Now imagine what happens as this wind-borne charge passes over a region with clouds, either fully formed or incipient (**Fig. 8.10**, *see next page*). Attraction ought to predominate. The positive charges should draw inexorably toward the negatively charged assemblage of vesicles, entering from above and from the side and partly neutralizing the clouds' negative charge (*panel i*).

Predictable events should follow. First, the cloud should begin descending (*panel ii*) because its reduced negativity diminishes its repulsion from the earth. Meanwhile, positive ions continue swarming toward the negatively charged cloud, lowering it even further. Those absorbed positive charges may also serve as attractors of nearby vesicles, the aggregation of which will increase cloud size (*panel ii*).

Clouds may also begin merging with one another (*panel iii*). Because of the free positive charges arriving with the wind, negatively charged clouds can coalesce. Such like-likes-like amalgamation can transform discrete clouds into the long, continuous stratus clouds typically referred to as "rain clouds."

Meanwhile, the freshly arrived positivity may also affect the cloud's luminosity. With more positive charges available to facilitate like-likes-like attraction, the negatively charged cloud droplets may draw closer to one another, fusing into larger droplets. Those larger droplets back-scatter more of the sunlight entering from above, leaving less light coming through below. This confers a darker appearance on the cloud, giving it the ominous look of a rain cloud.

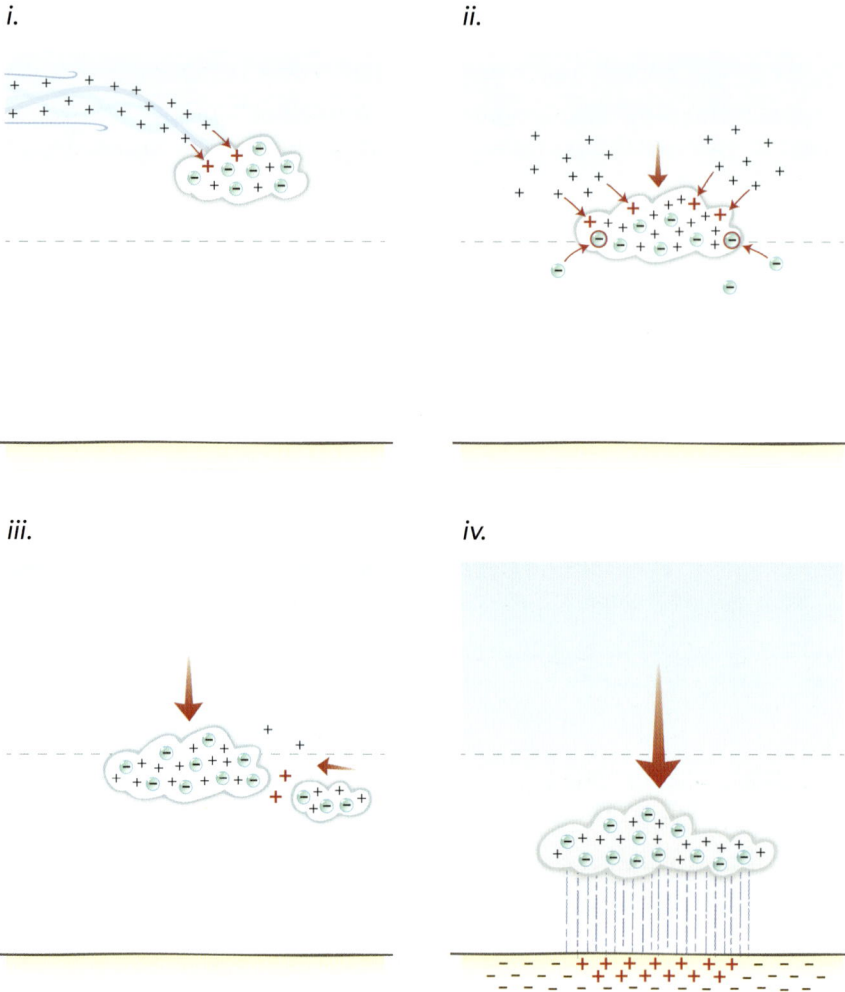

Figure 8.10. Events leading to rainfall. See text for explanation.

Finally, once the cloud has descended sufficiently, inductive forces may begin to prevail (*panel iv*). The cloud should begin inducing appreciable opposite charge on the earth's periphery below, while that periphery should in turn induce additional opposite charge on the cloud. The growing attractive forces ought to draw the cloud even farther downward. Beyond a critical point, the cloud's attraction to the earth should become sufficiently great that rainfall commences.

If that critical point is not reached, then no rain should fall, despite vast cloudiness. Hence, *the "decision" to rain should depend on whether the critical point of inductive takeover has, or has not, been reached.*

You might be well aware of that cloud-lowering criterion if you live near mountains. From my home in Seattle, I can see the Cascade Mountains to the east. When clouds lie well above those visible snow-capped peaks, it generally won't rain. But, as the peaks become shrouded in cloud cover, we commonly surmise that rain may be on its way; and often it is. Cloud lowering seems critical for rainfall, as the induction effect predicts, and as commonly observed.[w12]

Finally, why does rain fall in separated drops instead of as one large dump? To break free of the cloud, a droplet (vesicle) must experience a downward pulling force that exceeds that droplet's propensity to cling to the rest of the cloud. Think of pulled taffy. Such rip-off ought to happen at the bottom of the cloud, where proximity to the induced positive charge below creates the largest downward pull. Hence, bottom droplets ought to get plucked from the cloud one at a time.

Much of this inductive action, according to the hypothesis, may originate following the arrival of wind-borne positive charge. That explains why foul weather doesn't just appear — it often "moves in" from some distant locale. From where, exactly, is a question I'll address in a moment. Irrespective, those incoming winds may carry more than just a force pressing on your face; they may carry the positive charge responsible for initiating foul weather.

Finally, the incoming positive charge also brings semantic implications. If you happen to dislike rain, you might say that positive charge brings negative consequence.

Why Consistent Patterns of Weather?

Seattle summers can be pleasant and dry, while bleak skies and appreciable rain predominate during winter. Other regions of the earth experience comparable seasonal patterns, and the question is why. What brings alternating fair and rainy seasons to these regions year after year?

For explanatory purposes, I hold up the Pacific Northwest as an example. During summer, long days with tourist-drawing weather result from the earth's tilt. With gobs of sunshine, plenty of evaporating vesicles should rise as the day progresses (Chapter 7). If enough positivity appears as well, then puffy white clouds may dot the summer sky. If not, then clear blue prevails.

Winter presents more interesting patterns. Maximum solar intensity now shifts to the southern hemisphere, spanning from deep into that hemisphere at one edge, to somewhere just north of the equator at the other (**Fig. 8.11**). Within that belt, high-level positivity should build from sunlight. At the belt's edges, however, those positive charges should tend to escape to less well-endowed regions; they should spread both southward (producing the notoriously hostile southern seas), and northward, including toward the wintry Pacific Northwest.

SUMMER IN SOUTHERN HEMISPHERE

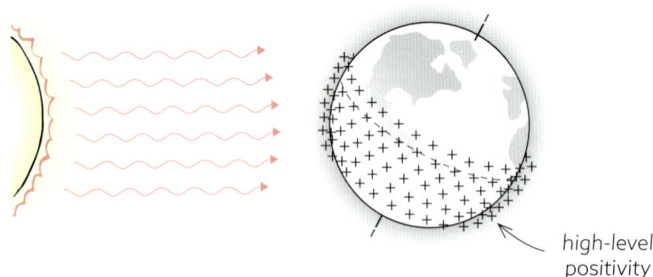

high-level
positivity

Figure 8.11. In the southern hemisphere, summertime builds high positive charge. That positive charge tends to escape to regions less endowed with high positive charge, both to the north and south.

That north-directed high-level positivity should arrive in the Pacific Northwest in abundance (**Fig. 8.12**). The region also experiences the prevailing westerly wind, which delivers moist air from over the Pacific (Chapter 4). Now, envision the convergence of those two systems. With positive charge arriving from the south and moist air from the west, the elements required for serious cloud formation are present in abundance. Seattle should experience the inevitable southwesterly flow of winter rain and gloom.

This seasonal analysis should apply at any place with geographical features comparable to those of the Pacific Northwest. One of those is Northwestern Europe, whose latitude and position just east of a great

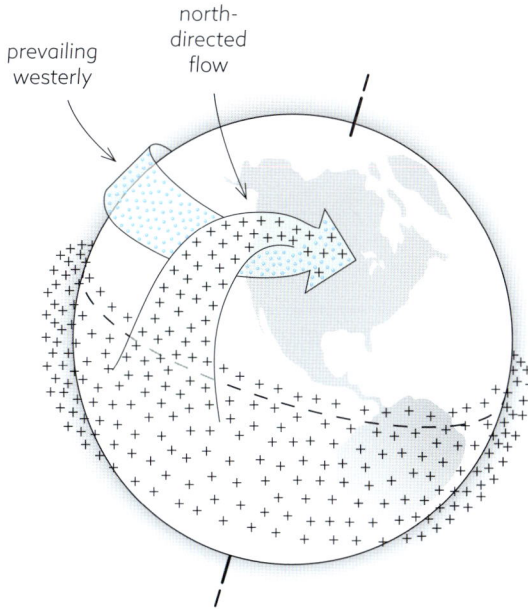

prevailing
westerly

north-
directed
flow

Figure 8.12. When winter comes to the Pacific Northwest (with summer below the equator), high-altitude positive charges from near the equator head north, mixing with moist air from the prevailing westerly wind flow. This convergence may bring rain to Seattle.

body of water resemble that in the Pacific Northwest. Both regions experience similar patterns of pleasant summers and mild, rainy winters. Plants that thrive in Seattle also thrive in London.

While such predictability might delude us into thinking that summer sunshine and winter rains prevail unceasingly in those areas, patterns can of course vary. The Pacific Northwest does enjoy occasional breaks of sunshine amidst the prevailing winter gloom, which can bring halfhearted smiles. On the other hand, knowledgeable locals can often spot rain's imminent return by using the Mount Rainier gauge: When clouds gain enough positivity to settle like a cap on that nearby peak (**Fig. 8.13**), locals know that rain is surely on its way.

Figure 8.13. Mt. Rainier, with cloud cap.

Fitzroy's Retort

The Fitzroy Retort is a device used for predicting weather, based on the character of an internal precipitate. It was named after Vice Admiral Robert Fitzroy, who accompanied Charles Darwin on his famous voyage on the HMS Beagle. Fitzroy was a meteorologist, intent on improving weather prediction, and renowned for inventing this classic device.

To construct the retort, certain chemicals are mixed to yield a precipitate inside a liquid. Depending on weather conditions, the precipitate changes shape, texture, and volume, sometimes dramatically (see figure). The retort remains sealed.

Sealing should confer immunity against the influence of atmospheric pressure changes; and, as the retort is generally kept in an environment with fairly stable conditions, neither should ambient temperature variations play a decidedly critical role.

Therefore, some other atmospheric feature likely induces the observed changes. Nobody to my knowledge has identified that feature, but one wonders whether subtle changes of atmospheric charge or electric field might bear responsibility for the stunning pattern variants, which help predict the weather.

What creates those variants from the prevailing pattern?

Weather blips may arise from variations of the sun's intensity.[7] If solar intensity and other sources of electromagnetic energy were to remain constant, then weather patterns should remain relatively stable. But those energies undergo seemingly unpredictable fluctuations as well as cyclic changes, which can demonstrably impact weather patterns.[8] Perhaps that's why sunny respites may punctuate winter's gloom, and why some winters will be more oppressive than others.

Weather may even be influenced by energy coming from distant stars.[w13] Stars radiate electromagnetic energy, just like our sun. Although of relatively low magnitude on Earth, starlight may nevertheless exert some influence on our weather. Stars also emit cosmic energy, in the form of protons and alpha particles. Those positively charged particles could help promote cloud formation in the same way as do other positive charges.[w14]

These seemingly unpredictable variations in incoming energy threaten to make weather predictions almost reliably inaccurate. Even subtle changes of incident intensity could seriously impact atmospheric charge distribution, which in turn, could significantly alter our weather. Who would have thought that the distant universe might bear such an intimate connection to our everyday lives?

Summary

The concepts presented in this chapter emphasize a previously unrecognized player in the drama of weather: electrical charge. As the sun beats down on the earth's water, positive charges build in the atmosphere. High positive charge alone presages fair weather. Some of that positive charge blown into a region replete with atmospheric moisture, however, brings something more notable: clouds and foul weather.

Cloud formation, according to the proffered hypothesis, requires two components: positively charged hydronium ions and negatively charged vesicles, the latter replete with moisture. When those oppositely charged components mix, clouds form. The clouds exhibit a billowy texture, a signature feature that may arise from the like-likes-like interaction, which maintains entities at some distance from one another. Hence, the vesicles maintain some separation. Separation confers a feathery appearance, rather than a liquidy one, as ordinary vapor condensation might produce. No zipper holds liquid water inside the cloud.

The altitude of a cloud ought to depend on the cloud's net charge. Clouds bearing high negativity should lodge high up in the atmosphere because of strong repulsion from the earth's negativity. Clouds with less negativity, *i.e.*, with relatively more positive charge content, should settle lower. The higher the fraction of positive charge, the lower the cloud.

When a cloud descends low enough, it begins to induce opposite (positive) charge on the earth just beneath.[w10] That opposite charge pulls the cloud further downward. Once the cloud descends to some critical threshold, the inductive pull may grow strong enough to draw the clouds' contents resolutely earthward — in the form of rain. Rain falls, I argue, because the earth *attracts* constituent droplets. It draws them downward — a notion supported by evidence of unexpectedly rapid rainfall.[6] Rain is apparently *pulled* toward the earth.

Rainy seasons occur when the two necessary components, moisture (vesicles) and positive charges (hydronium ions), appear consistently in the same locale. The moisture may come from vesicle-containing air rising from above oceans or large lakes. The positive charge may arrive windborne, from places with excess positivity. When seasonal situations dispose a region to a consistent feed of those two components, rain may occur frequently. Such areas include the Pacific Northwest and Northwestern Europe, both of which experience predictable winter clouds and rain.

Armed with the rudiments of a proposed weather paradigm, we can now move on to consider more exotic kinds of weather. The charge-centered concept, I believe, can offer straightforward explanations for some of the more impressive weather phenomena ranging from tropical storms all the way to typhoons and tornadoes. The next chapter explores how. I will argue that *locally* generated clusters of positive and negative charge can create a super-abundance of cloud-building constituents, leading the severest formations of foul weather.

CHAPTER 9

Thunderstorms

It was Philadelphia, and I was a poor graduate student at the University of Pennsylvania, trying to figure out how to furnish my new apartment. The solution seemed at hand. My friend's mother was remodeling, and for practically no cost at all I could furnish my bedroom, including the double bed I'd hoped might sustain some good use.

Only one obstacle: a looming thunderstorm. On the very day I arranged to purchase the furniture, the forecast was not optimistic. Thunderstorms appear commonly during summers on the east coast. You can expect them on a semi-regular basis. But for the most part, you go about your life without a whole lot of forethought.

So, I proceeded to rent the cheapest trailer I could find and attached it to the rear of my Peugeot 404. It worked. I proceeded to Ms. Pennock's home, made my purchase, and loaded up. Loading was easy because of the trailer's open top. But just as I began driving, the skies opened. The thunderstorm unleashed four inches (10 cm) of rain in the hour it took me to drive to my apartment. Rarely had I encountered such pounding rain.

Eventually, most of the furniture dried out, but the mattress never did recover. Even years later, with squeaks and groans at each body turn, the mattress let me know of its less-than-illustrious history. The lesson was clear: Respect the weather.

And, that we do, as we continue our exploration of weather events in this chapter. Having dealt with fair weather and ordinary rainfall, we begin exploring the more serious types of weather events, starting with thunderstorms. What creates those heavy rain squalls? And why are they accompanied by flashes of lightning and roars of thunder?

Thunderstorms' Features

If you surmise that I will build from the foundation established in the previous two chapters, then you are on track. I promised one central paradigm for all weather. I suggested that if adequate, that paradigm ought to account for events ranging from light drizzle to monster hurricanes. I've not forgotten that promise — a quest for paradigmatic simplicity.

Figure 9.1. A thunderstorm, in Eastbourne, UK.

Most readers will first wish to know the "prevailing wisdom." Thunderstorms have been with us forever, and by now you'd expect that atmospheric scientists surely would have figured out why we experience those flashes of lightning and crashes of thunder (**Fig. 9.1**). Can't we simply accept the established points of view and move on from there?

On the topic of "Thunderstorms," here's an opening snippet from Wikipedia:[w1]

> *Thunderstorms result from the rapid upward movement of warm, moist air, sometimes along a front. As the warm, moist air moves upward, it cools, condenses, and forms a cumulonimbus cloud that can reach heights of over 20 kilometers (12 mi). As the rising air reaches its dew point temperature, water vapor condenses into water droplets or ice, reducing pressure locally within the thunderstorm cell. Any precipitation falls the long distance through the clouds towards the Earth's surface. As the droplets fall, they collide with other droplets and become larger. The falling droplets create a downdraft as it pulls cold air with it, and this cold air spreads out at the Earth's surface, occasionally causing strong winds that are commonly associated with thunderstorms.*

And, just in case our friends across the pond in the UK hold differing views, here's what Britannica says on the same topic:[w2]

> ***Thunderstorm***, *a violent, short-lived weather disturbance that is almost always associated with lightning, thunder, dense clouds, heavy rain or hail, and strong, gusty winds. Thunderstorms arise when layers of warm, moist air rise in a large, swift updraft to cooler regions of the atmosphere. There the moisture contained in the updraft condenses to form towering cumulonimbus clouds and, eventually, precipitation. Columns of cooled air then sink earthward, striking the ground with strong downdrafts and horizontal winds. At the same time, electrical charges accumulate on cloud particles (water droplets and ice). Lightning discharges occur when the accumulated electric charge becomes sufficiently large. Lightning heats the air it passes through so intensely and quickly that shock waves are produced; these shock waves are heard as claps and rolls of thunder. On occasion, severe thunderstorms are accompanied by swirling vortices of air that become concentrated and powerful enough to form tornadoes.*

To summarize: Abundant updrafts and downdrafts seem to constitute the key points characterizing thunderstorm genesis. These features seem sufficiently well documented as to be considered "facts." We will find no reason to dispute them.

On the other hand, what's missing from the scenarios outlined above are answers to the "how" and "why" questions. You may dig further, but I don't think you'll find answers. For example: Why should like-charged droplets collide? And how could opposite charges separate enough to create intense flashes of lightning discharge?

So, we approach the thunderstorm from the vantage of the mechanism under development. We focus on the two rising species: the moisture contained in negatively charged vesicles; and the oppositely charged hydronium ions. Positive-negative; yin – yang. We don't discount any of the phenomena known to accompany thunderstorms. Instead, we see how they may integrate naturally into this emerging framework.

A Late Summer Afternoon in South Carolina

Whereas thunder-bearing storms may come at any time of year at virtually any place, in the US they show up most regularly in the eastern half of the country, particularly in late afternoons during warm summers and early autumns. If you've lived in the eastern US, you'll know what I mean.

Why at that place and during that time? What clues can we extract from that information?

Vast bodies of water surround the eastern US — the Great Lakes to the north, the Atlantic Ocean to the east, and the Caribbean Sea to the south. Water lies practically everywhere. In summertime, when evaporative processes shift into high gear, the atmosphere should be packed with vesicles and positive charges (Chapter 7). Especially by mid-afternoon, the air can grow oppressively hot and humid.

High humidity and high positivity should create the outcomes described in the previous chapter. Positive charges are attracted to the negatively charged vesicles, condensing them into clouds. Nothing remarkable, except in terms of quantity: Owing to the plentiful nearby waters and rapid evaporation evoked by summer heat, the two components critical for cloud formation should appear in abundance.

With plentiful moisture and positive charge, clouds should inevitably grow. Commonly, they begin as ordinary cumulus clouds. Existing clouds may then annex nearby cloud wisps or, they may merge with neighboring clouds through positive-charge linkages (think: like-likes-like) to create larger clouds. All necessary components are abundantly present.

Growth should proceed in the vertical direction as well (**Fig. 9.2**). With building components plentifully supplied, ordinary cumulus clouds may grow into towering cumulonimbus clouds that may extend way up into

Figure 9.2.
Majesty and threat.
A cumulonimbus cloud
with anvil.

the troposphere. They can grow tall. Those clouds may also darken, as positive charges fuse the cloud's vesicular droplets into larger droplets, scattering more light, and leaving the cloud darker.

And so it goes during a typical late summer afternoon in South Carolina. Abundant humidity and local charge can build huge clouds that may grow darker and more threatening by the minute.

To understand the culminating event, please recall Chapter 8's central argument. Bringing precipitation requires three consecutive steps (**Fig. 9.3**): First (*panel i*), the negatively charged cloud draws in high-altitude positive charge. Then (*panel ii*), because the addition of that positive charge diminishes the cloud's repulsion from the negative earth, the cloud descends. Descent brings the cloud into the effective range for inductive buildup of positive charge on the earth beneath (*panel iii*), leading to the pulling of individual droplets downward as rain.

The scenario just outlined could apply to any cloud system. What's different about thunderclouds? Why do they unleash torrential rains,

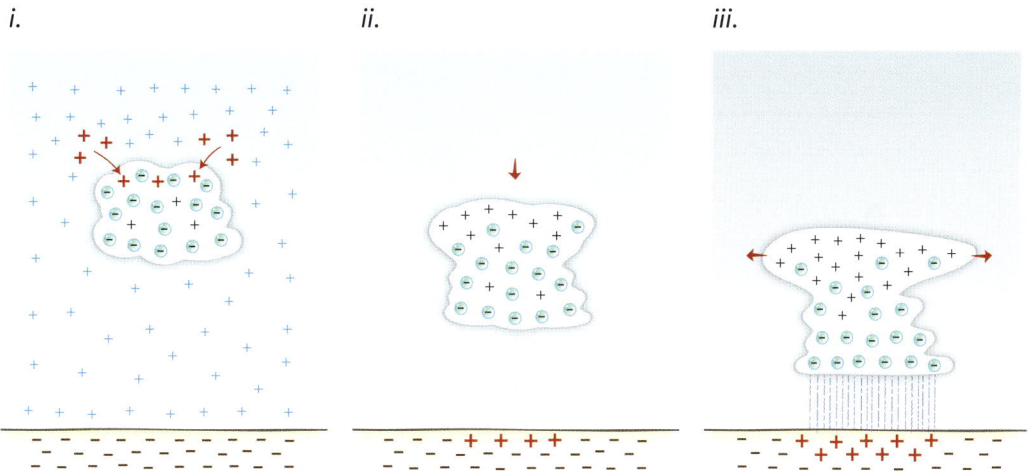

Figure 9.3. An incipient thundercloud draws positive charge from the atmosphere *(i)*. Partial neutralization lowers the cloud *(ii)*, eventually inducing ample opposite charge on the earth. Said charge draws the cloud further downward, increasing induction, and potentially bringing heavy rainfall *(iii)*. Excess positive charge in upper regions may create internal repulsion, resulting in the iconic lateral expansion, or "muffin top," as seen in **Fig. 9.2**.

powerful electrical discharge, and deafening thunder? Might such drama come from the copious amounts of moisture and charge contained within them? Could the difference between an ordinary rainstorm and a thunderstorm lie in the *amounts* of moisture and charge contained within?

I will consider the abundance thesis in a moment. Meanwhile, let me dispel the notion of an apparent exception: In the Pacific Northwest where I live, despite plentiful moisture theoretically available from the Pacific Ocean nearby, why do we see relatively few summer thunderstorms? The answer may lie in the rate of evaporation. The ocean temperature during the summer is on the order of 55 °F, while off Florida by contrast, the ocean temperature will rise to well above 80 °F.[w3] With the persistently low temperature in Northwest waters, evaporation ought to remain limited, and the abundance of evaporative products needed to produce thunderclouds should be relatively scant. So, towering thunderclouds, according to the developing paradigm, should seldom form off the Northwest coast during summer — as observed. The local situation does not contradict the abundance thesis.

Hence, we remain poised to explore the simple hypothesis: The towering cumulonimbus clouds that bring thunderstorms arise when the two essential cloud components are in plentiful supply — negatively charged vesicles and positively charged hydronium ions. More meat and vegetables produce a more bountiful stew.

Size Does Matter

Is quantity truly the critical factor? I will describe next how bountiful amounts of the two required constituents can come from primarily *local* sources. I will then explain how that abundance can produce the lightning, thunder, winds, and massive rain that we usually associate with thunderstorms.

Recall the common rainfall scenario. For the seasonal Northwest rainfall described in the previous chapter, the required positivity arguably comes from some distance — mainly from the belt of high charge well to the south (**Fig. 8.11**). Flowing toward both poles, those positive charges may feed any negatively charged clouds encountered along the way, setting the stage for rain. No reason why those flowing charges might not contribute to thunderstorm formation as well, but I will argue that for the latter phenomenon, other protagonists may take center stage.

Those players are locally generated.

To understand how and why, please recall the role of the blistering summer sun (Chapter 7). Intense solar energy propels massive evaporation. Evaporation brings negatively charged vesicles into the atmosphere in some quantity. Those plentiful vesicles can be detected because they scatter ample light, creating the familiar summertime haze; see **Figure 9.4**. The rising-vesicle feature also fits conventional thunderstorm understanding, that *"Thunderstorms arise when layers of warm, moist air rise in a large, swift updraft."* So far, no serious deviation from convention.

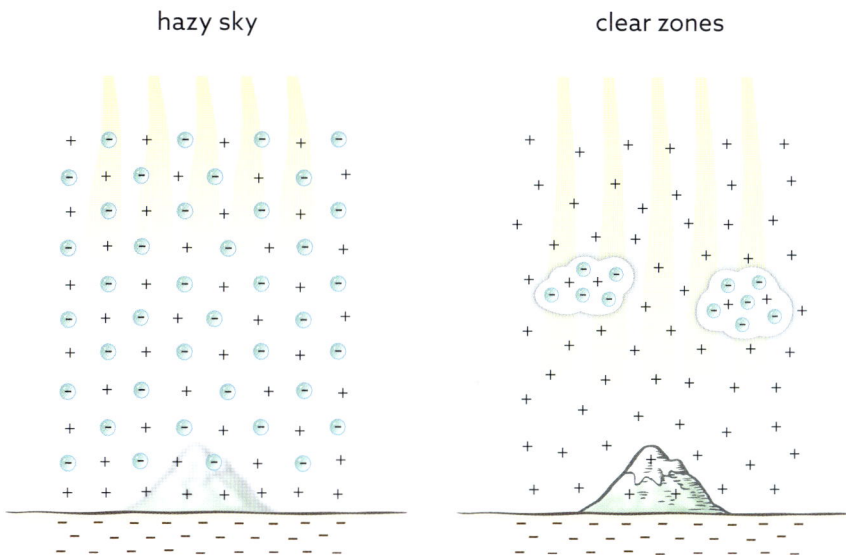

Figure 9.4. Abundant airborne vesicles scatter light, creating haze and rendering distant images such as the mountain peak fuzzy (*left*). Increasing positive atmospheric charge may condense those vesicles into clouds (*right*), leaving clear regions in between. Penetrating through those clear zones, sunlight creates even more local evaporative positivity.

Besides propelling the rise of those moisture-containing vesicles, the intense solar energy also drives positive charges to rise. Evaporation generates both components. Thus, the two critical components necessary for cloud formation should be abundantly available to perform their collective mischief.

The consequences?

Regenerative cloud buildup. As positive charges draw vesicles into increasingly closer proximity to one another, a residual ring-like clear zone opens around those condensing vesicles (*right panel*, **Fig. 9.4**). That clear space permits sunlight to pass through unscathed. The passing sunlight then drives additional evaporating vesicles and positive charges high into the atmosphere, augmenting the existing cloud and allowing it to grow taller, broader, and meaner. That's the "regenerative" feature: Cloud growth leads to more growth.

Because condensation creates those peripheral clear spaces for moisture and positivity to continue to rise, those clouds keep growing. Once they grow bulky enough to begin obscuring all penetrating sunshine, buildup may finally stop. By then, those clouds may have grown to monumental size — as cumulonimbus clouds do.

Still more growth may be promoted by contributions from outside the immediate vicinity. Depending on atmospheric conditions, swarms of high positive charge can come from practically any direction. Such swarms can feed those negatively charged clouds with additional positivity, augmenting thundercloud buildup even further.

The main point here is quantity. Through ample sunshine and regenerative processes, the contributions of copious moisture and positive charge can make the difference between modestly sized, garden-variety rainclouds and looming monsters.

Cloud shape may also differ in those behemoths. Since the positive charge feeding those monster clouds will often rise high into the atmosphere, said charge may approach the cloud from near its top. Hence, you'd expect the cloud's upper regions to bear ample positivity, an attribute that's confirmed.[w4] Further, since positive charges repel one another, you might also anticipate some upper level expansion, creating the thundercloud's characteristic anvil shape (**Fig. 9.3**).

More generally, thunderclouds loom uniquely tall and massive, features that arguably arise from regenerative cloud buildup.

Origins of Thunder, Lightning, and Wind

With their huge size, we may understand why thunderclouds can produce prodigious amounts of rain. But why also the characteristic lightning and the thunder?

Lightning discharges when the potential difference between two points in the atmosphere exceeds a certain threshold. Then, the medium between those points can no longer sustain that electrical potential and current flashes over. We saw an example in the Kelvin Water Dropper demonstration (Chapter 1). The same happens in the atmosphere.

In theory, the flash could occur anywhere in the cloud system. It could occur within the cloud, as atmospheric positive charges absorb at the top, creating a large potential difference between the cloud's top and lower regions. It could also occur from cloud to cloud, with the more negative region of one cloud flashing over to the less negative region of another. Or, it can occur from cloud to ground, with induction (*see* Chapter 8) creating a large potential difference that ultimately gives way to lightning discharge. All such variants are commonly observed.

In setting the stage for cloud lightning, a critical factor may be the rapidity of charge buildup. Separated charges within a body will ordinarily tend to recombine, driving the system toward bland uniformity. That feature may prevail during ordinary rainstorms, where clouds build at relatively modest speeds and separated charges have ample time to recombine. When cloud buildup is rapid and extensive, however, as in the thunderstorms' towering cumulonimbus clouds, accomplishing this equilibration might not be feasible. Then, release may come in the form of lightning discharge.

It's a bit like human relationships. The two parties may sometimes disagree. If those disagreements come few and far between, then time for accommodation may be possible. But too many coming at once will frequently lead to "lightning" discharge.

Arguably, then, the distinction between ordinary rain and thunderstorms may lie in the rapidity of cloud buildup. With slow buildup, ample time may exist for cloud charges to even out. With the rapid, regenerative buildup that characterizes thunderstorm clouds, time may be too short, resulting in lightning discharge.

Where there's lightning, there's also thunder. The prevailing explanation is that discharge current warms the ionized air, which then expands quickly, creating a shock wave, which we hear as thunder. In fact, any movement will create sound. One can imagine the formidable movements of mass that likely accompany any electrical discharge as creating a voluminous crack. So, thunder inevitably follows lightning.

What about the thunderstorms' characteristic winds? Recall the evaporative vesicles rising in the cloud's vicinity. Any such rising air must be accompanied by falling air — otherwise our atmosphere would vanish. With both rising and falling air in reasonable proximity, the aerodynamic circuit must be completed by horizontal winds; hence, those winds are a must.

Winds may also originate from above the clouds. With positive charges rising high in the atmosphere, downdrafts arriving from altitudes as high as Mt. Everest may bring positively charged (cool) air toward the negatively charged earth. Thus, winds are essential features of thunderstorms. Fierce winds are routinely mentioned in conventional thunderstorm descriptors; hence, little in this developing paradigm lies out of accord with those descriptors — only the underlying whys and the wherefores.

Then, there's the thunderstorm's heavy rainfall. As we've argued, the massive positive charge feeding into the vesicle-rich cloud should bring torrents of rain. Indeed, the intense supply of positive charge may make the difference between the fury of the ravaging thunderstorm and the tenderness of gentle rain.

Finally, there are the fears. Should you suffer the misfortune of getting caught where these bodies of negatively charged vesicles and positively charged air slide rapidly past one another, beware. You might find yourself enduring blinding flashes of lightning, deafening roars of thunder, terrifying speeds of wind, and torrents of rainfall — surely a daunting prospect for even the most intrepid among us.

And, if that's not enough, there's also the potential to encounter hail (see box). Assuming you are spared the unwelcome fate of having served as a human lightning rod, that doesn't mean you're safe. There's always the chance that giant hailstones might unexpectedly whack you in the head (**Fig. 9.5**).

Figure 9.5. Hail stones can be massive.

What Creates Hail?

Hail is nothing more than ice. Ice certainly forms when it's cold enough; that's common experience.

I've argued, however, that it's not low temperature *per se* that creates the ice, but arguably the cold-induced EZ growth (see Pollack[1]). As it grows, the EZ initially pushes positive charges out of its expanding lattice. When those positive charges grow intense enough, they massively invade the negatively charged EZ, turning its ice-like structure into ice.

The transition from EZ to ice can happen even at temperatures well above freezing, provided the needed positive charges are present in abundance. That has been demonstrated experimentally.[2] Ice formation does *not* require Arctic temperatures.

The same ice-forming progression can take place inside clouds. Imagine high concentrations of positive charge passing next to low-temperature cloud droplets. Since each droplet contains a negatively charged EZ shell, those proximate positive charges can invade the shell, creating ice. Such "ice droplets" would then link to one another by excess charges, building softball-sized hailstones. This charge-based understanding helps explain why hailstones can form even at summer temperatures.

Beware!

I've just outlined what I believe happens to create thunderstorms. Charges are central. Those tall clouds that we refer to as "thunderclouds" are full of electrical charges. Their massive size permits electrical events to be fully manifested. And with those electrically centered dynamics, we have the lightning, thunder, wind, and rain events that, together, classify as "thunderstorms."

When Will That Thunderstorm Come?

If thunderstorms arise when atmospheric moisture and positivity appear in hefty amounts, then we can make some further predictions about their likely timing and occurrence.

Whereas some of those storms may contain only a single isolated cell, or cloud unit (**Fig. 9.3**), often, thunderstorms consist of linked chains containing multiple cells. Widespread zones of high-altitude positivity presumably help feed those cells. Once the source of positive charge has been drained, the bank of cells will have run its course. The storms associated with each individual cloud unit should then dwindle more-or-less simultaneously. That outcome is common.

Also somewhat predictable is the endpoint's timing. The storm should begin to dwindle as its position on the earth's surface rotates away from the sun. Recall that it was the sun's energy that ultimately created the cloud. Storms do generally dissipate as the sun disappears over the horizon. Occasional storms may linger well into the night, presumably in the presence of unrelenting humidity and unusually prodigious positivity, fueled either by energy re-radiated from the earth or by cosmic-based positivity (protons and alpha particles). Most storms, however, run their course sometime during the evening, as the plug gets pulled on their major source of energy.

Predictability of thunderstorms depends on where the required components originate. If the sources are primarily local and consistently present, then those thunderstorms may occur with regularity, sometimes almost daily. That happens in south Florida during summer months, where ample sources of water lying on all three boundaries of the Florida peninsula supply reliable moisture. The pattern holds as well near other large bodies of water: Northern Lake Victoria in Uganda experiences some 240 thunderstorms annually, while northern Lake Maracaibo in Venezuela is not far behind with approximately 150. Local sources of moisture create predictable thunderstorm activity.

In locales where humidity or positivity come from more distant sources, by contrast, thunderstorm occurrence should be less regular. Examples include the Midwestern US, where environmental moisture is supplied by the Great Lakes — *i.e.,* from a distance. In this case, wind direction would determine the local quantity of atmospheric moisture, making the incidence of thunderstorms more difficult to predict.

Exceptions to the general rule of predictability may arise in mountainous regions, where earth charge can accumulate on jagged peaks. Concentrated mountaintop charge makes for intense electric fields between cloud and ground, precipitating lightning and thunder in places where standard thinking might not predict.

Beyond their often-predictable occurrence and consequence, thunderstorms may also bring relief. On hot, sticky summer days, respite may come from the cooler and drier air left in the storm's wake. Simple evaporation may help bring that relief.[w4] But relief may also come from high-altitude positivity — from the positive charge that helped create the thunderstorm in the first place. Any lingering mass of high positive charge should continue its draw toward the negative earth, even after the main event. That dry, cool, falling air from above can create the so-called "cool front" that often follows thunderstorm mugginess.

Such respite may be brief, yet one can almost hear those inevitable sighs of relief.

Summary

Thunderstorms may be little more than grown-up rainstorms.

Rainstorms build simply. According to the proposed paradigm, ordinary rain may come when positive charges show up in high enough quantity to "condense" negatively charged vesicles (humidity) into clouds. The higher the positive charge contribution, the lower the altitude to which the cloud will descend. Altitude matters. Once the cloud descends far enough to induce appreciable opposite charge on the earth below, inductive forces should begin pulling droplets from the cloud, creating rainfall.

Similar principles may create thunderstorms, although they are typically more intense. Their intensity comes from two features: plenty of local moisture and regenerative cloud-building behavior. The regenerative feature arises as condensation pulls vesicles together to form a cloud, leaving a ring-like, vesicle-free zone around each incipient cloud. That clear zone allows sunshine to penetrate, evaporating additional cloud components, which contributes to additional growth.

So long as local sources of moisture and charge amply persist, and vesicle condensation continues to open peripheral rings for sun-driven evaporation, the cloud should continue to grow. This regenerative feature assures the massively tall entities typical of cumulonimbus thunderclouds.

Those tall, electrically charged clouds should have plenty of opportunity to discharge. Any internal region with high negativity will tend to give way to less negative regions, whether situated inside or outside the cloud. Such discharge creates lightning, followed by the inevitable roar of thunder. Ordinary rainclouds don't manifest those phenomena, arguably because the charges build slowly enough to permit gradual evening out, eliminating the steep gradients that ordinarily bring electrical discharge.

So, while rainclouds and thunderclouds may seem like different species, the distinction may be largely quantitative. The thunderstorm's characteristic features may arise from the clouds' rapid charge buildup and massive size — features emerging because components are sourced locally. Thus, swift buildup happens because the contents come from localized sources rather than from distant, scarcer sources. Propinquity makes the difference.

While thunderstorms may sometimes pack enough punch to scare the daylights out of you, those storms are not weather's only cataclysmic manifestations. The next chapter explores weather events that can turn even more devastating: tropical storms, hurricanes, even tornadoes. I will argue that those weather events may all arise from principles that are similar to those suggested to create thunderstorms, differing for the most part only in intensity and degree.

Indeed, weather could turn out to be a lot less mysterious than presumed.

CHAPTER 10

Weather Exotica

Weather events can certainly arouse fear. I recollect an incident in my youth, driving westward across the country with my girlfriend. We had stopped for a picnic lunch on a rise overlooking a vast expanse of Midwestern plain. The day began gloriously. By midafternoon, the wind had picked up, and barreling toward us from the distant south raced a dark cloud as ominous as any I'd ever seen. This was tornado country.

The wind intensified practically by the minute. Paralyzed by fear of the oncoming peril, I sat terrified in the car while my girlfriend, showing no such panic, calmly darted out to gather the blowing lunch debris so that we could press on and outrun the looming monster. Not one of my more gallant moments, to be sure... but we did manage to survive that tornado.

Tornadoes play a central role in this chapter, as do tropical storms, hurricanes and the like. Exotic and potentially devastating, those weather events remain poorly understood. Here, we will consider the features that conspire to create them.

To do so, we will build on the charge-based foundation developed in the previous three chapters. I will argue that the set of principles developed in those chapters lead directly to these exotic weather phenomena, the outcomes differing depending on variations of moisture and charge.

Tropical Storms: A Beginning

As progenitors of hurricanes, tropical storms are the more muscular variants of thunderstorms. They commonly form in the warm waters near the equator. When ample sunshine and elevated water temperatures intensify evaporation during summer and early autumn, the local atmosphere ripens for mischief. The same conditions that lead to thunderstorms (Chapter 9) may also lead to tropical storms.

Indeed, those tropical storms closely resemble thunderstorms, often unleashing abundant wind and prodigious rainfall. The sea provides all the necessary ingredients: Practically endless supplies of vesicles and positive charges ensure ample buildup of towering cumulonimbus clouds.

These seaborne storms lumber slowly westward, driven by the trade winds (Chapter 4). Those winds drive both Atlantic and Pacific storms, which continue to pick up ocean-derived moisture as they travel westward. They may veer slightly northward or southward, but their main course commonly tracks toward the west.

Along their westward paths, tropical weather systems will often release their contents as localized storms, plaguing ocean-going vessels with strong winds and heavy rains. Vulnerable ships may capsize. Satisfied perhaps with their delivery of squall-based devastation, those weather systems will sometimes dissipate. Others may not. The surviving systems may build up to wreak additional havoc. Continuing to pick up moisture as they proceed further westward, those storms may eventually build into full-fledged hurricanes (also called cyclones or typhoons, depending on where they occur), unleashing considerable fury.

The action, however, begins with tropical storms, where endless local water sources and plentiful sunshine promote regenerative buildup — the same as for thunderstorms but even more marked.

Hurricanes and Their Iconic Features

Hurricanes, as you know, may pound the earth with torrential rains, generate colossal winds, and produce widespread coastal flooding. Eventually they peter out as their moisture source ebbs, either as they pass over dry land, or as they move toward the poles, where cooler temperatures limit evaporative feed. In a nutshell, that's the hurricane's biography.

More than merely strong storms, however, hurricanes show certain distinctive features (**Fig. 10.1**). Two stand out: First, the well-developed hurricane has a relatively cloud-free center, or "eye"; and second, the zone surrounding the eye spirals, dragging the storm's clouds behind.

So calm is this central eye that any person naïve enough to linger in its presence might think that the cataclysm has safely passed. A well-known story demonstrates the danger. Early in the last century, before the hurricane's structure had been elucidated, a hurricane of considerable strength passed through Florida. As the wind died down, people strolled onto causeways and beaches to check out the still-roaring surf. Then, the eye passed. Suddenly, the winds reappeared with a vengeance, drowning hundreds.

Figure 10.1. Satellite photo of a strong Atlantic hurricane.

What creates the hurricane's iconic eye, amidst the swirling, moisture-laden vortex? I searched early on for answers. Plenty of material describing the characteristics of the eye and its surrounding wall could be readily identified, but as for explanations of "how" and "why," the best I could find was "nobody really knows." So, we proceed to explore whether the principles under consideration here can bring some clarity.

It's helpful to start with a few relevant considerations:

• First, clouds tend to swirl. In the northern hemisphere, viewed from above, they commonly swirl in the counterclockwise direction. The reverse is true in the southern hemisphere. (I'll soon explain why these swirling patterns arise from the shape of the rotating earth.)

• Second, clouds can obviously produce rain. If they absorb enough positive charge, they can theoretically dump the entirety of their moisture, leaving no cloud at all — or at least a clear region within a cloud (an "eye").

- Third, an unforeseen attraction can help sustain the cloud's forward motion. Clouds pushing into a zone of positively charged air should transiently compress the positive charges ahead. Concentrated, those positive charges should more strongly attract the negatively charged cloud, sustaining its forward motion. The faster the cloud's forward motion, the stronger the effect. This inertia-like effect — a body in motion tends to stay in motion — can help sustain the westward course of the hurricane.

This bootstrapping effect may at first seem like the dubious "perpetual motion" machine, but it's not. The rescue comes from the fact that it requires energy. Energy (from the sun) separates atmospheric positive charges from the earth's negative charges and keeps them separated. So, the process is energy dependent: You don't get something for nothing.

Given those three features, let's talk about hurricane genesis.

Explaining the Hurricane's Iconic Features

It's a balmy September. Somewhere over the Atlantic, a cloud mass forms from ample moisture and positive charge (**Fig. 10.2**). The cloud blocks the sunlight from reaching the water directly beneath. However, the clear region around the cloud permits ample passage of sunlight, driving normal evaporative processes around the cloud. That ring-like pattern of evaporation produces a rising circle of moisture and intense positivity, ripe for interaction with the negatively charged cloud that it envelops.

Hence, abundant cloud-building components are locally sourced, as in thunderstorms (Chapter 9). And because September's waters are warmed from the summer's heat, those components should appear plentifully, setting the stage for a large cumulonimbus cloud. The situation is ripe for troublemaking.

Meanwhile, as our friendly cloud grows, the trade winds should nudge it continuously westward (*panel i,* **Fig. 10.2**). The moving cloud pushes against the positive atmospheric zone ahead, transiently compressing and concentrating its positive charges. At the same time, the cloud's rear edge experiences the opposite: Because the just-passed cloud had transiently blocked the sun, the buildup of atmospheric positivity just behind the moving cloud should largely have ceased. In this way, an

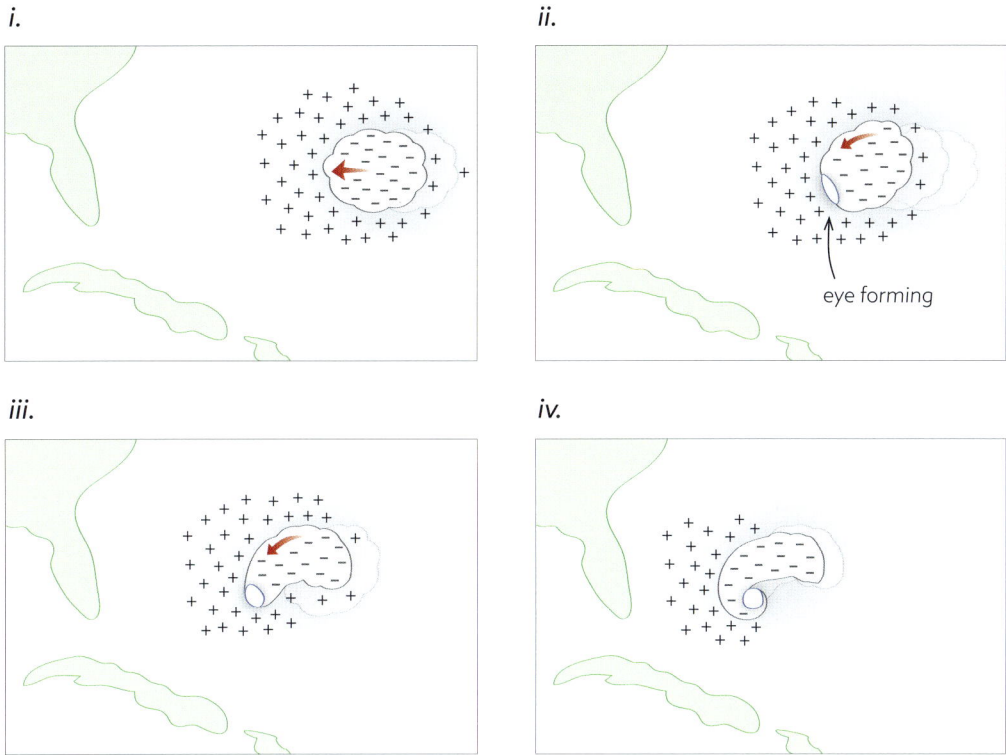

Figure 10.2. Hurricane cloud system forms as westward movement creates positive charge asymmetry, along with rotation. See text for explanation.

asymmetry develops: concentrated positivity ahead, with diminished positivity in its track (*panel i*).

With such asymmetric pulling force, the cloud's westward drive should persist. Even if the trade wind should transiently diminish its intensity, the moving cloud ought to keep moving. It should progress ever westward from the (energy dependent) bootstrapping effect of the compressed positive charges ahead.

But those compressed positive charges don't merely pull. Attracted to the cloud's negative charges, the concentrated positive charges can be expected to penetrate the cloud's leading edge. Of sufficient magnitude, penetration should create torrential rainfall, as commonly experienced at the storm's leading edge. Having spilled its guts, the cloud's western edge ought to largely vanish. Such a vanishing act creates an empty zone, free of moisture and free of charge — which, I argue, will become the hurricane's eye (*panel ii,* **Fig 10.2**).

To understand how the iconic hurricane features form, please recall that the cloud also rotates. In the northern hemisphere, it rotates anti-clockwise (see first bullet point in the previous section). Presumably, that counterclockwise rotation helps create the features inherent in the typical hurricane (**Fig. 10.1**). Here's how those features may build naturally.

Modest counterclockwise cloud rotation may arise from the earth's shape and its rotation. (For explanation, please see **Figure 10.7**.) Such modest rotational motion should begin inching the western clear zone toward the south (*panel iii,* **Fig. 10.2**).

Because those rotational forces are weak, they can create no more than a gentle rotational nudge. Once that rotation begins, however, more serious rotational impetus may come into play from the difference between the cloud's northern and southern edges. The northern edge should contain plenty of the cloud's ample supply of negative charge. The southern edge ought to contain less because of the emerging eye in the southwest: Its negative charge is largely depleted. Hence, the southern edge ought to draw less strongly toward the positive charge ahead than the northern edge. The north dominates.

Why should that difference drive rotation? Please recall the cloud's fundamental nature (Chapter 8): a cohesive entity that holds itself together. If so, then a forward attraction that's stronger at the cloud's northern edge than at its southern edge will drive rotation, like the induced spin of a ball. Cloud rotation should pick up speed.

Rotational action may in turn explain the hurricane's classic shape (*panel iv,* **Fig. 10.2**). As the emerging eye shifts progressively southward, then eastward, *etc.,* it eventually tracks a spiral course. In this context, action occurring at the cloud's edges should matter: The negative charges within those peripheral cloud regions will inevitably suffer intrusion from surrounding positive charges, acting to neutralize those peripheral regions. Modest neutralization creates relatively clear spaces within the cloud's edges. Those clear spaces should ensure distinct separation of trailing regions as the cloud continues to spiral, resulting in the hurricane's iconic shape (**Fig. 10.1**).

With some rationale in place for interpreting both the eye's genesis (vesicle depletion) and its spiral track, we have established a provisional foundation for understanding the hurricane's essence.

By the mechanism outlined in this section, an ordinary tropical thunderstorm may evolve into a hurricane. The transition should require nothing more than a locally abundant feed of moisture and positive charge, both of which can be found above warmed late-summer oceans. That abundance requirement carries potential consequence: It explains why hurricanes commonly begin their lives above vast open waters, and why they generally die out as they pass over dry land.

Why Low Pressure?

Hurricanes classify as extremely "low-pressure" systems — but why? What does pressure have to do with weather? As argued in Chapter 3, pressure may arise from the atmosphere's positive charges attracting to, and therefore pressing on, the negative earth. Plus charge clings to minus charge. Such pressure should be highest during fair weather, when ample sunshine heightens water splitting and drives positive charges high into the atmosphere. That tall column presses on the earth, yielding highest pressure (**Fig. 10.3**, *left*).

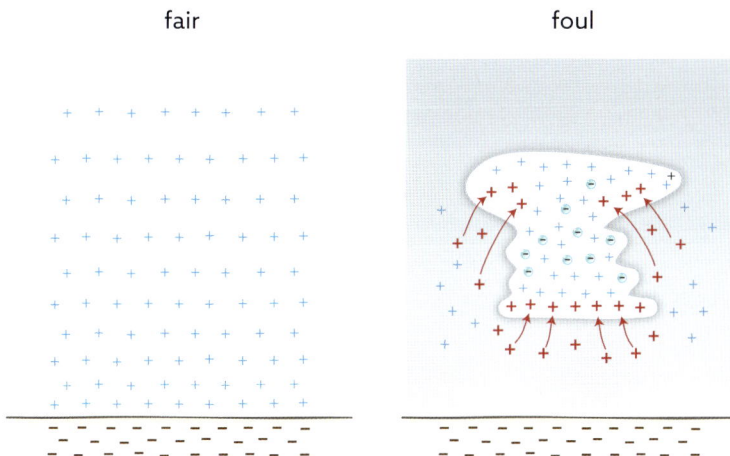

fair   foul

Figure 10.3. Surface pressure should be lower in foul weather than fair weather because positive charges around the cloud tend to get sucked into the cloud, preventing them from rising, stacking, and thereby pressing on the negative earth.

During the foul weather of a hurricane or tropical storm (**Fig. 10.3,** *right*), rising positive charges may never make it that high. Many of those ascending charges will get quickly drawn off into the massive, negatively charged cloud system, never building to the altitudes characteristic of fair weather. Without those many positives pushing down, the pressure experienced below ought to be relatively low. The eye should exert the lowest pressure of all, since that region ought to be virtually charge-free. Any positive charge attempting to rise inside the eye will quickly get drawn off into the enveloping negatively charged cloud, minimizing any positive-charge buildup in the eye. Hence, the more intense the hurricane, the lower should be the eye pressure.

In sum, hurricanes and their iconic features may arise from ordinary aspects of the environment. With an abundant supply of sun-driven moisture and positive charge, and with both components rising locally from the warm oceans beneath, strong electrostatic forces can turn an ordinary tropical thunderstorm into a full-fledged hurricane, with the structure shown in **Figure 10.1.**

Figure 10.4. Hurricanes generate rough seas.

Oceanic Swells, Storm Surges, and Flooding

Hurricanes bring consequences beyond just wind, rain, lightning, and thunder. Among them is the generation of towering oceanic swells. Treacherous seas seem so tightly intertwined with hurricanes that we may inadvertently fail to ask the deeper "how" and "why" questions about their origin. Sometimes, those questions may bring unexpected answers.

You'd think that the hurricane's fierce winds alone ought to suffice for driving and sustaining those immense waves. While such linkage may seem a no-brainer, cloud-borne electrical forces could play a surprisingly important role. Let me explain.

Once a hurricane cloud descends sufficiently close to the earth's surface, induction enters the equation. The negatively charged cloud induces opposite charges in the water beneath. Thus, cloud contents draw strongly toward the water, while the water draws strongly toward the cloud. We simulated this inductive effect in the laboratory by placing a charged rod above a beaker of water. Immediately, the water rose like a mound beneath the rod (**Fig. 10.5**). If the rod was positioned low enough, the water would rise to touch it; in that case, the rod immediately discharged, the mound fell, and the cycle could repeat, again and again. Clearly, charge forces can be strong enough to impact the water beneath.

i.   *ii.*

water surface

Figure 10.5. Effect of a charged glass rod, ~1 mm wide, approaching the water (black) beneath. *(i)* No rod. *(ii)* As the charged rod descends, the water surface rises because of electrostatically induced opposite charge.

Given that kind of attraction, imagine a hurricane-style cloud with its strong charge, descending on an oceanic body of water. If the charge differential between cloud and water is sufficiently high, then the attracted water ought to rise like a tent. Such rise may be substantial, as the induced charge will ordinarily concentrate atop the rise; so, the pulling force can create a swell of considerable magnitude. Blown by the wind, the swell could then become a substantial wave.

Even more substantial lift may be achieved as fast-moving clouds pass over said wave and inductive processes continue to operate. With such persistent lifting forces, the hurricane's wind-driven clouds could eventually spawn those towering waves that have the capacity to overturn ships and bring massive coastal storm surges. Responsibility for those profound effects would then rest not so much on the wind's mechanical forces, but largely on the moving clouds' electrical forces. Those forces confer *lift* on the waves.

Such charge-based interactions may also help explain the coastal flooding that commonly accompanies hurricanes. Again, we reflexively assume that the hurricane's winds fully explain the flooding. That presumption seems natural. However, those winds do not always blow toward land; sometimes they blow oppositely. This immediately raises questions about the wind-driven flooding hypothesis.

Another possible explanation for coastal flooding derives, again, from the influence of electrical charge. So long as post-hurricane clouds linger over coastal waters, cloud charge ought to induce opposite charge in the waters beneath. That should sustain the rise. When hurricane Irma hit Houston in 2017, for example, flooding persisted in some areas for up to six days after the hurricane winds had subsided.[wl] Two explanations seem plausible: Water-filled sewers is one of them. A second possibility lies in the aforementioned charge forces. So long as overhead clouds linger, the same electrical forces argued to create hurricane waves could likewise sustain coastal flooding.

Beyond sustaining flooding and building hurricane-level waves, electrical forces could bear responsibility for something less exotic: ordinary water waves. Invoking wind as the driver of those waves comes practically reflexively, for wind commonly accompanies waves. Yet, asking the inevitable "how" and "why" questions easily leads to frustration. For example, how does ordinary wind create the upward force required to create the wave? Whether any such force could come from atmospheric or cloud charge is an issue that may be worth considering (albeit not here, as we must press on).

Tornadoes: Creatures of Anomaly

Beyond tropical storms and hurricanes, large, dark clouds can sometimes spawn cataclysmic entities known as tornadoes, sometimes called twisters. You've probably seen photos (**Fig. 10.6**). Twisters have been known to lift cows and trains — even classrooms full of children. How is this possible?

Figure 10.6. A funnel cloud touching down in Orchard, Iowa, on June 10, 2008

In considering the tornado's origin, one needs to explain at least its three main features: the tornado's characteristic funnel shape; its twist; and its capacity to raise objects from the ground toward the cloud above. To identify the accepted explanations for these attributes, I commend the reader to popular sources. I am, frankly, at a loss to find any standard interpretation of the tornado's main features that I could convey in a concise, understandable way.

Tornadic clouds commonly exhibit electrical discharges, and I hope to show that an electrostatically based explanation may offer some promise — and with a little twist (if you'll pardon the pun), that the principles suggested to govern thunderstorms and hurricanes may also apply to tornadoes.

Twisters typically funnel out from dark cumulonimbus clouds that come close to the ground (**Fig. 10.6**). This close proximity implies that the cloud can induce substantial opposite charge on the ground beneath, which, in turn, should create a high electric field between cloud and ground. Your hair may stand on end.

I'm not the first to suggest a linkage between tornadoes and electrical charge. Half a century ago, the prominent atmospheric scientist Bernard Vonnegut argued strongly for such linkage.[1] Others have recognized the role of strong electric fields in tornadoes, and have proposed various mechanisms based on their presence.[w2,w3] Thus, while mainstream models continue to focus on thermodynamic variables such as pressure and temperature, some scientists have recognized the potential explanatory power of electrical charges, especially in explaining why objects may be drawn upward and why lightning discharges appear commonly in tornadoes. I'm not alone in pursuing this line of thinking.

A central consideration is the electrical field direction. As we've discussed, clouds bear negative charge (which keeps them afloat), while positive charge gets induced immediately below (see Chapter 9, **Fig. 9.3**). That ground-borne positive charge, I have argued, can create rainfall by pulling negatively charged droplets from the cloud proper. In the tornado, the action is curiously opposite: It's not the droplets that get pulled downward, but the loose earth, sand, debris, *etc.*, that get pulled upward. Those particles break loose from the ground, moving up and creating the iconic tornado structure. The action is up instead of down. Why?

Logic suggests the influence of some feature opposite to the usual one. What might that feature be?

While no cloud should bear net positive charge — otherwise it would collapse onto the negative earth — regions within clouds, especially the huge cumulonimbus clouds that commonly spawn tornadoes, *can* contain zones of net positivity. Charge maps confirm their existence.[w4] Often, those zones lie near the clouds' bottoms. How those positive zones might arise remains unsettled, although they could conceivably develop during cloud merger: A positive atmospheric region could get entrapped between two existing clouds, leaving a pocket of positivity in the otherwise negatively charged cloud.

Could such zones of positivity play a critical role in tornado genesis? Such zones are not usual; nor are tornadoes.

During tornadoes, researchers report electrical discharges of "anomalous" nature: *i.e.*, a negative ground gives way to an upper zone of positivity.[2] That's opposite the usual direction. In fact, during intense storms in tornado country, stunning videos show upside-down lightning.[w5] Thus, said discharge presumably involves negative earth discharging onto some pocket of cloud positivity — leading directly to the following question: Could any such zone of positivity figure in spawning the tornado?

In dealing with those positive cloud zones, consider again the phenomenon of induction. A positively charged cloud zone, if substantial enough and low enough in the cloud, ought to induce appreciable negative charge immediately beneath. Such ground negativity could matter. In the same way that positive ground charge can pull negative cloud droplets downward (Chapter 8), the opposite could also hold: A positive cloud zone could pull negatively charged ground debris upward, potentially creating the tornadic rise. The rise of debris and the fall of rain could be polar opposites.

In such context, think of the tornado's limited size. Could its narrowly confined upper margin correspond to the similarly confined zone of cloud positivity? Might such correspondence help explain the tornado's iconic shape?

In short, could we be on the right track?

Tornadic Pull: Ever Upward

Gravitation keeps things in place. If negatively charged debris gets drawn upward, such debris would need to overcome the force of gravity. However, gravitation is extremely weak (Chapter 1); so, only a modest amount of negative earth charge should be required to facilitate the rise toward a positively charged cloud region.

Such modest amounts of earth charge should be easily achievable because of multiple contributors. One of those contributors lies beyond the two already cited, *i.e.*, beyond the induced negative charge (see previous section) and the EZ-based negative charge (Chapter 2). I refer here to "triboelectric" charge. Firmly established by physicists and one I will deal with in detail in Chapter 14, the triboelectric mechanism ("tribo" means "friction") refers to charges developed by materials rubbing against one another. One prominent feature: Any object passing through (or rubbing against) air acquires negative charge. The faster the object moves, the higher the acquired negativity, which can be substantial.

That triboelectric contribution ought to enhance the debris' upward draw. As they begin their upward course, the negatively charged particles should acquire additional negative charge, virtually assuring completion of their upward traverse toward the positively charged cloud region. Gravitation should pose no serious obstacle.

OUT-ON-A-LIMB
METER

This augmentation effect lends additional credence to the inductive draw mechanism. Once earthly debris gets moving, it should keep moving all the way up to any positively charged cloud region —this is virtually assured. With this added feature, reasonable basis would seem to exist to formally hypothesize the mechanism to which we have alluded: *Localized zones of cloud positivity may be necessary conditions for spawning a tornado.*

To appreciate the dynamics, envision the following scenario. A positively charged region at the bottom of a cumulonimbus cloud passes over the ground. That passage induces negative charge just beneath, incrementing the earth's usual negative charge. Soon, the increasing electric field between cloud region and earth begins drawing negatively charged particulate matter upward from the ground, pulling it toward the positively charged cloud zone above. Meanwhile, particulate matter gains additional negative charge from the triboelectric effect as it moves through the air, thereby enhancing the upward pull. Everything moves up.

Such hapless fate applies to practically anything unlucky enough to have been caught in the tornado's path. If your hair stands on end, then the rising debris could include you — drawn upward along with the dirt, sand, or whatever else might be free to fly.

In this scenario, the tornado's action may better resemble the pull of a lofty magnet than the push of a windstorm. Air and debris get drawn upward by a force from above.

While tornadoes evidently levitate earthly materials (rendering a tornado visible), a fair question to ask is why, instead, don't those earthly materials draw the positive cloud region downward? In fact, they do — to some extent. Often, the visible tornado begins with a small, funnel-like structure descending from the bottom of the cloud, toward the ground. Only then does the ascending debris begin rising, thereby sustaining the visible tornado.

In terms of whether matter rises up from the earth or falls down from the clouds, it's helpful to think in terms of scale. Consider the cumulonimbus cloud vs. say, a refrigerator. Which one will give way to the other?

A food-filled refrigerator may weigh as much as a tenth of an elephant, but the cloud may weigh as much as a million elephants. Considering the cloud as a "solid" may not be entirely accurate, but the cloud does hang together as a cohesive unit. That behemoth should not easily yield to the more diminutive refrigerator. So, it's the fridge that gives way. The refrigerator gets drawn upward electrostatically, along with everything else on the ground left unsecured.

Hang on for dear life!

Twister Rotation

While the tornado's uplifting action seems provisionally explained by electrical forces, what about its rotation? What accounts for the tornado's characteristic twist?

In theory, the twist could initiate from the parent cloud's rotation. Rotation, in turn, arises as a natural consequence of the earth's spin and sphericality, a non-intuitive feature that is worth explaining. While cloud rotation itself may constitute no more than a partial answer to the question of tornadic rotation, explanation of the phenomenon does tend to stimulate those grey cells.

Consider the wind velocity at different heights above the earth. At ground level, the air moves at roughly the same speed as the spinning earth, otherwise you'd feel a perpetual wind. Relative to the cosmic frame, however, that near-earth velocity amounts to roughly twice the speed of a jet plane (Chapter 6).

Far above the earth on the other hand, the air velocity should be negligibly low: The thinned-out layer of air ought to hardly move at all as it smoothly melds into the edge of cosmic space. According to the cosmic frame, therefore, air velocity from the earth's surface to the outermost edge of the atmosphere should range from very high to very low.

I hope you're with me.

Now here's the critical point. That vertical gradient ought to be steepest above the equator, where the earth spins fastest. Near the poles, where the earth-spin velocity is negligibly small, the difference between earth-surface velocity and upper zone velocity should practically vanish; so, the vertical gradient ought to be least steep at the poles. Evidently, the magnitude of that vertical velocity gradient depends strongly on latitude.

Given that understanding, imagine a cloud-level horizontal slice through the atmosphere, say partway between the earth's surface and the atmosphere's outer edge. At that elevation, horizontal wind velocities should likewise depend on the latitude — fastest just above the equator and progressively slower toward the poles. Such velocity difference at cloud level implies that clouds must twist. In the northern hemisphere, the twist direction is counterclockwise (**Fig. 10.7**).

Hence, *cloud rotation should arise naturally from the earth's shape and spin.*

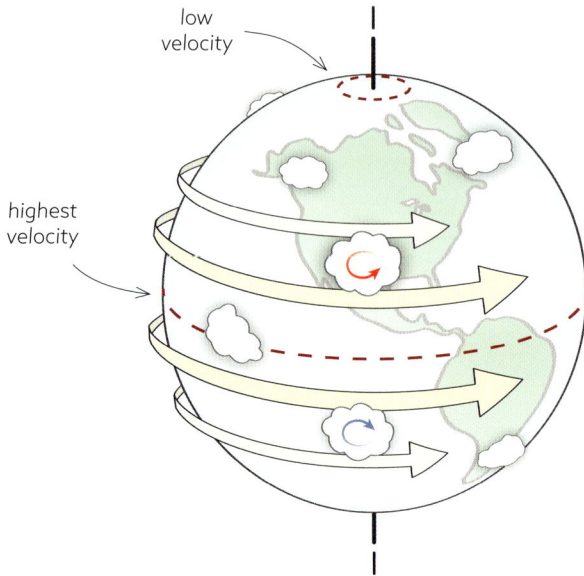

Figure 10.7. Clouds in the northern hemisphere rotate counterclockwise (red arrow) because the air moves faster above the equator than above the poles. In the southern hemisphere, rotation is opposite (blue arrow).

Scientists refer loosely to this phenomenon as a manifestation of the "Coriolis" effect. For small clouds, the rotational speed ought to be barely noticeable, and I don't mean to suggest that the Coriolis effect explains all of tornadic rotation. For clouds spanning a larger area, however, the effect can be appreciable. All of this, just because the spinning earth has a spherical shape.

Any natural cloud rotation arising from the Coriolis effect can be enhanced by local winds. For example, a mass of dry (positive) air flowing southward on the west side of a cloud would accelerate the rotation. With such enhanced rotation and proper inductive conditions, the emergence of twisters becomes easier to envision. Indeed, twisters do often descend from the storms' southwest,[w6] as expected from the airflow scenario just described.

On the other hand, you might wonder whether even such enhanced Coriolis-based cloud swirling would have enough oomph to drive tornadic rotation. To get some idea, consider some fairly arbitrary numbers. Suppose a large tornadic cloud rotates once per minute; and suppose further that a large twister clings beneath. The twister would then also rotate once per minute, just like the cloud. While such rotation is not particularly impressive, suppose further that the tornado has a radius of 500 meters (~1,650 feet). Since that tornado spins *as a cohesive unit,* the wind velocity at the outer edge of that tornado computes to something close to 200 km per hour (~125 miles per hour). That could demolish your home.

On the other hand, the numbers above may be arbitrarily favorable to the argument. Most tornadoes are narrower than one kilometer, and the clouds from which they descend may well rotate at speeds less than once per minute. Still, those tornadoes may twist ferociously. How could this be?

Charge-Induced Rotation

To understand how tornado-rotation speed could accelerate well beyond cloud-rotation speed, we return to the forward-motion mechanism elaborated for hurricanes (**Fig. 10.2**).

Like the hurricane, the tornado doesn't just rotate; it also advances. The tornadic advance compresses the atmospheric positive charges ahead. Those compressed charges then pull on the tornado's negative charges, perpetuating the advance. Continuing tornadic advance compresses additional positive charges, *etc.* Through this bootstrapping process, the tornado progresses forward.

Might that process also augment the tornado's twist? Here again, the central actor is the assembly of compressed positive charges ahead

(**Fig. 10.8**). As in the hurricane scenario, some of those compressed charges will inevitably combine with the negative charges at the advancing tornado's leading edge. Such action creates an asymmetry: The tornado's retreating edge, now partially neutralized, should be left less negative than its advancing edge. The dominant pull on the advancing edge should quicken the twist. Hence, the rotational speed of the tornado is by no means limited by the relatively feeble parent-cloud rotation. Rather, the charge-asymmetry phenomenon should appreciably enhance that speed.

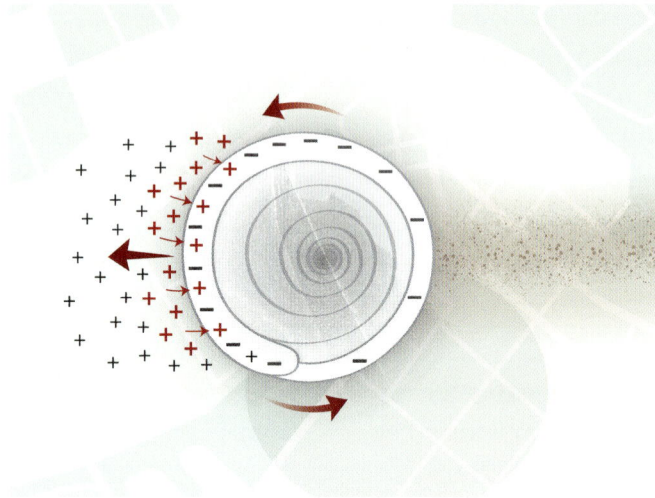

Figure 10.8. Mechanism of tornado twist. Positive charges ahead of the tornado attract negative charges on the tornado's periphery. The two combine, diminishing peripheral negativity of the receding edge. The tornado twists because negative charge on the leading edge (top) exceeds negative charge on the receding edge (bottom).

Key to this entire argument is the fact that the twister structure ultimately hangs together; *i.e.*, particles and debris remain loosely stuck to one another. Absent such cohesiveness, the twisting structure would immediately fly apart.

Cohesiveness of that sort can be produced in at least two ways:

- The cloud-to-ground electric field could vertically align the materials' molecular dipoles, positive at one end and negative at the other, which then cling together;

- Any positive charges lying between the negatively charged particles could create cohesion by the like-likes-like principle.

Either way, the particles cling to one another, creating a structural entity that spins like a top. Tops spin fastest at their edges, potentially explaining the extreme wind velocity experienced as the advancing tornado strikes.

So, while we can perhaps understand why the tornado's constituent particles can stick together, why does the tornado itself assume its classical funnel shape (**Fig. 10.9**)? One relevant consideration is the extent of the positively charged cloud zone. If the top of the tornado originates from such a zone, then the extent of that zone ought to put a lid on the tornado's upper diameter. Its size will evidently vary, depending on the particular cloud. The ground diameter, typically narrower, may be set by the region where the induced ground charge

Fig. 10.9. Tornadoes commonly exhibit funnel-like configurations.

is highest. A tendency does exist for electric fields to focus on narrow places, keeping the bottom diameter limited in size.

But there may be more to shape determination. Yes, as the loose terrain lifts to create the visible tornado, constituent particles do tend to stick together. However, any such cohesiveness must have limits. As the ascending particles twirl, centrifugal forces can be expected to drive those particles increasingly outward, to the extent they can. That action should help create the iconic funnel shape, the upper diameter being limited by the extent of the positive cloud zone.

Dust Devils and Waterspouts

Dust devils (figure, left) resemble small tornadoes. They commonly occur in dry places such as Arizona, New Mexico, and eastern California. Given their resemblance to the tornado, we surmise that the underlying basis could be similar — conditions of unusually high atmospheric positive charge drawing negatively charged dust and debris from the ground. Indeed, when inclement weather happens to move in, dust devils do sometimes convert to tornadoes, underlining the likely similarity of their genesis.

Closely related to the tornado and the dust devil is the waterspout.

Vacationing with his family in Genoa, Italy, Evgeny Drokov photographed the dramatic waterspout pictured below.

Waterspouts emerge from the ocean's surface, rising toward the cloud and twisting much like tornadoes. Here, the "debris" is water, droplets of which contain separated charges.[3] Electric fields can easily polarize those charges to create adhesion. Forces similar to those responsible for creating tornadoes very likely create waterspouts — idiosyncratic creatures both, yet plausibly understandable when considered in terms of electrical charge.

Tornadoes run their course. Some may persist for an hour or more, but most dissipate within minutes. You could imagine any of a number of factors that might exhaust the tornado. According to the hypothesis presented here, primary among those factors may be the exhaustion of the cloud's positive zone. Once those positive charges neutralize, the electric field required for the tornado's sustenance should vanish, and so should the tornado.

Improved Weather Forecasts?

With our provisional understanding of the genesis of weather events, the question arises: Could that proposed understanding help improve weather forecasts?

Some years ago, I served as an advisor to a task force of the National Science Board, the body that governs the US National Science Foundation. The Board had been trying to introduce more transformative science into the Foundation's grant portfolio. During the many hearings and meetings, I had the good fortune to meet and chat with one of the leaders in the atmospheric science field. I found myself shocked to learn that fundamental research on the physics of weather had practically vanished from the scientific scene; almost all weather science now deals with computer simulations.

Of the many variables used in those weather-prediction simulations, atmospheric pressure and temperature, dominate. Predictions have certainly improved with the increase of computational power; nevertheless, just because the weatherman says it will rain tomorrow, it doesn't mean it will. Perhaps we'll never know for sure. On the other hand, if temperature and pressure are not the most critical variables determining weather, then algorithms based mainly on those two variables would be unlikely to ever achieve robust predictive capability.

Consider, for example, the role of atmospheric pressure. Low pressure generally presages foul weather, while high pressure corresponds to fair weather. Correlation, however, need not imply cause. And indeed, the predictive value of pressure remains uncertain. One reason is the small difference between the levels of pressure distinguishing fair weather from severe foul weather, a difference amounting to only a few percent of the average pressure. To imagine that a small divergence can make the difference between dazzling sunshine and a monstrous hurricane may seem something of a stretch.

Temperature, in theory, would appear to hold more promise as a predictive variable because of its larger natural range. However, even that is not so clear: Serious storms can occur in sultry sunshine as well as in subzero frigidity. Further, anticipated airflow patterns do not reliably arise from temperature differences: The denser, cooler air atop a mountain does not routinely slide down to displace the lighter, warmer air beneath (Chapter 3). This apparent "paradox" can resolve within the charge-based paradigm: The force of the electrical charges on the air may greatly exceed standard gravitational forces pulling on the air (Chapter 1). Thus, any prediction based solely on the gravitational pull of cooler, denser air versus warmer, lighter air may miss a critical part of the big picture.

Pressure and temperature are certainly involved in weather, but the implicit presumption that these variables play the most pivotal roles creates ample opportunity for forecasts to go seriously astray. They may easily miss the mark.

Whether charge-based predictions will eventually prove more accurate remains to be seen. Perhaps one day, though, we will know with increased certainty whether to leave home carrying an umbrella.

Section Summary and Conclusions

In the preceding four chapters, I have attempted to better explain the origin of weather. Instead of relying on traditional variables such as pressure and temperature, I examined weather in the context of electrical charge. Charges can produce monumental forces (Chapter 1), forces so large that it would seem imprudent to ignore them when considering the enormous power exhibited by weather events such as thunderstorms and typhoons.

The proposed paradigm contains two interacting components: negatively charged vesicles (moisture), and atmospheric positivity. Cloud formation arguably requires both. And clouds, after all, are the forerunners of weather. To know weather is to know clouds; and to know clouds, I argue, it's necessary to understand the roles of those two oppositely charged species.

Of those two species, the negative charge comes from the vesicles evaporating from the earth's water. Under appropriate circumstances, you can actually see them evaporating. Those vesicles arguably hold most of the atmosphere's moisture.

The complementary positive charges also derive from the earth's water, in the form of hydronium ions. Those positive ions may enter the atmosphere in two contexts: (*i*) directly from the water, where their mutual repulsion creates pressure to force their escape into the air; and (*ii*) initially clinging to the emerging water vesicles, and then getting released as those vesicles rise and disperse.

Thus, negative and positive components both derive from the earth's water, initially co-existing in the atmosphere. Naturally, they attract. When local concentrations of both species grow high enough, the positive charges begin seriously interacting with the negatively charged vesicles, condensing them into clouds by the like-likes-like mechanism.

A critical feature of that mechanism is that the condensing vesicles do not ordinarily coalesce; typically, they maintain some separation. That separation confers the cloud's fluffy character — not at all like the liquid water that you'd expect from ordinary condensation. Clouds bear no resemblance to water-filled bathtubs.

A cloud's height in the sky should depend on its level of negativity. The fair-weather cloud will often sit moderately high because its appreciable negativity abundantly repels the earth's negative charge. If it absorbs positive charges, the cloud will become less negative overall, and should therefore descend.

As a cloud descends close enough to the earth's surface, induction should begin to take hold. The cloud's net negative charge should induce positive charge on the region of the earth lying just beneath. Since opposite charges attract, the cloud's contents should then be pulled down as rain, electrostatically *attracted* to the earth. Thus, pounding rain doesn't merely fall; it gets *drawn* to the earth.

To understand the dynamics of rainfall, a distinction needs to be made between the charge of the cloud as a whole and the charge of constituent droplets within the cloud. Cloud droplets (vesicles) bear net negative charge. In the induction scenario, the ground immediately below the cloud ought to develop positive charge. If intense enough, the pull of that positive ground charge should dislodge individual negatively charged vesicles from the cloud, explaining why rain comes a droplet at a time, and not as a massive water dump.

What happens in and around the cloud matters a lot. Thus, relatively weak positivity arriving from afar may merge with negatively charged clouds to produce ordinary seasonal rains. Not very much drama. When atmospheric positivity intensifies, clouds can become mean: If enough positive charge joins the negatively charged cloud, the diminished repulsion from the earth will lower the cloud. When sufficiently low, the cloud may induce enough opposite charge on the ground to trigger flashes of lightning along with other thunderstorm manifestations. Drama increases.

More serious drama may take place when positive and negative charges meet over vast expanses of warm water, where evaporative charges are particularly bountiful. Intense positive-negative interactions can then create severe tropical storms, even hurricanes. The charges themselves play a central role in creating a hurricane's iconic features such as its rotation around a central eye. The same for tornadoes.

In each of the above-mentioned weather events, the action comes from some mixture of positive charge and negative moisture, combinations of which can sometimes scare the daylights out of you.

And if hurricanes fail to inspire enough fear, then the job can certainly be done by tornadoes, especially if one of them comes barreling directly toward you. For most, it's utter panic. Tornadoes form under unusual conditions, arguably when a confined, near-bottom region within a tall cumulonimbus cloud bears positive charge. If large enough, that positive zone should induce negative charge on any ground-borne material below, raising any and all loosely secured objects — hopefully not including yourself.

—

Any reader seriously interested in the issue of weather would do well to read any of the many standard textbooks or web offerings on the subject. The merits of prevailing views can then be compared against the merits of the view offered here.

Personally, I find the textbooks' myriad complexities troubling. A central theme seems wanting. On the other hand, nature could well operate in ways far more complex than this author believes, and I leave it to the reader to decide which approach may offer a more promising pathway toward truth.

Perhaps one day as that truth emerges, our understanding of weather may advance to the point where inaccurate forecasts become rare or non-existent. I hope so — even if we run the risk of finding ourselves at coffee breaks with nothing to talk about.

Surely, it would be a small price to pay for that understanding.

SECTION IV

Gravitation

Gravitation is the force we experience relentlessly — but do we really comprehend its underlying nature?

Gravitation is said to arise from the attraction of masses — *e.g.,* the mass of your body attracted to the mass of the earth. That formulation would appear to suffice. After all, the fat goose does weigh more than the scrawny chicken. Observations of that sort confirm our expectations; they reinforce the view that for explaining the phenomenon of gravitation, the simple attraction of masses should work just fine.

Yet, we are left with the usual "how" and "why" questions. *Why* should masses attract? Posing that question commonly elicits blank responses. The reigning presumption seems to be that masses simply *do* attract, and that's all there is to it. For many, that genre of response has seemed sufficient. Physicists skirt the issue by relying on abstract mathematics. Others find themselves mildly frustrated, hoping for a more satisfying answer to that perennial scientific question: *Why*?

Beyond that philosophical concern lies an issue that is potentially more serious. The mass-based formulation may work well in most circumstances, but not all. Recognized anomalies abound. We may ignore those anomalies, hoping that someone will eventually bring resolution; or, we can step back and reconsider the merits/demerits of our current understanding, to see whether some refinement may perhaps help bring more complete understanding. Here, we opt for that latter course.

Mass obviously plays some central role in gravitation — we routinely witness its impact. By no means am I proposing to dump mass into the scientific toilet. On the other hand, masses are composed of atoms, and atoms are built of charged particles. Exploring whether the most prominent among those particles might prove more consequential for creating the gravitational force than ordinarily considered would constitute a departure from conventional thinking, but one that is modest rather than radical.

Thus, the issue: could such a deviation from conventional thinking help us arrive at a more intuitive and satisfying understanding of why we are drawn to the earth?

CHAPTER 11

Can Pigs Fly? On the Origin of Gravitation

Dropped objects fall to the earth.

Why they fall is something few stop to consider. Children simply accept it. Adults with education may opine that those objects fall because of gravitation, *i.e.*, because the mass of the earth attracts the mass of objects above. They've integrated the views of the experts, happy to pass on that understanding.

But *why* should masses attract?

To this date, nobody really knows. Einstein sought a unified understanding of the forces governing the entire physical realm, including gravitation; but beyond his vaunted identification of the curvature of space-time, and subsequent mathematical constructs embedded in the abstract notions of quantum mechanics, this holy grail of science has never been found, or even described in an intuitively understandable way. We accept that masses attract, leaving it to the experts to argue among themselves how it really works.

What We (Think We) Know about Gravitation

Until recently, I was as detached from the question of gravitation's origin as most. With limited time available to pursue the subject, I blithely accepted gravitation as one of those axiomatic features of nature. It was as simple as that.

But an unexpectedly jarring presentation drew me out of that complacency. Let me elaborate.

Physicists have embodied the gravitational pull into a universal gravitational constant, G, often called "big G." Knowing the value of G and the magnitudes of any two masses, you can easily calculate the force of attraction between those masses. Any competent Physics-101 student can do it.

As a fundamental constant of nature, G should be invariant, just like *pi*. Physicists accept that notion. I had little reason to think otherwise until I heard Rupert Sheldrake argue that G, in fact, *does* vary from time to time and place to place. When Sheldrake presented his evidence at a scientific conference, you could have heard a pin drop.

I followed up by reading about his findings in more detail in his eye-opening book, *Science Set Free*.[1] I was stunned. To obtain his evidence, Sheldrake pursued an unorthodox approach. He went to a London library that kept handbooks of scientific constants. Those handbooks are periodically refreshed with updated information. When a new volume appears, fastidious librarians commonly discard older volumes to save shelf space. This particular library, however, kept editions going as far back as 1973. So, with proper dusting, Sheldrake could easily find earlier-reported values of G. Plotting those values over time yielded the plot reproduced in **Figure 11.1**.

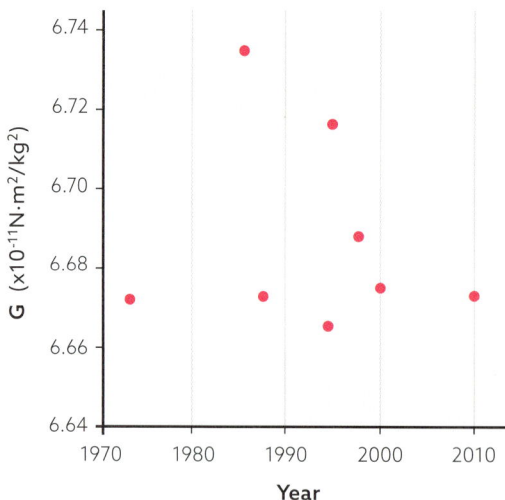

Figure 11.1. Values of G obtained at different times between 1973 and 2010. After Sheldrake, 2012.[1]

The measured values of *G* evidently do not remain constant. They vary over time. Clearly, some of that variation must arise from measurement imprecision, leading to the obvious question: how much?

To establish the values of such constants, several prominent laboratories are charged with making the relevant experimental measurements. That takes time and effort. The measurements are then vetted by professionals, highly trained to evaluate the levels of uncertainty. Handbooks report the mean values among those laboratories, along with the overall level of uncertainty, which in the case of the high-precision instruments developed to measure gravitational forces, amounts to a few parts per million.

Here's the problem: Those parts-per-million uncertainties are too small to make a difference. The reported uncertainty values are in fact 40 times smaller than the difference between the highest and lowest points on the graph. They hardly matter. So, if those professionals are correct in their uncertainty computations, then the graphed differences of *G* cannot arise out of sloppy experimentation. Although the graphed differences are relatively modest, they nevertheless imply something unexpected: *G* might indeed vary over time. According to conventional physical thinking, that should not happen. Constants should remain constant.

Shown the results, the London librarian retorted that something *had to* be wrong with the plot, because... well... everyone knows that *G* is constant!

If this evidence truly reflects reality, then scientists are in a pinch, for how could a fundamental constant vary? Constants remain constant. If *G* does vary over time, then a constant it is not, and proper understanding of gravitation will perhaps require the identification of some more fundamental variable that does remain constant.

That masses attract is certainly true: Voluminous evidence supports that notion, as does common experience. It's real. On the other hand, that experience does not exclude the possibility that some variant of the standard mass-centered paradigm may be worth exploring. Thus, rather than mass *per se*, the attraction could arise from some force intimately related to mass. If so, then perhaps some sense could be made of Sheldrake's otherwise unsettling observation.

Sheldrake's demonstration does hint that something could be amiss with current thinking about gravitation. On the other hand, the obvious question arises: Just how important is that lone demonstration? Could a single anomaly suffice for unseating a long-standing paradigm that most have come to accept as foundational truth? How do we deal with this paradox in a definitive way?

The approach I have taken is to ask whether Sheldrake's demonstration is the only one to raise eyebrows, or whether additional gravitational anomalies exist. If others exist, and especially if they are serious enough, then we may do science a disservice if we sweep them under the rug and press on. I think you understand.

Marching Directions

Where, then, is this chapter headed?

As you might have predicted, I shall first document multiple gravitational anomalies, some of them serious. To deal with that issue, I will then explore a variation on the common gravitational theme: whether the gravitational force may arise not from the masses themselves, but from the protons and electrons that make up those masses. The distinction may seem subtle; but the concept of charge-based electrical forces could have meaningful implications for ultimate understanding. We will consider whether those elementary charges can fill some explanatory potholes.

Along the way, I will describe several long-standing clues implying gravitation's possible linkage to electrical phenomena. Such a connection is hardly new. An electrical linkage has long been implicated by multiple scientists, including several whose names you likely know. Robust interest persists to this day. Hence, while the introduction of electrical charge forces us into a gravitational paradigm that might be viewed as beyond humdrum, neither is it particularly radical. Others have been there before.

I'll say in advance that the paradigm to be developed in these pages remains incomplete. Not all the relevant evidence is in. Thus, it's not my intent to prove the new paradigm adequate (if ever anyone could "prove" any hypothesis to be adequate). Rather, my goal is to present the core idea with reasonable justification, so that any interested party may follow up with deeper and more probing quantitative analysis.

On the other hand, the mechanism I'm about to describe does provide a simple understanding of why the earth pulls on all objects in its vicinity, and, also, why the sun attracts the earth and its sister planets. Thus, reason exists to believe that the developing paradigm may have some promise.

I'm also confident that, in its simplicity, the proffered mechanism can be understood even by those lacking advanced degrees in physics.

SOME WEIRDNESS

Gravitational Anomalies

Do Sheldrake's concerns about big G stand alone? Or do other issues raise additional doubts?

For describing phenomena that fail to fit common expectations, scientists use the word "anomalies." The underlying theory is fine, it's implicitly presumed, but a few anomalies remain to be reconciled. Loose ends need tying. Those fraying snags form the grist for many physicists' theoretical mills, challenging them to come up with satisfying explanations.

Many anomalies resolve easily. Some reconcile through nothing more than some grinding of the cerebral millstone — filling some missing gaps in logic. Others turn out to have arisen from observational errors, while still others resolve through minor theoretical modification. Certain anomalies, on the other hand, require major theoretical patching to cover gaping conceptual wounds.

The field of gravitation recognizes an appreciable number of those unresolved anomalies. Here are a few, some more serious than others.

- Perhaps the best-known gravitational anomaly concerns the so-called "missing mass." Since mass radiates energy, the amount of radiated energy is commonly used to compute the amount of mass. The procedure lends convenience, especially for masses that are inaccessible. For years, it's been physicists' standard tool. More recently, that same procedure has been applied to galaxies, with troubling results.

Galaxies revolve about their galactic centers (**Fig. 11.2**). According to the gravitational laws formulated by the German astronomer Johannes Kepler (1571 - 1630), the period of revolution should be calculable

Figure 11. 2. A representative galaxy. Note the structural similarity to the hurricane — perhaps hinting that both derive from similar kinds of force.

Dark matter conjuring, as demonstrated by Darth Sidious

Figure 11.3. Dark matter: reality or expedient of convenience?

from the galaxy's mass and its distance from the galactic center. Those numbers generally work, but not always. For galactic rotation, the observed periods require mass-correction factors that can amount to some *ten times* the measured mass. The failures are not trivial.

The physics community has responded by postulating the existence of a type of mass that supplements the measured mass — one that doesn't radiate any energy at all. This is the so-called "dark" matter, "dark" because of its failure to radiate. To account for the correct revolutionary period, then, one just adds the right amount of undetectable matter to the detectable matter. Voila! Problem solved.

While the theorized presence of dark matter has proven convenient for matching expectation with observation, dark matter's existence remains a supposition. If dark matter's existence is ever *proven*, then the mass anomaly could vanish in an instant. On the other hand, definitive proof continues to elude physicists, leaving open the possibility that dark matter might not exist at all. In that case, the edifice of standard gravitational theory will continue to suffer a troubling defect — a profound failure of the mass-based formulation to give the right answer (**Fig. 11.3**).

- A second gravitational issue is the so-called Pioneer anomaly. The Pioneer 10 and Pioneer 11 satellites were launched into space in the early 1970s. Their orbits should be predictable from their masses. But those predictions have proved erroneous: These satellites have inexplicably slowed down relative to NASA's calculated expectations. Each year, they fall behind their projected travel by an additional 5,000 km (~3,100 miles). Physicists continue to debate the origin of these long-standing anomalies. Evidently, standard gravitational considerations do not suffice, even though the supposed predictability of satellite orbits is frequently touted as evidence for the conventional paradigm's success. The reality seems less certain.

- Gravitational anomalies also exist on the earth. One such anomaly is the gravitational pull's season-to-season variability. In the geophysical research station 1.4 km (0.86 miles) below Gran Sasso Mountain in Italy, scientists have confirmed that gravitation pulls more strongly in the summer than in the winter.[w1] If mass is all that matters, then why should summer gravitation exceed winter gravitation? Does summer sunshine create more mass? Or what?

Besides those seasonal differences, gravitation also varies over the course of the *day*. An MIT group measured the values of the gravitational constant, G, hourly, around the clock over a seven-month period. G varied with time, with a periodicity of 24 hours.[2] Two measurement methods gave the same result. A supposed constant that is fundamental to the nature of the universe should not vary over the course of the day, or the season. Yet, the standard gravitational constant apparently varies over time, and in such a provocative way.

- The weight of a mass is generally taken as a measure of the gravitational pull on that mass. Unless that mass is augmented or eroded, its weight should remain invariant. However, weight changes reportedly do occur when certain substances react in sealed flasks. The reported weight changes are modest, but reproducible; and, proper control experiments designed to show no change indeed show no change.[3,4,5] Evidently, the value of G can vary as substances react inside those sealed flasks.

In a similar vein, weight may also vary when a piece of metal is brought close to the metal being weighed.[6] Why the presence of nearby metal should impact an object's weight is not evident, at least within the conventional framework. It's as though you weigh more when your friend stands nearby, staring at the scale. Within the accepted gravitational framework, such behavior would be classified as anomalous.

Gravitational Anomaly or Hoax?

Beyond the gravitational anomalies listed in this chapter, certain physical sites boast local gravitational distortions. If real, those distortions add further to the list of acknowledged anomalies. If not, they may at least offer some measure of needed amusement.

One site of apparent gravitational distortion lies near Santa Cruz, California. Visitors to this well-known site experience a range of anomalies, including disorientation (pets avoid the area), instability, *etc*. Water apparently runs uphill. The site is touted as an area of gravitational distortion.

Connected to these apparent gravitational anomalies are some remarkable optical aberrations, almost too extreme to qualify as real (figure, left panels). Indeed, the hefty threads of commercialism rampant at such sites have earned them reputations for charlatanism. You might even hear the Santa Cruz site described as "the world's biggest hoax."

Curious, I visited the Santa Cruz site twice. Each time, I took photographs similar to the half-century old marketing photos shown in the figure (left). I took my snapshots (right) at a spot less crowded with tourists because the optical distortions there were less striking. I set my camera on a tripod and photographed two people positioned equidistant from my camera.

Because background tilt and apparent ground tilt (though much less than implied by the curving stone wall behind) can confuse the mind, I measured the subjects' heights objectively, from the images. I did this by counting pixels from head to foot. The number of pixels, *i.e.*, the subjects' apparent heights, definitely varied depending on their position. However, I judged the differences too modest to convince, absent needed repetition.

I plan to return to explore further, focusing on one of the more popular zones (left) where the position-dependent differences are dramatic enough to routinely elicit popping eyes and dropping jaws. It's worth the trip.

Following my picture-taking jaunts, I discovered something surprising: someone else had already explored the same phenomenon.[8] The author visited several sites in the Western US. At each one, including the Santa Cruz site, he could systematically confirm that the apparent height depended on position. The author went on to quote reports that the distortion magnitude could even vary over time — some days more than others.[8] Whether those optical distortions are real, or merely a matter of perception, remains to be determined. Hefty skepticism abounds from all quarters. Should the distortions turn out to be products of our imagination,

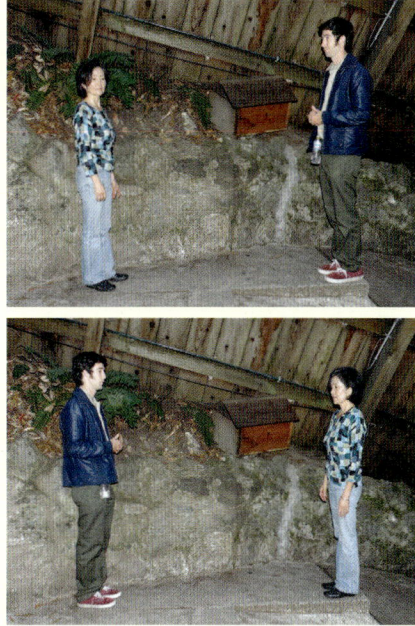

(a)

(b)

Old postcard images illustrating local optical aberrations at the Santa Cruz "Mystery Spot," (a). Images taken by the author at a site situated about ten meters from the one at the left, (b).

then they will have nevertheless served a useful purpose: providing some needed diversion from serious world events. Should they turn out to be real, then they may tell us something notable about gravitation — i.e., that something lying just underground and therefore connected to gravitation's source might alter local refractive properties of the air above, thereby bending the light and creating those distortions.

To most readers, optical phenomena would seem as distinct from gravitation as sardines are from strawberry jam. Please recall, however, that gravitation is known to bend light: Einstein predicted that in 1905, and observations eventually confirmed it. Therefore, should those optical distortions turn out to be real, it may well be that they can tell us something meaningful about the nature of gravitation.

- Gravitational pull varies at different positions on the earth's surface. A strong dip of G occurs off the coast of India in the Bay of Bengal, as well as at other sites such as Hudson Bay in Canada, and certain Himalayan regions. These variations are neither secret nor obscure — NASA has officially mapped them.[w2] In theory, small variations of G could arise from differences in local subsurface composition. Indeed, such variations provide clues for detecting and exploiting the presence of subsurface oil. Compared to the huge mass of the earth, however, any such local density variation ought to have no more than a trifling effect on G. Nor is it clear why any such local effect should manifest over massive bodies of water such as Hudson Bay or the Bay of Bengal. Spatial variations of G evidently exist, and within the conventional paradigm, the reasons for those differences remain unclear. They classify as anomalous.

- Finally, gravitation can vary during an eclipse. We don't expect eclipses to modify the mass of the earth; yet reports show a transient decrease of local gravitational pull during solar eclipses.[7] Further, the pull's nominally vertical direction undergoes minor tilt at the edge of the eclipse zone, but not at positions even slightly removed. Apparently, something beyond just earthly mass is impacting the gravitational pull, another hint that the standard gravitational paradigm does not account for all features of gravitation.

Implications for Pervasive Anomalies

The anomalies detailed above present challenges to standard gravitational theory. Some are subtle. Others, such as the need to invoke massive amounts of dark matter to accommodate theoretical expectations, seem more substantial. While expedients such as dark matter could indeed turn out to be part of the natural fabric of our physical universe, the alternative is also possible: *i.e.*, the need to invoke so radical a concept may point to something seriously amiss with our understanding of the rudiments of gravitation.

For now, those anomalies, together with the one identified by Sheldrake, challenge the notion that gravitational theory is one of those neat and tidy areas of science in which all evidence fits. It does not. We may prefer to think that it ought to fit, but that is evidently not the case.

Thus, room exists for fresh thinking, and as you'll see in a moment, I'm not alone in harboring that sentiment.

Whether fresh thinking can resolve those multiple anomalies is an issue I will address at the end of the subsequent chapter. I will re-visit the list of anomalies, evaluating the extent to which any revised thinking resolves those anomalies.

A ROLE FOR ELECTRICAL CHARGES?

Clues for a Possible Electrical Contribution

Of the plausible paradigmatic options to consider, one that seems worth contemplating is a charge-centered mechanism in which gravitational force arises from the charges contained *within* the mass. That way, the charges remain firmly linked to mass, retaining mass's obvious dominance in the gravitational paradigm. In the remainder of this chapter and the two that follow, I will attempt to show that a simple charge-based mechanism can go a long way toward accounting for the main features of gravitation.

But wait. If this line of thinking has any merit at all, then you may well be inclined to ask: How come nobody has thought of it before?

In fact, some have. It may surprise you to learn that, now and again, distinguished scientists have proposed that gravitation might arise from electrostatic or electromagnetic fields. Michael Faraday, the great 19th century English physicist and the inventor of the electric motor, was perhaps the first to think along such lines. Later, Einstein tried in vain to establish some link between electromagnetic and gravitational forces. And, following an entirely different approach, Feynman attempted to do so as well.[9] Even today, attempts to relate electrical or electromagnetic forces to gravitation persist, as any search through the literature will demonstrate. An active forum for these presentations has been the *Electric Universe*. Up to recently, hundreds have been gathering annually to discuss the role of electricity in the cosmos and on the earth. That discussion includes Wallace Thornhill's impressive electric dipole-based theory of gravitation.[w3] And more: While in the final throes of preparing this book for publication, I came upon a proposal closely similar to the one I'll be suggesting.[w4] Evidently, I'm not alone in thinking that electrical charges may hold more than passing relevance for gravitation.

Given such precedent, seeking a linkage between electrical phenomena and gravitation should not seem frivolous. On the other hand, even those earlier luminaries could not achieve a satisfying understanding of that linkage. Such failure implies that mere mortals like us might have a hard time of it — unless, possibly, those giants might have missed some non-obvious but crucial fact that only today we are able to recognize.

Coming to my own approach to the possible link between charge and gravitation began from two hints. The first involved weather (Chapters 7 – 10). There, we saw not only how repulsive charge forces could keep clouds separated from the earth, but conversely, how attractive charge forces could draw raindrops *toward* the earth. If inductive forces can pull raindrops toward the earth, then why might they not pull you? Or your uncle? Why couldn't inductive forces rule the roost?

The second hint came from noting the similarity between the mathematical expressions for electrostatic attraction and gravitational attraction: Those equations are practically identical, differing only by a constant.

Thus, the value of the force, F, between two charged bodies (assumed, for convenience, to be points) is given by:

$$F = K(q_1 \times q_2) / r^2 \qquad (1)$$

where K is a proportionality constant, q_1 and q_2 are the charges, and r is the distance between those two-point charges.

Similarly, for the gravitational attraction, the attractive force, F, between point masses is given as:

$$F = G(m_1 \times m_2) / r^2 \qquad (2)$$

where G is the universal gravitational constant, m_1 and m_2 are the relevant masses, and r is the distance between their centers.

Equations *(1)* and *(2)* have precisely the same form — possibly a fortuitous coincidence, or perhaps a clue that swapping charge for mass (and choosing the appropriate constant) could give a similar result.

Of course. we know that a simple swap of that kind cannot suffice. Mass must matter. Attractive forces are greater for larger masses than for smaller ones. Recall videos showing astronauts leaping playfully, high up from the surface of the low-mass moon (**Fig. 11.4**). Evidently, mass does matter. Therefore, while some temptation might exist to begin afresh with the notion of charge-based gravitation, any proposed electrostatic mechanism would need to link charge with mass, at least loosely. That's the course we will follow.

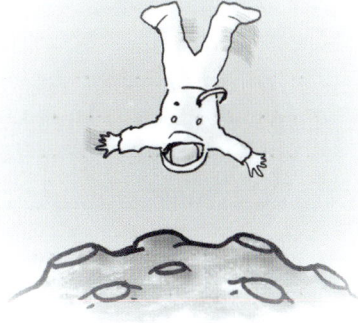

Figure 11.4. Astronauts jump higher on the moon than on the earth. Evidence for mass-based gravitation?

Down to Earth with Gravitation

To begin our pursuit, we ask the following question: What if standard mass-based gravitation were not a fundamental force, but merely an ill-understood interpretation of an electrostatic force? If that were true, then said electrostatic force would need to explain at least the most obvious gravitational basics: (*a*) Gravitation always attracts, but never repels; (*b*) the gravitational force (considered by physicists to be a distinctly "weak" force) is more feeble than ordinary electrostatic forces; and (*c*) the gravitational force generally increases in strength with greater mass.

Does any such charge-based force exist?

Consider induction. We encountered inductive forces in the context of weather. You may recall **Figure 8.9** where we considered how cloud charge could induce opposite charge on the earth immediately beneath, pulling the cloud contents downward (as rainfall). In the case of gravitation, the inductive pulling force would arise from a physically closer connection: an imagined mass lying directly on

Figure 11.5. Inductive charge separation in a block lying on the negative earth.

the surface of the negative earth (**Fig. 11.5**). The earth's negative charge should induce positive charge on the bottom of the block, causing an attractive force to develop between the block's lower surface and the earth.

Inducing positive charge on the block's lower surface presumably involves charge rearrangements throughout. Thus, while the block's positive charges are pulled downward, the complementary negative charges should be pushed upward. Genuine charge separation may occur. Hence, while it may be convenient to think of the induction process as creating atomic dipoles (**Fig. 11.6**), the actual effect may extend well beyond that feature; it may involve the actual separation of charges, as depicted in **Figure 11.5**.

In principle, the same should happen whether the block sits directly on the earth, or lodges somewhere above. Only the strength of the effect should vary.

In this formulation, the attraction to the earth comes not from one force but from the difference of two forces. The positive charges at the bottom certainly get pulled downward. Meanwhile, the negative charges on top repel the earth's negative charge, creating an upward, anti-gravity force. But the two opposing forces are unequal: The distance from the earth to those upper negative charges always exceeds the distance to the corresponding positive charges, making the repulsive force slightly weaker than the attractive force. So, the net force always pulls the mass downward, toward the earth, as we expect from gravitation.

Figure 11.6. The earth's negative charge may reorient the charges in a mass's structural units into dipoles.

And it's also expectedly weak because of its differential nature, *i.e.*, it arises from the difference between two slightly dissimilar forces.

Even if the earth's surface happened to be positively charged instead of negatively charged, the result would be the same. Induced dipoles will always create a net attractive force. Thus, induction satisfies condition (*a*) in our list of necessary features. The force is always attractive and never repels (**Fig. 11.6**).

The net attraction may nevertheless be more evident from **Figure 11.5**, which emphasizes the charges accumulating at the top and bottom of the block. Attraction exceeds repulsion because the respective distances of positive and negative charges to the earth differ more obviously. In reality, those differences may still be relatively modest; but, then again, the gravitational pull is acknowledged as "weak" — many, many orders of magnitude smaller than the pull between charges in simple electrostatic attraction (Chapter 1). So, the rather weak force anticipated by the induction mechanism satisfies condition (b) in our list of requirements, that gravitation is considerably weaker than electrostatic forces.

So far, induction seems to show a modicum of promise.

What about condition (c)? Does mass matter in this gravitational paradigm? Larger masses of the same material contain more atoms, hence more separated charges, and more downward force. So larger masses must weigh more. Thus, this model predicts that objects with greater mass should generally experience stronger attraction to the earth, thereby satisfying condition (c) in our list. The greater the mass, the greater the gravitational pull.

The weight concept deserves a few additional words. In the proposed formulation, weight is given by the strength of the earthly attraction. As such, weight ought to depend on the degree to which the charges of constituent atoms can separate.

Consider, for example, conducting materials (e.g., many metals). In such materials, outer-shell electrons can shift rather freely — that's what confers the high conductivity. In insulating materials, by contrast, the outer-shell electrons cannot move as freely. This difference should explain the tendency of conductors to weigh more than insulators. Generally, that's the case: Metals typically weigh more than insulators. What counts in determining weight, however, is not just the outer-shell electrons but all electrons; hence, we don't necessarily expect all conductors to weigh more than insulators (and some don't). Nevertheless, the electrostatic paradigm offers a vehicle for understanding the concept of density: It lies in the ability of charges to shift.

In sum, the charge-based paradigm shows at least some modicum of reasonableness. Because of the earth's negative charge, the positive end of the dipole should always lie nearest the earth's surface, ensuring that gravitation always pulls but never pushes (a). The pulling force is expectedly weak because it depends not on charge per se, but on a difference of two charges (*b*). And, the pulling force increases as the number of separated charges increase, meaning that mass does count (*c*). Hence, the proposed paradigm conforms to all three criteria.

Why the Mass's Shape Hardly Matters

How might we expect the shape of the mass to play into the gravitational pull? Should a vertically oriented pole weigh more (or less) than one of the same mass that is horizontally oriented?

To address the shape issue, we must first deal with the location of the earth's charge. We've considered that charge to be fixed on and beneath the earth's surface (Chapter 3). But the natural question arises: Could the earth sustain any such shell of negative charge indefinitely? Positive charges from the atmosphere should flow toward those negative charges, theoretically neutralizing the earth's negativity, at least near its surface. The same could also happen with objects sitting *on* the earth's surface: Their downward-positioned positive charges could also compromise surface negativity. Any such charge recombination, if serious enough, could impair the proposed gravitational pull mechanism, potentially releasing us from the earth's tether.

Some such charge recombination seems inevitable. The fusion of negative and positive charges should create a zone of near-neutrality on the earth's very surface (Chapter 3), leaving the negative charges mainly beneath that surface, perhaps meters to kilometers beneath. Nevertheless, the earth's negativity should largely endure (as electric-field measurements confirm). And, with solar energy perpetually recharging the earth, the proposed electrostatic "gravitational" potential should likewise endure, notwithstanding some surface neutralization. We're not about to float.

The sub-surface locus of the earth's negative charge carries a practical implication: It helps explain why objects of the same mass and material should weigh much the same, irrespective of their shape. Let me explain.

To understand why, imagine two cubes of mass glued together. Suppose that in one case those masses lie next to one another, while in the other case, one rests atop the other. In the stacked situation, the upper mass is relatively farther from the earth's charge, so the effectiveness of the gravitational pull should be smaller; hence, that top mass should theoretically weigh less than the one beneath. However, any such weight difference ought to be trivial if the earthly charge is centered well below the earth's surface. Relative to that substantial distance, any difference in height will be miniscule. Hence, the weight should not vary a whole lot whether the glued cubes are vertically stacked or positioned side-by-side. In terms of weight, mass shape should be virtually irrelevant.

The same for us. When we step on the bathroom scale to receive the bad news, the scale should register much the same weight whether our arms dangle or project outward. The distribution of mass should be largely irrelevant.

For resolving personal weight issues, then, it's either dieting or perhaps moving to a planet with less charge. Contorting your body will not do the trick.

Cavendish's Balls: The Torsion Pendulum

In the charge-based gravitational framework, an issue that lingers is one that you may not immediately appreciate, though physicists certainly do: the lateral gravitational force. Gravitation acts in all directions, not merely up and down. Two masses set at the same height on a table will attract one another. The distinguished English physicist Henry Cavendish (**Fig. 11.7**) carried out elegant experiments two centuries ago that confirmed such attraction.

Cavendish's motivation for pursuing that question is worthy of note. It began in the United States. Surveyors Charles Mason and Jeremiah Dixon were surveying land to demarcate the boundary between Pennsylvania and Maryland (a boundary

Figure 11.7. Lord (Henry) Cavendish 1731–1810.

that later became part of the so-called "Mason-Dixon line"). Learning of this endeavor, Cavendish worried that the gravitational pull of the nearby Allegheny mountains might deflect the plumb bob used for that survey, thereby distorting measurements enough to cause considerable error. Hence, he set out to determine just how serious the lateral gravitational attraction might be.

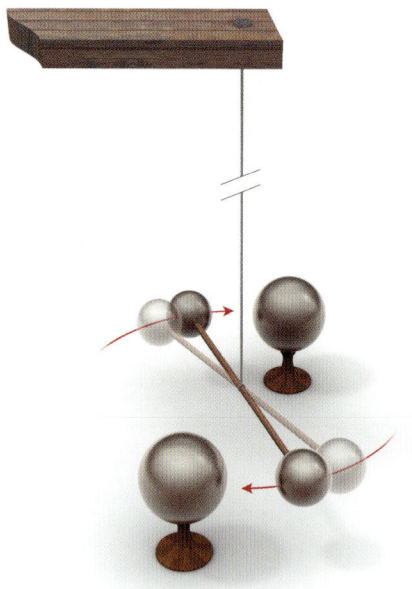

Figure 11.8. Cavendish's torsion pendulum.

To demonstrate that attraction, Cavendish built a clever device. Essentially a barbell, the so-called "torsion pendulum" hangs from its midpoint by a long metallic wire (**Fig. 11.8**). The barbell can rotate freely in the horizontal plane, twisting the suspension wire as it turns.

Cavendish used lightweight wood to make the strut that connected two heavy lead spheres. He then positioned two additional, larger lead masses near the suspended ones, so that any mass-mass attractions would act to twist the barbell.

It worked. Over a period of minutes, the masses moved appreciably toward one another. Although weak, gravitational attractions definitely occurred in the horizontal plane.

Could such lateral attraction be accommodated within the induction-based framework? Imagine a simple situation: two lead masses placed close to one another. (Cavendish used four, to achieve proper balance.) How might the arrays of charges in one mass interact with the arrays in the other mass? Could attraction between those charge arrays account for the observed gravitational pull?

Within each lead mass, we may envision numerous atomic dipoles. For simplicity, suppose we approximate each such dipolar array as a single dipole. The dipole of one mass then faces the dipole of the other mass (**Fig. 11.9**). In the (symmetrical) situation of *panel i.*, nothing much should happen, except perhaps for a miniscule repulsive force between charges of opposing dipoles.

Suppose, however, that some minor asymmetry exists, as represented by modest dipolar offset (*panel ii.*). Some asymmetry would be hard to avoid. (In fact, it should occur more readily than the special case of symmetry.) Head-to-tail attractive forces should then develop between opposing

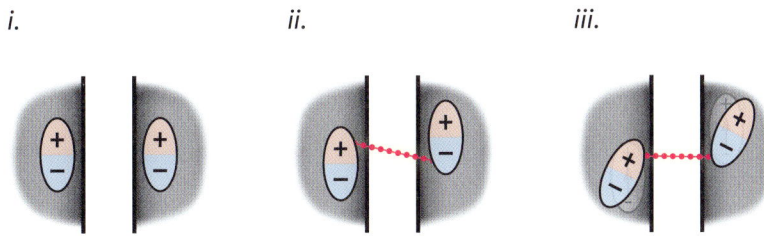

Figure 11.9. Schematic view of dipoles (viewed from above) that make up the masses in *Figure 11.8*.

dipoles (*panel ii.*), which may then induce dipolar rotation (*panel iii.*). As the dipoles rotate, charges get closer together, so their attraction should increase. Inevitably, the positive end of one dipole should come as close as possible to the negative end of the other, maximizing the attraction.

By this mechanism, laterally separated masses should inevitably attract — as Cavendish demonstrated. Such attraction should occur whenever two masses come into sufficient proximity. Induction does the rest. It enhances the attractive pull.

Hence, we can appreciate why "gravitation" can act laterally. The lateral attraction seems explainable from the charges that make up the masses, just as those same charges arguably explain gravitation's vertical pull.

A natural question is whether the value of the gravitational constant measured in the horizontal direction agrees with the known value prevailing in the vertical direction. Should orientation matter? In the electrostatic framework, the gravitational constant, G, reflects the efficacy of the inductive process — how readily charge can be induced in a nearby object. It's a physical feature, unrelated to orientation. Thus, horizontally obtained values of G should agree with vertically determined values. The effect should be direction insensitive (as anticipated equally in the mass-based paradigm).

Of course, the magnitude of the vertically directed gravitational force will ordinarily exceed that of the horizontally directed force because of the earth's huge mass and charge, but the gravitational constant, G, should remain insensitive to direction. In this respect, mass-based and charge-based paradigms behave similarly.

Testing for Charge Involvement

If gravitational forces arise as a consequence of the earth's net negative charge, then canceling that earthly charge should allow masses to float. Gravitation should vanish. Likewise, objects acquiring enough net negative charge should defy gravitation by repelling the earth and rising upward. Testing these predictions would seem straightforward, so the question arises: Why hasn't anyone tried it?

They have, albeit indirectly and facing inevitable obstacles.

Thus, attempts to deal with cancelation of gravity are routinely met with a dismissive suspicion. We "know" in advance that canceling gravity should be impossible; hence, anything that suggests otherwise smacks of delusional thinking, and quickly gets labeled as "pseudoscience." Perhaps this explains why, of late, few academic physicists have dared to study anti-gravity phenomena seriously.

Nevertheless, we do hear about anti-gravity experiments carried out with "lifters." A typical lifter consists of an ultra-light tubular or geometric frame built around a vertical axis (**Fig. 11.10**). A pair of vertically separated conductors runs around the frame, tightly secured to it. Connected to those two conductors through long flexible wires, a power supply imposes a large potential difference between the conductors.

When the voltage is cranked up sufficiently, on the order of 10 kV - 30 kV, the assembly lifts, floating above the earth. It may hover indefinitely, moving randomly to and fro in a manner reminiscent of a flying kite. When the voltage is turned off, the lifter plummets back to the earth. Somehow, the presence of electrical charge creates lift.

Naturally, we were prompted to build lifters of our own, and two different versions were constructed in our laboratory. They rise up, rarely failing to impress visitors. Just as intriguing to watch are web-based videos of other lifters in action.[w5, w6] At one time, gravity-canceling devices of this ilk were the objects of serious study. The well-known engineer and inventor, T. Townsend Brown, acquired several patents in the 1960s for his

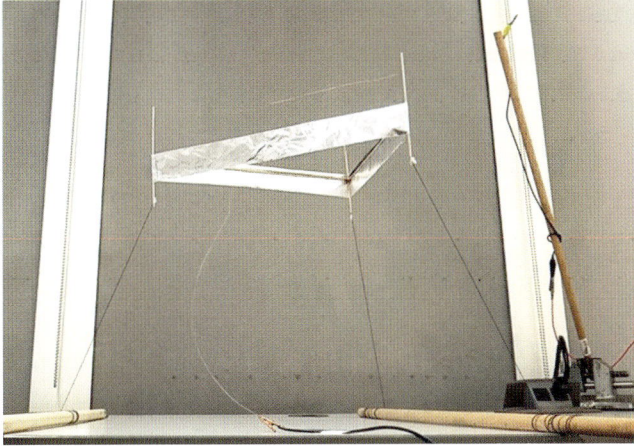

Figure 11.10. Lifter, floating above table. Angled guy wires constrain lifter from excessive elevation. Electrode pair consists of aluminum foil (bottom electrode), and above it, a thin, parallel wire (barely visible) held atop vertical separators. High-voltage source wires minimally visible at bottom and to the right.

"electro-kinetic" devices — machines that used electric fields to levitate. Since nobody could explain precisely how they worked, the explanation got pigeonholed as the "Biefeld-Brown effect." Name assignment notwithstanding, still, the underlying mechanism has remained somewhat obscure. In some way, the imposed electric field creates anti-gravitational lift.

An extensive review by Paul LaViolette[10] reveals a rich history of many successful anti-gravity devices, along with detailed descriptions of the sometimes-prominent engineers and scientists who helped develop them. Such gravitation-defying devices are surprisingly common. Evidently, engineless devices can fly.

Regarding the material above, an academic might say that plenty of "preliminary evidence" shows that electrical devices can cancel gravity. On the other hand, the term "anti-gravity" smacks of UFOs, along with the aliens who might inhabit them, enabling the quick dismissal of this sort of discourse as foolish fantasy. Such ready dismissal keeps otherwise interested scientists from seriously investigating these phenomena.

It's not just lifters that use charge to rise up. In subsequent chapters, I will present evidence that many common flying objects can arguably gain loft by acquiring enough negative charge to repel the earth. The list ranges from paper planes and frisbees, all the way to birds, insects, and even gliders. The underlying mechanism is much the same as argued for those billowy white clouds (Chapter 8): simple repulsion from the earth.

While all such observations imply that gravitation (and anti-gravitation) may well involve the earth's charge, they do not specifically address the possible role of induction. We consider that next.

Testing the Charge-Induction Mechanism

Possible tests of the charge-induction model for gravitation can be envisioned, although some of those tests may be less than straightforward.

One might argue that induction is automatic: If the earth is negatively charged and Faraday induction is real, then any object situated on or above the earth must be subject to inductive attraction; the presence of such attraction cannot be escaped. The challenge is to determine whether the size of the attraction matches what we expect from gravitation, and that exercise remains subject to multiple uncertainties, hence difficult to reliably pursue at the moment.

In terms of experimental tests, perhaps the most obvious one consists of placing an electrostatic shield around a mass. Canceling the earth's electric field, according to the proposed theory, should cancel gravitation; the enveloped mass ought then to float. While this maneuver could constitute an important test of the proposed paradigm, no shield that I'm aware of, including the so-called "Faraday cage," can yet achieve this kind of isolation, for the shielded chamber itself may experience charge induction. The same for a suit of mylar. Hence, we may need to look elsewhere. Perhaps, one day the shield problem could be solved.

Another conceivable test involves placing the leads of a voltmeter respectively at the top and bottom surfaces of a mass. The separated charges illustrated in **Figures 11.5** and **11.6** imply that the voltmeter ought to detect a potential difference. But not so fast. Voltmeter readings depend on a flow of charge into the meter; otherwise, the meter senses nothing. In the present context, that flow of charge would require the mass's dipolar atoms or molecules to get ripped apart — a demanding scenario. Therefore, the voltmeter test may be inapplicable, although this genre of experiment could be particularly revealing if one could devise a proper way to test the proposed top-bottom separation of charges (**Fig. 11.5**).

A third and perhaps more promising testing route would measure dipole orientation. The Cavendish experiment implies that placing a huge lead mass next to an object should reorient at least some of that object's dipoles. In the proposed gravitational context, that would correspond to a weight change. The relevant test has, in fact, been carried out, with a surprising result: an object's weight does change when a metal mass is brought close by.[6] If the object's dipolar orientation could be tracked during such a weight-change experiment, then this might constitute an effective approach to testing the induction mechanism.

In sum, direct tests of the inductive mechanism should be possible, although challenging. Meanwhile, we are obliged to rely principally on existing experimental data, and on the degree to which they fit expectations.

Wrapping Up

Should the electrostatic approach to gravitation eventually prove adequate, then simplifying consequences could accrue. Think of those floating clouds, for example. In Chapter 8, we considered their height to be set by a balance between two forces — the repulsive force between cloud and earth and the attractive force of gravitation. That balance, it was suggested, sets the clouds' height. On the other hand, if the gravitational force itself arises electrostatically, then the need to deal with any such distinct, mass-based gravitational force would be obviated; all relevant forces would be electrostatic. One essential mechanism could lead to a more parsimonious understanding, not only of cloud dynamics but also of the dynamics of any, and all, objects situated on or above the earth.

Our understanding of gravitation-related phenomena could thus gain fresh depth if gravitation were to arise not from mass itself, but from the charges contained within the mass. The distinction between those two attributes may seem subtle, but the impact for understanding could be potentially far reaching — as I hope to show in subsequent chapters.

Summary

Masses attract. Despite the widespread presumption that mass-based attraction bears direct responsibility for the phenomenon that we refer to as "gravitation," scientists have yet to reach a fundamental understanding of what prompts those masses to draw toward one another. The "why" question remains unanswered.

For most, gravitational attraction is largely axiomatic. We learned it in middle school, and the concept has felt comfortable ever since — one of those certainties of natural science that we have incorporated into our very being. It's become part of us. The last thing we want is to divest ourselves of that satisfying understanding.

Yet, concerns arise. While many observations fit the mass-based paradigm, others fit less well, and still others don't fit at all. The failures get routinely dismissed as anomalies. But those anomalies demand explanation. The presence of enough unresolved anomalies ought to jar us out of our complacency, prompting us to re-think that central mass-based paradigm. It may feel comfortable, but can we be sure of its adequacy?

The alternative paradigm that we've begun considering does not discount mass, but centers on charges. The earth is negatively charged. That negative charge should induce an opposite charge on the closest surface of any object on or above the earth's surface. That creates attraction. The resulting attraction ought to pull that object toward the earth. That pulling force may be no different from the force we know as gravitation.

I remind the reader that this concept is hardly radical. Masses are made up of charges, and, rather than the masses themselves as the prime movers, we focus on the charges that make up those masses. This does not challenge the general concept that masses attract; it merely provides a conceptual basis underlying that attraction. Within that framework, we can understand *why* those masses attract — possibly, from the simple attraction of opposite charges.

Gravitation not only affects earthly bodies; it also has cosmological implications. The sun attracts the earth as well as the other planets of our solar system. We next address whether those attractions can be explained by the same electrostatic forces proposed to explain gravitation on the earth. Might those same principles suffice?

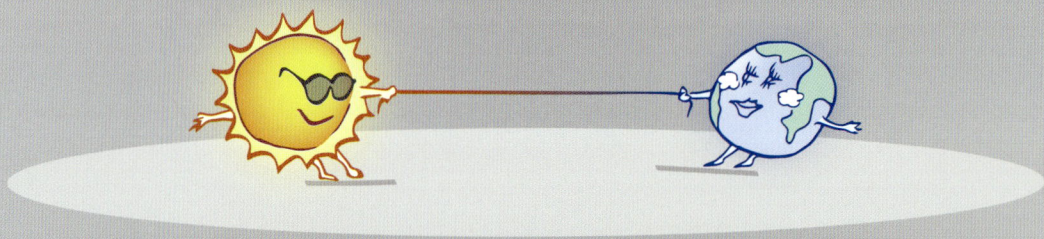

CHAPTER 12

What Ties Our Planet to the Sun?

Gravity connects us to the earth's surface. Our mass gets drawn toward the earth's mass supposedly because "masses attract." Some questions about that long-standing paradigm were raised in the previous chapter, and we began exploring an alternative mechanism for gravitation based on the charges lodged within those attracting masses.

In the proposed mechanism, charge reigns as king. The negative charge of the earth induces positive charge on the bottoms of any masses situated on or above the earth's surface. The result is an attraction — owed to the earth's negative charges pulling on those induced positive charges above. Attraction prevails, as gravitation warrants. And, since larger masses contain more charges, those larger masses get pulled more strongly than smaller ones. They "weigh" more, as common experience says they should.

Figure 12.1. Hannes Alfvén, (1908 – 1995).

Having dealt with on-earth gravitation, let us now turn our attention to the earth itself. The sun attracts the earth as well as the other planets of the solar system, keeping them from flying off into the far reaches of our galaxy. Could the charge-centered paradigm deal effectively with those larger-scale gravitational phenomena?

If electrical forces in the cosmos strike you as unlikely, or even bizarre, allow me to challenge that notion. Many decades ago, physics Nobel laureate Hannes Alfvén (**Fig. 12.1**) opined[1]:

"Certainly we have seen plenty of evidence of electrical phenomena out in space. Within the last few decades we have discovered several important electrical effects in the heavens: strong stellar magnetic fields such as could only be caused by large electric currents, radio waves emanating from the Sun and from many star systems, and the energetic cosmic rays, which are electrically charged particles accelerated to tremendous speeds."

Even Tesla had something to say on the subject of cosmic electrical phenomena.[w1] Dismissing Einstein's curved space-time formulation of the origin of gravitation as "impossible," Tesla went on to build his own "dynamic theory of gravity" based on electromagnetic phenomena.[w2] Tesla hypothesized that the sun and all stars emit "primary solar rays" that in turn produce the secondary radiations that underlie gravitation. Soon after announcing that his theory had been worked out in detail, Tesla died; so, regrettably, no formal mathematical description of his electrical theory has ever appeared.

In Chapter 11, I mentioned that a community of scientists and technologists, under the name, *"Electric Universe,"* has met regularly to present and discuss the critical importance of electricity in the cosmos. Picking up on the electrical gravitation theme, one of the topics under discussion in this group is the possible electrical origin of gravitation. The group receives plenty of criticism for their unconventional outlook; yet, having attended several of their conferences, I've found much to stimulate my own thinking. Judging from the size of their conference, interest in cosmic electrical phenomena seems to be robust.

Continuing along these unconventional lines of thinking, I will attempt to demonstrate the relevance of electrical phenomena in planetary gravitation. We begin with the earth's gravitational attraction to the sun. Could electrical phenomena dominate?

Why the Earth Loves the Sun: Consider the Solar Wind

Coming immediately to mind once again is the phenomenon of induction. The distance between earth and sun, however, may be too vast to lend credence to that option. Two more plausible charge-based options can be envisioned: (*i*) Sun and earth are oppositely charged, in which case simple electrostatic attraction might explain the gravitational pull; or (*ii*) those two bodies bear the same charge, in which case Feynman's like-likes-like attraction (Chapter 7) could stand as a possible candidate for explaining the gravitational pulling force.

Do any of those options show promise?

Regarding the polarity of the sun's charge, nobody has yet managed to get to that glowing body to check. But one prominent feature of the sun's output bears relevance because it may contain net charge: the so-called solar wind. Evidence for the presence of solar wind on earth is manifested as the northern and southern lights (more about that below).

This "wind" originates from the outermost part of the sun's atmosphere, the corona. From there, it blows all the way to the far reaches of the solar system, beyond even Pluto. That long-ranging wind manifests as a somewhat wavy sheet, measuring about 6,000 miles thick as it passes nearest the earth.[w3] Curiously, the sheet intersects all of the solar system's planets (**Fig. 12.2**), raising the question of whether such intersection may be more than coincidental.

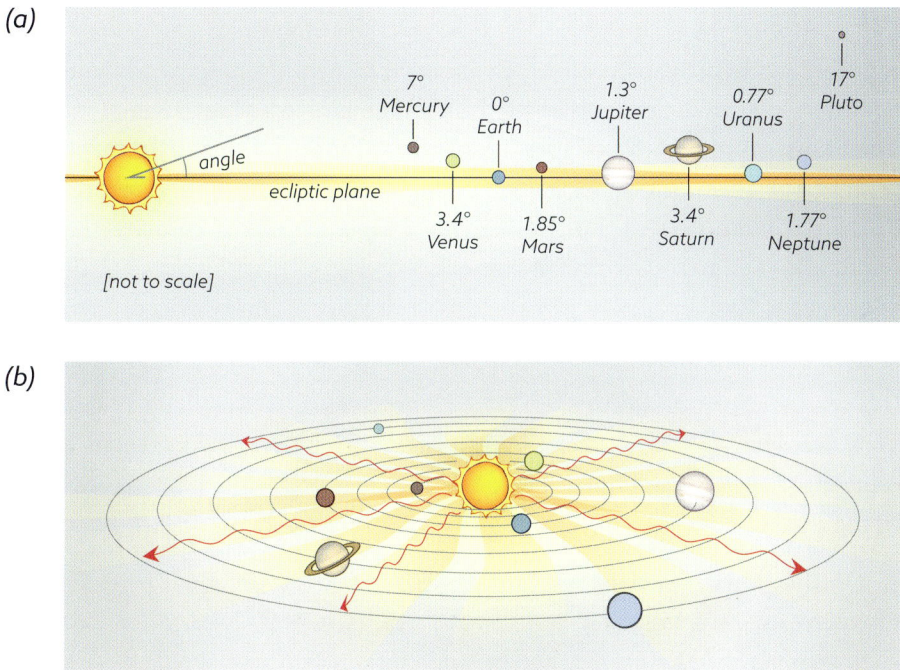

Figure 12.2. (*a*) Planets lie largely in a single plane. The "ecliptic plane" denotes Earth's plane of revolution around the sun. The various angles in the figure refer to the respective deviations from that plane, which are small. Pluto's more substantial deviation is one reason why it has been designated a non-planet. (*b*) The solar wind intersects all of those planets.

The solar wind is sometimes described as a current sheet, with charges flowing from the sun to myriad places beyond.[w4] The character of that flowing charge may carry relevance. Thus, we would like to know the following: its charge polarity; and whether those charges bear responsibility for propelling the wind to the far reaches of the solar system.

For driving the solar wind, the prevailing explanation is that the propulsive force comes from the extreme temperature at the wind's origin. That explanation could suffice for explaining propulsion in the immediate vicinity of the sun's corona; however, it may lose relevance in explaining why the wind can reach beyond Pluto, where temperatures lower than even negative 200°C would freeze even cold-hardened Eskimos. Could a temperature-driven wind propel itself in so frigid an environment?

Another possibility is that the wind is driven by repulsive charges. The solar wind appears to consist mainly of protons. Routinely measured by ion-spectrometer-containing spacecraft such as Ulysses and ACE, the solar wind's composition is well established: approximately 95 percent ionized hydrogen (protons, positively charged); 4 percent ionized helium (alpha particles, two positive charges); and less than 0.5 percent other ions. Since positive charges repel one another, the release of those positive ions from the solar corona could create extreme pressure that drives the wind flow — not unlike the charges proposed to drive the wind flow on Earth (Chapter 4).

Such massive positive charge release makes it easy to understand how the solar wind could propel itself with blazing speed — less than five seconds for transiting the distance between Los Angeles to New York. Imagine! Proton flow could also account for the wind's label as a "current sheet," for the flow of positive charge amounts to nothing less than current.

On the other hand, many scientists presume that an equal number of electrons must lie in the solar wind mix to achieve neutrality. But such *presumed* neutrality, debated on and off for a century,[w5] could pose a problem: It would eradicate any possibility of a charge-based propulsive force. With positive charges alone released from the solar corona, on the other hand, repulsion-based propagation is natural. Some propulsive force is necessary. Without it, any resistance encountered along the way would quickly compromise an unpropelled wind, which, in fact, broadcasts over *billions* of miles.

Support for the positive-charge paradigm comes from comparison with the energy emitted by stars. The sun, after all, is a star. Cosmic rays from distant stars reach all the way to the earth. We know that they consist mainly of protons (90%) and alpha particles (9%), *i.e.*, positive charges.[w6] If the energy emitted by those remote stars bears similarity to the energy emitted by our own star, then the net positive charge hypothesis gains strength. No need to invoke electrons to "neutralize" the solar wind.

Additional support for the net positive-charge paradigm comes from a different solar wind feature: its associated magnetic field.[w7] Currents produce magnetic fields. If the wind generates a magnetic field, then the flowing currents imply that charges should not experience neutralization. The wind ought to contain an excess of one polarity or the other, and the experimental evidence implies that that species should be positive.

While these arguments may enjoy a modicum of reasonableness, some challenge to the notion of positive solar wind comes from two potentially inconvenient hitches. The first is the wind's characteristic plate-like shape. With positive charges alone, you'd expect the wind to expand in all three dimensions as it emerges from the sun's corona, rather than confining itself to a (wavy) sheet-like configuration. Some restraining agent seems necessary, particularly at the sheet's surfaces, and a speculation is the incorporation of random negative charges drawn from the environment. While not the dominant charge species, those negative charges could nevertheless combine with the positive wind at its periphery, preventing said expansion. Incorporation of some electrons at the flat boundaries does not necessarily negate the wind's net positive-charge characterization but does provide for at least some mechanism to confine the wind into its signature, sheet-like configuration.

The second hitch: If the sun continuously releases positive charge, it should become

OUT-ON-A-LIMB METER

increasingly negative — *i.e.*, the sun would experience progressively increasing net negative charge. In theory, that could happen. The sun could continuously build negativity, eventually burning out (not soon, we hope) when that huge body becomes sufficiently charge unbalanced. Stars do burn out.[w8]

Notwithstanding the burnout option, maintaining the sun's neutrality would seem more natural. Sustaining neutrality implicitly demands that the sun release negative charges along with the positive. Reasonable arguments can be adduced for a mechanism that releases the needed negative charge, but in a different context — and I will reserve those arguments for a subsequent book in which I consider the nature of electromagnetic energy.

For now, we stick with what measurements and logic imply: a solar wind that is positively charged (**Fig. 12.3**).

Complementing that positive charge is the earth's negative charge (Chapter 2). In this scenario, the positive solar wind arriving from the sun should attract the negative earth — a feature that would appear to place option (*i*) as discussed earlier, in which the sun and earth are oppositely charged (simple electrostatic attraction), at center stage.

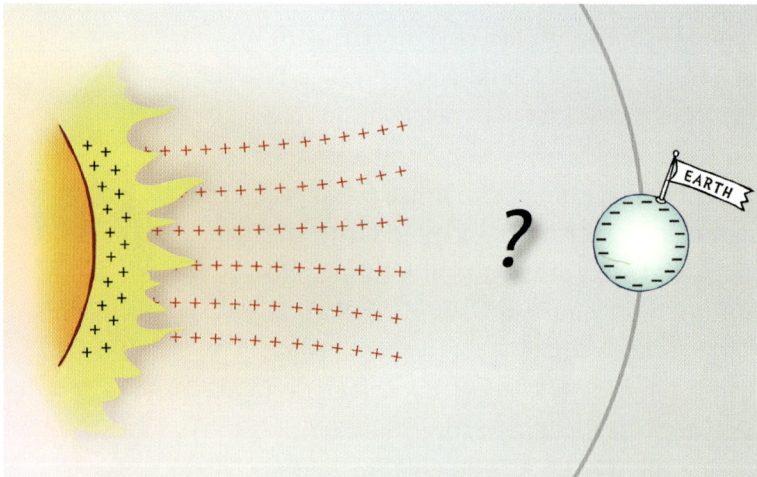

Figure 12.3. Provisional scenario (option *i*) in which the positively charged solar corona attracts negative charge on the earth.

More on the Solar Wind

Next, I will present arguments as to why the solar wind may be particularly effective in pulling the earth toward the sun. But first, I must mention another earth attractor: atmospheric positivity.

The earth's atmosphere contains plenty of positive charge (**Fig. 12.4**). Atmospheric positivity builds from the sun's energy (Chapter 4). The figure emphasizes what we have already seen: Atmospheric positivity develops on the earth's sunny side. That positive charge attracts the negative earth, inching it sunward. That sunward pull constitutes an attractive force, not unlike gravitation, drawing the earth toward the sun.

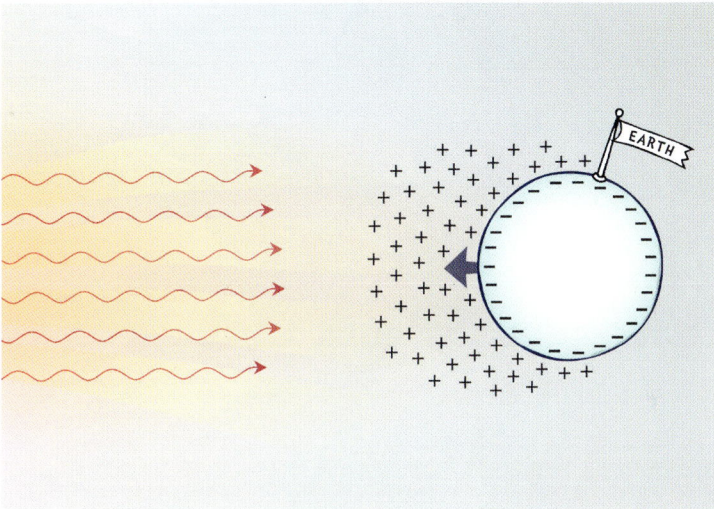

Figure 12.4. The sun's radiant energy builds positive charges in the earth's atmosphere (Chapter 4). Those charges pull on the negative earth.

Hence, the sunward pull on the negative earth may come from two bodies of positive charge: (*i*) atmospheric; and (*ii*) solar wind. The atmospheric pull is straightforward, though presumably weak. The solar wind's pull, on the other hand, is more complex and deserves further consideration. In particular, we need to address the question of why the solar wind's arriving positivity doesn't merely push against the atmosphere's positive charges, thereby repelling the earth.

Most of the solar wind does stop at the atmosphere's outer edge (**Fig. 12.5**). However, its two "polar cusps" deflect around the atmosphere's mass of positive charges, penetrating all the way to the earth's surface and intersecting obliquely near each pole. Because the positive-negative separation is thus small, the pulling force may be substantial. It may overwhelm the wind's push against atmospheric positivity.

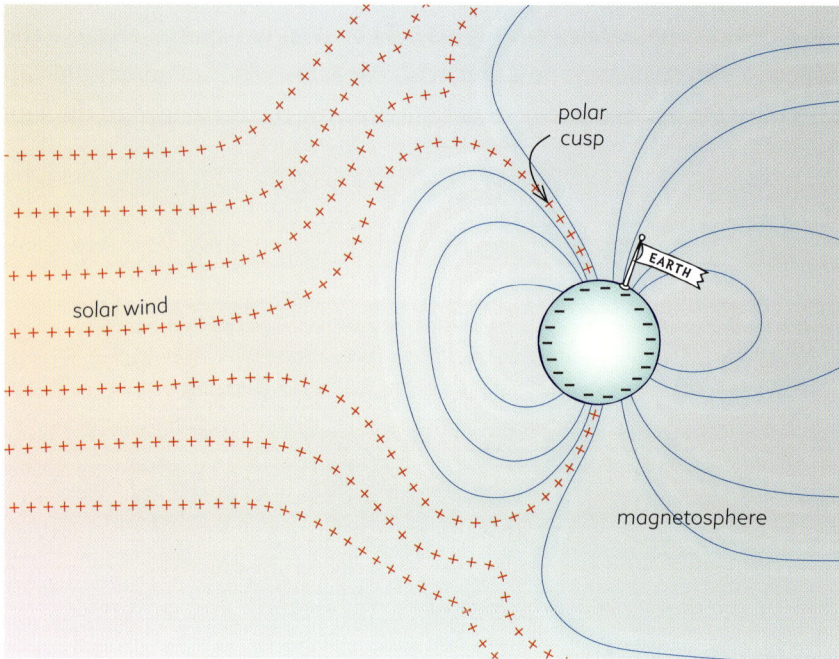

Figure 12.5. The deflected solar wind arrives at the poles of the earth, as commonly understood.

Those polar cusps may be unfamiliar, although you may know of something related: the northern and southern lights (more below). Those light shows presumably arise from the concentrated flow of current passing through those near-earth polar cusps. They light the sky with glorious spectacles.

The polar cusps have become well recognized by scientists, but the reason for their iconic shape has remained uncertain. We can now understand. The incoming solar wind's positive charges should get deflected by

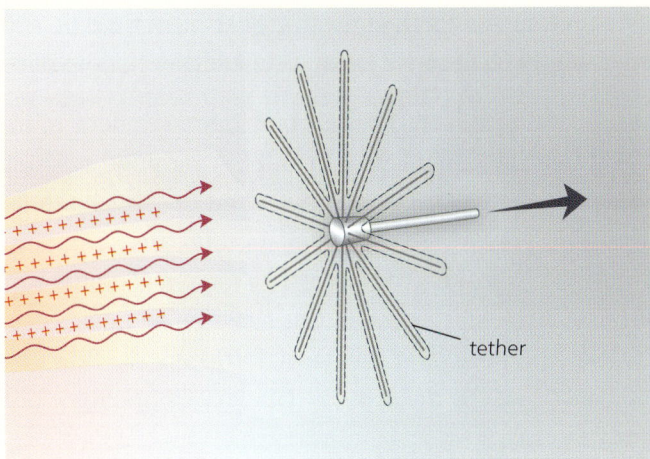

tether

Power From the Solar Wind

Given the solar wind's presumed net charge, you might anticipate some practical exploitation. It's happened. An example is the "electric sail" attached to a space vehicle, which won the 2010 Finnish Quality Innovation Prize.[w9] That device amounts to nothing more than a tethered bunch of long, wires emanating from the vehicle, the wires bearing positive charge.[2] The wires move consistently *away* from the sun, dragging the attached vehicle along with them. According to the inventors, the propulsive force arises from solar wind "pressure." One wonders: Could that so-called pressure arise from the wind's positivity, repelling the wires' positive charge and driving the wires away from the sun?

the atmosphere's positive charges, causing the wind's charges to veer off to either side and finally toward the poles (**Fig 12.5**). The poles, in turn, should willingly accept those incoming charges. Sunlight is minimal there, so the buildup of atmospheric positivity should likewise be minimal. With scant atmospheric positivity, the dominant feature becomes the earth's negative charge. That negativity should draw in those positively charged solar wind tails, explaining the solar wind's characteristic shape.

It's worth emphasizing the solar wind's oblique intersection with the earth (**Fig. 12.5**). The polar cusps intersect at some angle. Hence, a component of the wind's attraction to the earth lies in the direction of the sun, thereby pulling the earth sunward. That pull could be substantial since positive wind and negative earth ultimately lie close to one another.

As an aside, the term solar "wind" may open the door to misunderstanding. Although we might imagine a blowing wind pushing on the earth, that's not exactly so. The positive wind may push repulsively on the atmosphere's positive charges; but its tails, reaching close to the earth and charged oppositely to the earth, should *pull* on the earth. They ought not push. And, since charge forces can be so powerful (Chapter 1), that attractive pull could be substantial. Indeed, the tails of the solar wind could constitute the majority of the electrostatic force pulling the earth toward the sun.

That solar wind pull, along with the atmospheric charge pull, should draw the earth sunward. Whether the combined pull constitutes *the* gravitational force, keeping the earth from veering off from its orbit, can only be deduced by proper quantification, but the relevant numbers remain uncertain. Hence, the thesis must remain speculative, as you've likely already surmised.

The solar wind represents a prominent feature of the cosmos. It has nevertheless been considered little more than a sidebar to the thesis of cosmological physics. Here, I am suggesting that this rather conspicuous entity may play a more prominent, if not central, role in the earth's draw toward the sun.

Are the Solar Wind Tails Multifunctional?

Solar wind tails may be important not only for contributing to the gravitational pull on the earth, but potentially also for explaining some other earthly features. Current flow is one of them. Those positively charged winds flowing into the negative earth constitute current flow into the earth. This flow may be the source of the so-called "telluric" currents known to flow through the earth's crust and mantle. Those currents flow mainly from poles toward the equator, as might be expected from a solar wind origin. Further, they flow on the daylight side of the earth, where the positive charges insert, returning on the earth's dark side. Thus, solar wind charge may well account for those long-mysterious earthly currents, whose functional significance remains to be explored.

Incoming current entering the earth near the poles may also explain the tourist-attracting spectacle known as northern and southern lights (**Fig 12.6**). Those oft-spectacular auroras commonly show up in polar zones, where their appearance is thought to derive from the arrival of charged particles (current) from solar eruptions. Eruptions could well influence the solar wind-tail currents; however, the currents themselves could plausibly suffice for creating those awesome light displays, much like lightning, but sustained.

Figure 12.6. Northern lights, viewed from Iceland.

Finally, the solar wind tails may enhance the earth's rotational stability. Think of an old-fashioned globe held by a semicircular frame. In the same way that the grip of the frame constrains that globe's rotational axis, the grip of the solar wind may constrain the earth's rotational axis (**Fig. 12.5**). That grip might help maintain planetary stability. It might perhaps even explain the earth's characteristic tilt, whose angle does vary slightly over time as we might expect from anticipated fluctuations in the strength of the wind's charge source.[w10]

In sum of the sections above, abundant positive charges lie skyward from the sun-directed face of the earth. Those positive charges should pull on the negative earth, drawing it toward the sun. That pulling force could conceivably explain at least some of the sunward attractive pull that we currently attribute to mass-based gravitation. It might also account for various idiosyncratic features of earth physics. But principally, it may explain why planet earth is kept from flying off into the dark recesses of the cosmos.

Revisiting Anomalies

We now switch gears. The previous chapter listed a set of gravitational observations that appeared to conflict with conventional theory. As we close our discussion of planet-earth gravitation, we need to return to the lingering question: Does the newly proposed, charge-based framework resolve those anomalies? Let us consider them, one by one.

Dark matter. Physicists introduced the concept of "dark" matter to compensate for mass that seemed missing within the conventional gravitational framework. That "missing" mass was not trivial — it could amount to as much as ten times the detectable mass. The charge-based framework makes it unnecessary to invoke dark matter because force need not correlate directly with mass (although it often does); the primary variable is the charge contained within the mass. Hence, the concept of "missing mass" becomes irrelevant in the proposed schema. No need exists to invoke dark matter.

The Pioneer anomalies. In the new framework, satellite orbits like those of Pioneer 10 and 11 do not depend specifically on mass; so, once again, mass-based anomalies lose much of their relevance. The charge-based framework does allow for orbital computations. However, the results of any such reckonings depend on variables not yet well quantified — *e.g.,* the distribution of charges around the earth and the polarizability of satellite materials. Once those variables have been definitively mapped, comparing predicted and observed orbits should be possible.

Seasonal variation of gravitation. Summer's gravitational force exceeds winter's gravitational force, implying a seasonal change in a supposedly fundamental constant. In the conventional gravitational framework, this stands as a curious anomaly. In the new framework, summer's penetrating sunshine should produce greater atmospheric charge than winter's weaker sun. The higher local positive charge could create more local gravitational pull toward the negative earth, potentially resolving this anomaly.

Solar eclipses. Blockage of sunlight should impact local atmospheric charge; therefore, the eclipse should dynamically modify local gravitational force — as observed.[3]

Spatial variation of gravitational pull. Gravitational forces ordinarily pull vertically. However, lateral force components could arise if subsurface charge distributions are distorted by local variants in the earth's subsurface materials. Those components have been documented.[3] Such lateral force components could potentially drive water and cars uphill — as anecdotally reported in multiple areas but reflexively dismissed as optical illusions.

In sum, electric field effects could add a fresh twist to the fabric of understanding. Phenomena that have seemed "anomalous" by standard gravitational considerations could become predictable features of charge-based gravitation.

In this context, I refer again to the growing movement toward recognizing the centrality of cosmic electrical phenomena, pioneered by the Electric Universe group and its experimental voice, the Thunderbolts Project.[w11]

Relativity-related Anomalies

Beyond the aforementioned list of anomalies, I now refer to certain gravitation-related anomalies that had seemed resolved by general relativity theory. Only following their resolution did Einstein's theories finally gain broad acceptance. I mention these anomalies because even though the recent report of gravity waves [w12] has been argued to show consistency with relativistic predictions, the charge-centered gravitational framework may obviate the need for any relativistic considerations.

How so?

The charge-centered gravitational framework emphasizes the central role of electric fields. Enter the Kerr effect.[w13] The Kerr effect refers to the refraction of light caused by imposing an electric field. Applying a field bends light, in much the same way as does a camera lens or the air above a hot road (**Fig. 12.7**).

Figure 12.7. Changes in refractive index bend light, sometimes creating mirages.

From such electric-field mediated lens effects, unexpected features may follow. One of them is refraction. Any light ray passing obliquely through a region with an imposed electric field ought to deflect. In fact, rays of starlight do deflect when passing close to a galaxy or a planet (so-called "gravitational lensing"). Relativity explains the phenomenon gravitationally: It asserts that such bending arises out of a warping of space-time (a gravitational concept that I've tried, unsuccessfully, to understand). That may be. On the other hand, electrostatic considerations raise a potentially simpler explanatory option. If electric fields exist around planets, then the Kerr effect may offer an alternative explanation for the observed bending.

I'm not the first to raise questions about the origin of gravitational lensing. Don Scott's fascinating book, *The Electric Sky*,[4] discusses the issue at length, considering lensing as a function of charge (plasma). Related material can be found also in the popular Thornhill-Talbott piece, *The Electric Universe*.[5] Conceivably, the bending of starlight (or any light) could arise from simple electrostatic effects.

Related questions arise for those so-called altered-gravitation sites (Chapter 11, Box), where impressive dimensional aberrations can be witnessed. If gravitation involves electric fields and those fields suffer local distortions, then local optical refractive changes will follow. Observed dimensions might then vary with position. The obvious question arises: Could the weird height aberrations that so many tourists witness derive from these electrical phenomena? Measurements of local electric fields could answer the question, should someone with the relevant knowhow be tempted enough to pursue such measurements.

These considerations shed some light on the assumed relevance of general relativity. Recall the history. Early resistance to that theory melted away only when relativity seemed to quantitatively explain the observed bending of starlight around the sun during a solar eclipse. Before then, many scientists considered relativity theory inadequate, or, as one distinguished Princetonian put it, "a great and serious retrograde step." It was the relativistic interpretation of the bending of light that sealed relativity theory into the envelope of accepted fundamental science. If the bending arises instead from local electric field gradients, then general relativity theory might prove unnecessary. Gravitation-based anomalies could resolve in ways that obviate the need for that theoretical construct.

Summary

This chapter shifts emphasis from gravitation *on* the earth to gravitation *of* the earth. What pulls the earth toward the sun?

The natural presumption is the traditional one: The mass of the sun attracts the mass of the earth. However, the previous chapter offered arguments suggesting that the mass-based paradigm for gravitational phenomena was, at the very least, incomplete. Charges seemed important. Whether charge-based forces might apply as well to gravitation of our planet seemed worthy of consideration, and it turns out that such lines of consideration are not new: For some time, scientists have pursued, or are pursuing, electrically centered gravitational paradigms.

In any such gravitational paradigm, a likely protagonist may be the solar wind. Conspicuous throughout our solar system, that "wind" carries charge from the sun's corona all the way to regions beyond Neptune and Pluto. It intersects all planets. Multiple arguments imply that the wind may well carry net positive charge toward the earth, creating an attraction that could draw the negatively charged earth toward the sun.

That draw may gain strength from the solar wind's cusps. Like tentacles extending from the body of wind, those cusps project directly toward the earth. Proximity puts positive and negative charges close to one another, amplifying the attractive force that we may reflexively interpret as "gravitation."

Apart from the wind, a second component of that attractive force may come from the positive charges residing in the atmosphere. Those charges build on the earth's sunny side. By attracting the negative earth, the positive atmospheric charges could constitute a secondary component of the gravitational pulling force.

Hence, reasonable arguments point to the possibility that charge forces could govern gravitation *of* the earth, just as they may govern gravitation *on* the earth.

Finally, we dealt with the anomalies identified in the previous chapter, *i.e.*, the list of features that failed to fit the standard mass-based paradigm in a natural way. Do they resolve within the electrostatic framework? While firm conclusions rest in part on data not yet available, many of the issues arising in the mass-based framework become non-issues in the electrostatic framework, where charge prevails over mass.

Continuing along these unorthodox lines, the next chapter ventures to explore gravitation of other of our solar system's planets. Could charge play a role there as well?

CHAPTER 13

Gravitational Odds and Ends: The Cosmos

Having dealt with gravitation on familiar terrain, we move on to terrain less familiar. Could the electrically based paradigm proposed to govern earthly gravitation possibly extend to the cosmic realm?

The term "cosmic" commonly refers to something beyond easy reach. Planets are indeed distant. Gathering evidence on those remote bodies can be formidable, hence this chapter is thin on evidence. Nevertheless, by combining existing facts with logical treatment, we arrive at an electrically based formulation for explaining gravitational forces throughout our planetary neighborhood, one that I hope is worthy of your consideration.

Gravitational Options Beyond Earth

Like planet Earth, our sister planets make their respective journeys around the sun. They revolve. Revolution requires not only an orbitally directed force, but also a sun-directed force — otherwise those orbiting planets would fly off into the cosmos like a chip off a spinning flywheel.

We first address that sun-directed force. If electrical forces govern the Earth's attraction to the sun, then similar forces might govern our sister planets' attraction. Consider a pull on those planets from sources that are the same as those dealt with in the previous chapter: atmospheric charges and solar wind charges.

First the atmosphere. Atmospheres on remote planets are not something we envision reflexively. Nevertheless, they are broadly present. Best recognized is the atmosphere of Venus. But it's not just Venus: An Internet search quickly reveals some sort of atmosphere on each and every planet of our solar system. Atmospheres are pervasive.

And, as does the atmosphere of Earth, those atmospheres appear to bear charge. Spacecraft report lightning discharges not just on Venus,[1] but also on Saturn,[w1] Jupiter,[2] and Mars.[3] And those charges may be substantial: So massive are the dust devils on Mars that fears have arisen that astronauts' safety might be compromised by charged dust penetrating into space suits. For the aforementioned planets, at least, atmospheric charge is something to be reckoned with, just as it is on our own planet.

Those charges, in theory, could give rise to sunward attraction of planets, just as argued for planet Earth in the previous chapter. If so, then atmospheric charge could contribute to gravitational pull in a similar way on all planets of our solar system.

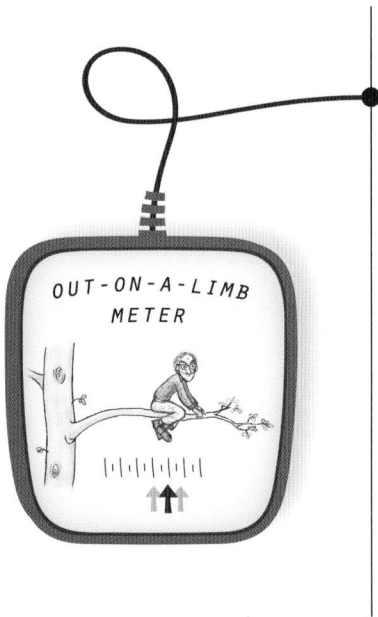

An obvious question is the source of those planetary atmospheric charges. On our planet, a central role is arguably played by water. Water splitting creates positive atmospheric charge. Could the same phenomenon apply on our sister planets? With increasing frequency, we hear reports of water lying practically everywhere throughout the cosmos.[w2] It's seemingly universal. Hence, water-derived charges remain an option. Once those atmospheric features are better established, it may be possible to decide whether the atmospheric charge option really makes sense as a universal component of the sunward pull, not just on Earth but also on our sister planets.

Besides the possible pull associated with atmospheric charge, consideration needs to be given to the second gravitational candidate: the solar wind. Recall that this charged, sun-derived "wind" does not terminate on Earth; it blows well beyond our planet (Chapter 12), reaching all the way to Uranus, Neptune, and (if you feel nostalgic and still consider it a planet) even Pluto. Curiously, the wind intersects *all* of those planets. In the same way that the tails of the positive solar wind arguably pull on the negative Earth (**Fig. 12.5**), similar tails could likewise tug on our sister planets. Why otherwise might the wind so consistently intersect all of those planets if not for some common functional role?

In support of the commonality argument, consider the following phenomenon: All planetary orbits around the sun are elliptical. When planets come closest to the sun (periapsis), orbital velocities are higher than when their orbits take them farthest from the sun (apoapsis). Such consistent speed differences, sometimes referred to as "Kepler's Second Law," apply to *all* planets. Without exception, orbital speeds vary with the planets' distance from the sun. Such consistency of dynamics, once again, implies that the driver is likely to be the same for all. Could that driving mechanism be the solar wind, operating not just on Earth but similarly on all planets?

I will deal more with the solar wind in a moment. For now, let me suggest that it's not just Earth, but also the other planets, on which the solar wind may create gravitation-like attractions. Throughout our solar system, solar wind and charged atmosphere may be central protagonists in the drama of sunward pull.

Solar Wind Pervasiveness

More about the solar wind, a potentially underappreciated force of nature. Consider a curious feature of our solar system: All planetary orbits lie in essentially the same plane (**Fig. 12.2**). Could such consistency happen by chance? Or more likely perhaps, could something constrain those planetary orbits?

An obvious candidate is the solar wind. Described as a wavy sheet of charge, that "wind" extends to the far reaches of the solar system, intersecting all the solar system's planets. If the solar wind grasps those planets as it grasps our Earth (**Fig. 12.5**), then that clutch could constrain the planets to lie in that particular plane. That is, the so-called wind could serve as a central organizing force for our entire solar system.

By contrast, the prevailing explanation for the near-planar distribution goes like this: Whatever created the planets in the first place set them into that planar arrangement, and that feature has remained conserved ever since. In other words, it started that way and stayed that way. That's certainly possible, but absent some physical constraint, you'd think that a few *billion* years would be enough time for the planets to follow their natural tendency to meander off track. Hence, some constraining mechanism seems likely, and the most obvious candidate must surely be that known connector: the solar wind.

In a similar vein, consider yet another consistent feature of planetary dynamics, the direction of revolution (**Fig. 13.1**). Oddly, all planets revolve around the sun in the same direction (the same as the sun's rotational direction.[w3]) Consistency of orbital direction implies yet again that the driver ought to be common to all planets. In plain view is that pervasive agent, the solar wind. If the wind were somehow to drive all the planets in common fashion, then directional uniformity might be a natural expectation. Otherwise, why the consistency?

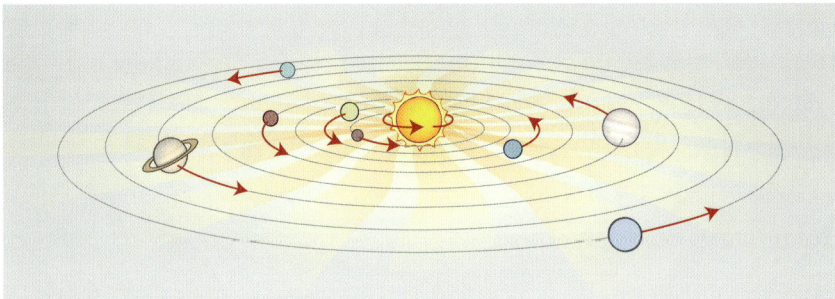

Figure 13.1. All planets of the solar system revolve in the same direction— same as the sun, and within the confines of a disc.

Supporting rationale therefore exists for the solar wind's central involvement in much of planetary dynamics. Not only may that wind help constrain the planets into their near-planar distribution, but it may also play a role in the consistency of their orbital directions. Indeed, the solar wind may be thought of as a physical extension of the sun — an octopus with radiating tentacles — projecting out to all planets and extending its grip to govern diverse features of all planetary motions.

The Revolutionary Drive

We return to the issue of forces. Having addressed the forces likely to pull the planets toward the sun, we now consider the forces propelling travel around the sun. All planets revolve, with periods ranging from weeks to decades. While revolution has been recognized since the time of Copernicus, the forces driving those orbital motions have remained somewhat obscure. Movement requires energy. So, we proceed to the heart of the matter: What energy, or force, propels the planets to orbit ceaselessly around the sun?

This question bears similarity to one addressed earlier (Chapter 6): Why do planets endlessly spin? Spinning and revolution both require energy for overcoming the inevitable drag and frictional forces that conspire to bring those motions to a halt. Those retarding forces may seem inconsequential (most scientists think there's barely any drag in the vacuum of space), but, over the eons, even miniscule losses add up. Something must keep the planets revolving around the sun.

Planets might keep revolving simply because of their immense inertia — once in motion the planets should stay in motion. Initially, I had dismissed that explanation; it seemed like an expedient conjured to skirt the obligation to deal with the nature of the driving force. I'm now less certain. In the next section, I will show that, far from having the bland nature that we commonly assign to inertia, a force that looks, feels, and even smells like inertia does require a continuous supply of energy. As such, some kind of "inertial" force could indeed hold relevance for sustaining the planets' orbital motions.

Before dealing with that phenomenon, let me consider an altogether different potential driver of orbital motion, one arising from the asymmetry of atmospheric charge. Said asymmetry applies to Earth, but the arguments could apply as well to other planets. To illustrate, please see **Figure 13.2**, *panel a* on the following page, which shows a view looking down toward the earth's north pole.

Recall that the earth spins from west to east. Thus, at any site on earth's daylight side, the region to the east will have been exposed to sunlight longer than the region to the west (Chapter 4). With more evaporative gain, the atmosphere to the east should therefore contain more positive charge than that to the west. Such asymmetry should create a skewing of atmospheric positive charges toward the eastern sky (**Fig. 13.2a**).

Such off-center distribution carries with it an implication. It means that the pull of atmospheric positivity against the negative earth should be angled. The angled force vector can be broken into its component parts in the standard way (**Fig. 13.2b**). The major component orients toward

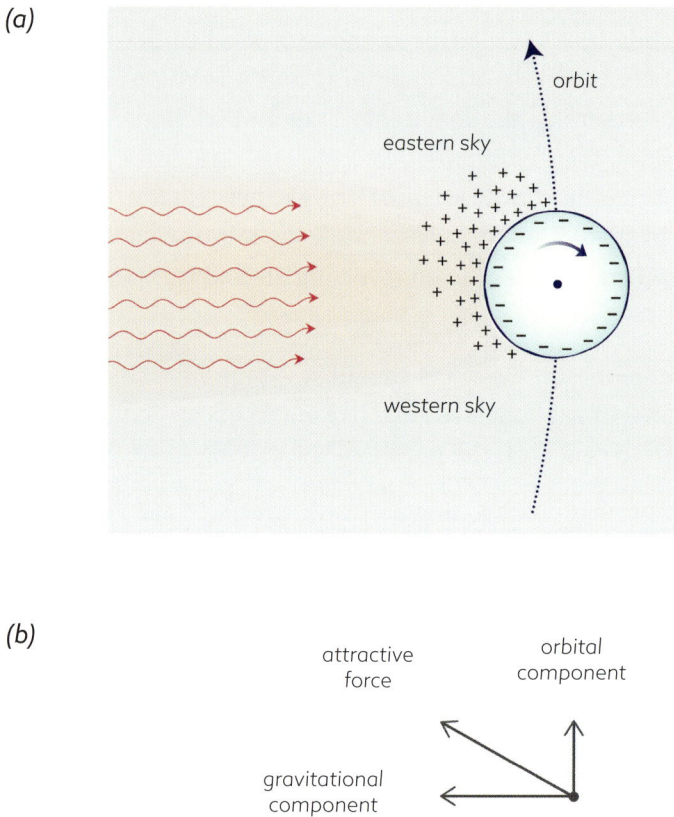

(a)

(b)

Figure 13.2. Asymmetrical distribution of atmospheric positivity. (*a*) View from above the north pole. The eastern sky carries more positive charge than the western sky. (*b*) That asymmetry creates an angled force vector, whose components include a sun-directed gravitational component and an orbital component. The latter may help drive orbital motion

the sun, producing the gravitation-like sun-earth attraction that we just discussed. The minor component orients perpendicular to that, *i.e.*, in the orbital direction. That latter force component could be significant: It should help propel the earth's orbital motion around the sun.

A charge-based orbital force may seem exotic, but you can see much the same demonstrated in space. A video[w4] shows charged water droplets orbiting around a charged Teflon rod under zero-gravity conditions. Closer drops precess faster, distant ones more slowly — much like the planets. Astronaut Don Petit, the presenter, asserts that this is *not* the same as gravitation; it's merely electrostatic. But, if gravitation *is* electrostatic, then those observations fit naturally. They demonstrate that revolution *could* be driven electrostatically.

Hence, the orbital drive of Earth and perhaps the other planets could arise at least in part from the skewing of atmospheric charges. Skewing creates a force component in the orbital direction. On the other hand, the full story of orbital drive may involve an additional force component, as we will see next.

Earth "Inertia" as a Second Revolutionary Driver

Besides implicating the skew of atmospheric charge in maintaining planetary revolution (**Fig. 13.2**), I alluded earlier to a second orbital driver — a kind of "inertial" force. Briefly introduced during our discussion of hurricanes (Chapter 10, **Fig. 10.2**), that force once again involves charges.

To set the scene, please recall Newton's first law of motion. Bodies at rest tend to remain at rest; bodies in motion tend to remain in motion with the same speed and in the same direction, unless acted upon by an external force. That defines "inertia." In fact, the phenomenon is frequently called "Newton's Law of Inertia." I will attempt to show how a quasi-inertial force could arise naturally from a body moving through a sea of charge, and how such a force could plausibly help sustain revolutionary motion of the planets.

Throughout our solar system, positive charges abound. Solar wind, atmosphere, and cosmic energy combine to create a vast sea of positivity enveloping our planet and possibly our sister planets as well. Let us envision a simplified model: a planetary body surrounded by a uniform sea of positive charge (**Fig. 13.3a**). What can this model tell us?

Suppose some force has already set that planetary body in motion. Motion continues, say, toward the left (**Fig. 13.3b**). As it moves leftward, the planet should compress the charges in its forward path, while leaving a void behind. Any such charge disparity abolishes the existing balance: The concentrated charges on the left increase the leftward pull, while the void on the right diminishes the rightward pull. So, the leftward-moving body should continue to move leftward. Once in motion, the body ought to keep moving — similar to the law of inertia.

(a)   *(b)*

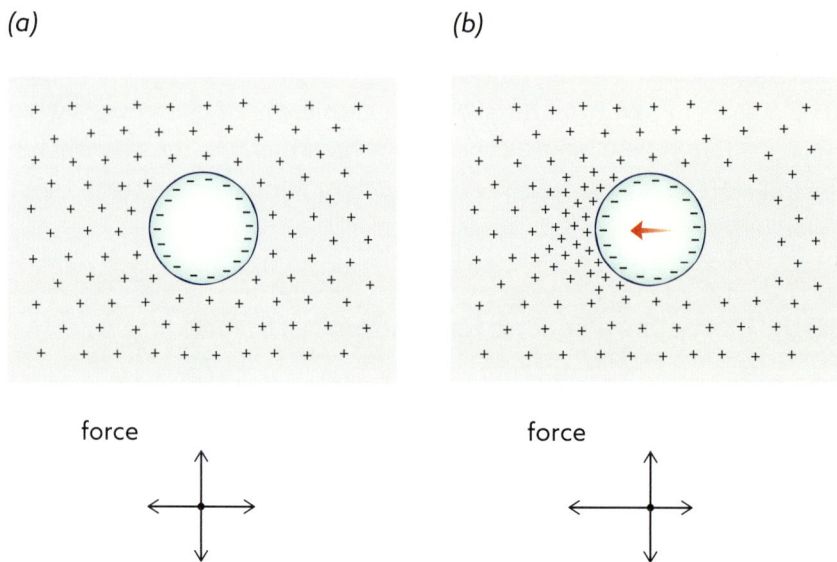

Figure 13.3. (*a*) A negatively charged planet immersed in a uniform sea of positivity. (*b*) Leftward movement will transiently unbalance the charges, creating additional leftward force and thereby perpetuating the leftward motion.

Naturally, any such imbalance can occur only if the body moves fast enough. If it moves too slowly, then the positive charges will have had ample time to redistribute and regain uniformity (which they will eventually do anyway). The higher the speed, the higher the imbalance, and therefore the higher the "inertial" force — very much like Newton's inertia.

A possible snag in this argument was flagged by my elder son, Seth. He suggested the following (politely, of course): Might the forward compression act as a kind of damper? It's like pushing against a spring; the spring pushes back, so nothing is gained.

His comment initially stopped me in my tracks. While I was proud of my son for coming forth with that insightful comment, additional thinking reminded me that his argument skirts a relevant feature: the effect of induction. The compressed positive charges ahead should induce negative charges on the body just behind (**Fig. 13.4**). Those increased negative charges behind, in turn, ought to induce additional compression of the positive charges ahead, *etc.* So, the act of pushing doesn't merely compress; it separates additional charges, which then amplify said body's tendency to advance. One might liken that advance to a bootstrap effect, assuring that the movement will persist.

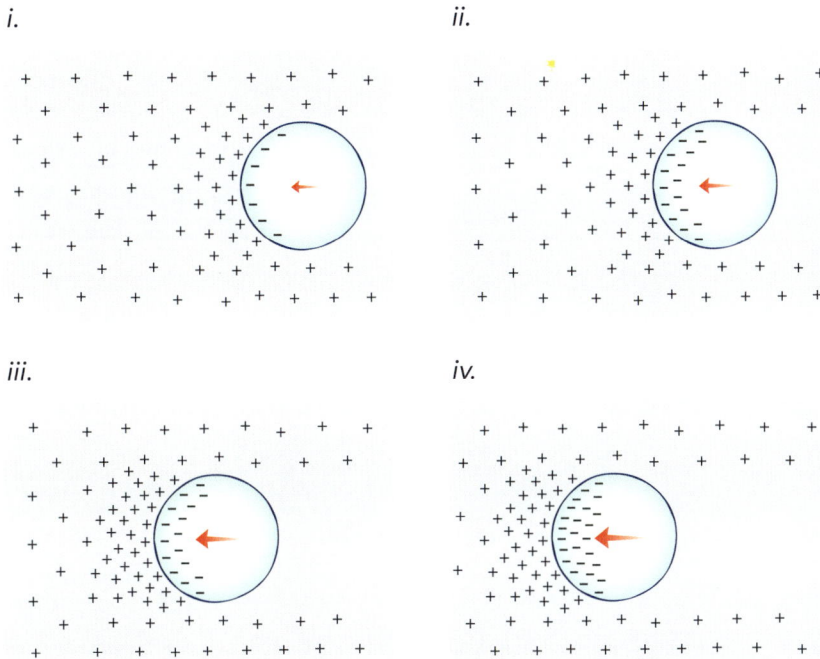

i.

ii.

iii.

iv.

Figure 13.4. Bootstrapping mechanism for perpetuation of orbital motion. Compression of positive charges ahead (*i*) induces negative charge on planet (*ii*), which induces additional positive charge ahead (*iii*), perpetuating the effect (*iv*) and keeping the planet moving in the orbital direction.

With such a bootstrapping mechanism in place, you'd surmise that a body might theoretically accelerate to infinity. However, that should not happen. Because they repel, like charges cannot concentrate indefinitely. In any realistic setting, this accelerative mechanism should, therefore, remain within sensible bounds. Whereas induction can theoretically promote continued, even enhanced motion, speed should remain within realistic limits.

While this quasi-inertial mechanism can arguably perpetuate motion, it may also act to prevent motion from starting up in the first place. Bodies at rest, Newton asserted, tend to remain at rest. In the considered framework, that reluctance to get going should occur because bodies at rest act to maintain the symmetrical charge distribution around them (**Fig. 13.3a**). While initiated movement tends to compress the charges on one side or the other, those positive charges do repel one another, posing some resistance to compression. Such resistance acts against motional startup — which helps explain why bodies at rest tend to remain at rest, and why, therefore, the rest position can be relatively stable.

But if a planet is already passing through a sea of positive charge (**Fig 13.3b**), the quasi-inertial force should ensure that the moving planet continues to move in the same direction. If some component of said force lies tangent to the planet's orbit around the sun, then the planet should keep moving in the orbital direction. By this feature, revolutionary motion is thereby sustained.

Meanwhile, the complementary sunward pull should add to the mix, creating all that is needed for maintaining the planet's continuing motion around the sun: gravitational pull, plus orbital drive.

Charge-based Inertia as a General Feature?

I hope you'll indulge a brief digression to consider the question of inertia in a more general context. Could the quasi-inertial mechanism just described apply more broadly?

I had thought initially that I might have nailed the long-elusive mechanism of inertia. Nobody had yet explained what Newton professed and what experiments had amply confirmed. I wondered: Could the charge-based mechanism suffice in general? I posed the proffered mechanism to a friend whose thoughtful, even-handed responses I trusted, and his positive reaction left me practically giddy with excitement. Then, reality struck. I soon remembered that if inertia works in a vacuum, then charge-based inertial forces would seem unlikely as a general mechanism (if, indeed, vacuums are genuinely free of charge).

Additional skepticism arose because this persisting-motion concept seemed at first to produce something for nothing — motion without energy. Any such unlikely feature would surely drive the final nail into the coffin of charge-based inertia. However, the energy issue can be easily brushed aside: It's the energy of the sun that creates the positive charges in both the solar wind and the atmosphere. That energy requirement separates the proffered mechanism from the dubious realm of "perpetual motion." You don't get something for nothing — the sun supplies the energy.

So, while the quasi-inertial mechanism might not operate on the universal scale that I had initially envisioned, it might still have application in situations where charges are present. The energy requirement is not necessarily an issue.

To put that concept into a more familiar context, think of the popular American pastime, baseball. A baseball thrown forcefully toward home plate continues along its path toward the batter. We reflexively attribute that persisting motion to the ball's inertia. However, the persisting speed could arise at least in part from charge imbalances between the ball's front and rear environments. Please recall, the atmosphere is populated with positive charges throughout (Chapter 2). Compressed positive charges ahead of the fast-moving ball should induce negative charges on the front surface of the ball. The resulting attraction should create a persisting advance force, similar as suggested for the planetary body. We refer to such continuing forward motion as arising from inertia, but could the inertia, in turn, arise from charge compression?

This same genre of explanation may help explain baseball's famous "curve ball" pitch (**Fig. 13.5**). How could a ball thrown in one direction

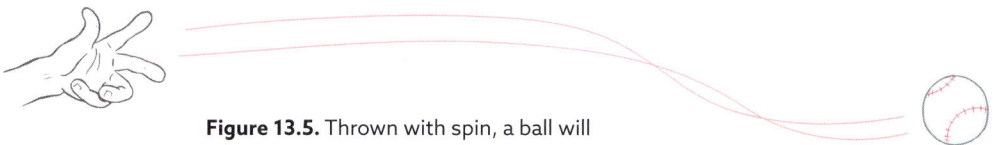

Figure 13.5. Thrown with spin, a ball will naturally curve because of induced charge forces.

eventually "decide" to veer off path to another direction? It's perplexing, to say the least. But, consider again the atmospheric charges lying ahead of the ball. Those charges should compress, inducing additional negative charge on the front of the ball, which, in turn, creates additional positive-charge cling in front. If the ball is thrown with some spin, then those clinging positive charges should rotate sideways. Enough of those positive side charges should exert a lateral pull on the ball, edging it sideways and confusing the hapless batter.

Inertia is recognized as a universal feature of nature. So familiar is the effect that most of us pay scant attention to its origin. Physicists do. They recognize that the inertial mechanism remains elusive. Think of it: Why should a moving body *necessarily* keep moving? I don't suggest that the issue resolves with the considerations of charge. However, I do suggest that at least some inertia-like effects could plausibly originate on that basis. The prospect seems worthy of exploration.

The Moon

In theory, the principles adduced to explain the dynamics of planets should apply equally to lunar dynamics. As a large dead rock, however, our moon contains practically none of the needed "atmospheric" charge (**Fig. 13.2**, *panels a and b*). Hence, at least one charge-based component is seemingly absent. Perhaps different principles might govern the motion of the moon?

One certainly hopes not.

If the arguments presented earlier in this chapter carry some measure of truth, then charge involvement in lunar dynamics would seem likely. The moon lies only 30 earth-diameters away from our planet. Because of that proximity, the moon should sense the closest of earth-associated charges: The earth's atmosphere is one, but the solar wind may be more pervasive and, in this regard, likely to be more significant. The positive charges contained within that vast sheet can be expected to induce negative charge on the moon's surface. Measurements concur: They reveal negative charge over the surface of the moon, with negative electrical potentials as high as thousands of volts.[w5] So, like the earth itself, the moon seems to bear negative charge.

For the earth and moon to remain coupled to one another, those two negatively charged bodies must attract. One possible way is through the like-likes-like mechanism (Chapter 7). That mechanism requires in-between positive charges, which could come from the solar wind. That wind lies in the same (ecliptic) plane that contains both the planets and, to a first approximation, also the moon. (Everything, it seems, lies in or near that same plane.) Hence, the solar wind could be the agent supplying the positive charges needed to sustain the like-likes-like attraction. It could serve as the critical feature holding the moon near the earth.

This suggested attraction mechanism may help explain why we always see the same face of the moon (**Fig. 13.6**). The face we see ought to be the one most attracted by the intervening positivity, *i.e.*, the face whose terrain features maximize the potential for charge induction. We don't know exactly what those features might be; however, we do know that the physical appearance of the moon's near and far faces differ strikingly.[w6] Some aspect of that difference might maximize the inductive pull on one side relative to the other. Any such charge difference is probably small, for the moon does exhibit some wobble, allowing us to sometimes catch a glimmer of a small portion of the far side.

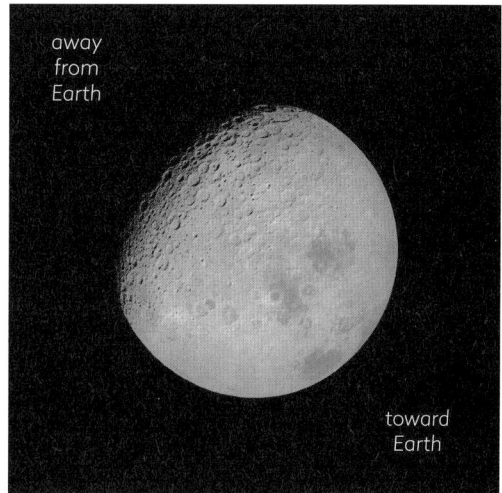

Figure 13.6. The cratered side of the moon always faces away from the earth.

The notion of consistently seeing the same side of the moon is reinforced in **Figure 13.7** (*see next page*). Imagine the earth as a point. Tie a string from that point to the surface of a moon-like sphere. To mimic the moon's orbit around the earth, twirl the sphere around the earth point. Adding some elasticity to the string helps, because the moon's orbit is slightly elliptical; essentially, this twirling dance models why everyone, everywhere, sees the same side of the moon.

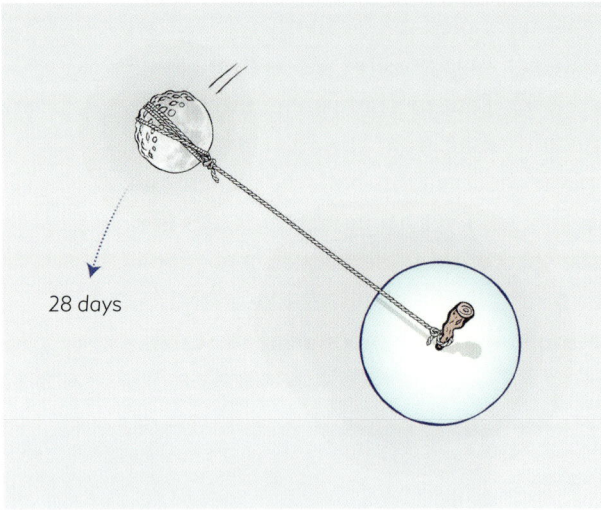

28 days

Figure 13.7. String and ball analogy, depicting the lunar trajectory around the Earth.

Implicit in this model is the absence of any lunar rotation. The moon does not spin on its own axis — otherwise we'd see all faces. The absence of spin fits with our discussion of planetary rotation (Chapter 6): If rotational drive requires atmospheric charge gradients and the moon has no atmosphere, then the moon should not rotate. And indeed, it does not.

These same dynamics may apply to moons other than our own. Jupiter's moon, Europa, according to NASA, has only an "extremely tenuous" atmosphere,[w7] while Saturn's moon Rhea contains only a thin, "wispy" atmosphere.[w8] Essentially, those moons are atmosphere free. Devoid of atmosphere and hence free of the charge-based force suggested to drive them round and round, those bodies become non-rotating moons instead of rotating planets. In theory, this principle should apply pervasively: Bodies without substantial atmospheric charges should not rotate — we classify them as moons.

While our own moon also does not rotate, it does revolve. It circles the earth in almost the same plane as the planets. To keep it revolving, some driving force should exist. That force could come from the above-proposed "inertial" mechanism. If the moon carves a path through a sea of positive charge, then that trajectory could create the same kind of quasi-inertial force proposed to help keep the earth revolving around the sun. It would keep the moon orbiting the earth and treating us to those awesome rising-moon vistas that can enrich our souls.

Is the Universe Expanding?

A final consideration in dealing with solar-system dynamics is the concept of the expanding universe. Probably you've heard the term. The universe's expansion is inferred from the "redshift" of starlight to longer wavelengths (**Fig. 13.8**). That wavelength shift is analogous to the train-whistle (Doppler) effect: As a sound source moves away from you, you hear its pitch drop. From a similar shift of the starlight reaching our eyes, we infer that the stars must be moving away from us, just like the train. Hence, the cosmos is thought to be expanding.

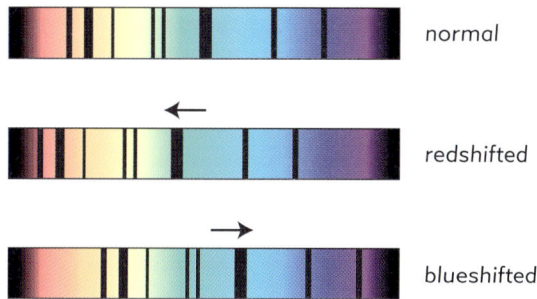

Figure 13.8. Common elements emit light in a signature-like series of narrow wavelengths, or spectral lines (top). If those lines shift toward longer wavelengths (middle), the light is said to be "redshifted." A spectrum that is redshifted may signify that the light emitter is moving away from the observer, analogous to the Doppler effect. Hence, light from distant celestial bodies that is redshifted may imply an expanding universe.

Imagine our precarious future if indeed the universe continues to thin out into nothingness.

An altogether different interpretation of the seemingly expanding universe could emerge if electric fields are present. Wavelength depends on optical refractive index. Since the refractive index changes in the presence of electric fields (the Kerr effect), any fields present throughout the cosmos could impact the observed wavelength. That being the

The Tidal Enigma: Why Two High Tides per Day Instead of One?

Tides are created by the lunar pull on the earth. The earth's liquid water, after all, is free to move, depending on what's pulling. Hence, the sea can rise and fall from the varying pull of the moon.

Tide-like effects demonstrated in our laboratory lend support to the involvement of electrical forces. Placing a charged rod just above a water-filled chamber causes the water immediately beneath the rod to rise up in a mound (see Fig 10.5). In a similar way, the moon's charge could plausibly lift the water on Earth. The aforementioned laboratory demonstration lends credence to such a proposition.[w9]

While this chapter's theme does not lie with earthly tides, I can't resist dealing with an aspect of the tidal phenomenon that has proved particularly enigmatic: why two tide cycles per day instead of one? If the moon pulls on the earth's water, then, at a given earth-surface location, you'd expect only one high tide each time that site comes closest to the moon. That's once per day. Yet, high tides generally come twice per day. How come?

Experts can't agree. Proffered explanations range from variations of the lunar gravitational potential field, all the way to the water's inertia, pulling the water away from the backside of the earth to create the second high tide. Diverse views abound. No consensus exists.

A potentially simple explanation may lie on the supply side. The rise of water depends on two competing factors: the gravitational draw from the moon, and the supply of water available to meet that demand. To illustrate, think of an isolated basin full of water. When the moon lies just above the basin, that water ought to rise, but fails to do so. That's because there is no water supply from which to draw. Similarly for lakes: no serious supply, hence no tides. And similarly for the eastern Mediterranean, where the only available (oceanic) supply must come through the narrow, nine-mile wide, Strait of Gibraltar. With the supply so restricted, the difference between high and low tide must remain minimal, as it does.

On the other hand, when the supply is as vast as a large, open ocean, tidal swings can be substantial. Seattle waters can experience sizable swings because the adjacent Pacific Ocean carries a practically endless supply of water. Those waters can deliver abundantly.

But why the second high tide, occurring just when the moon pulls on the opposite side of the earth? That secondary tide may not rise as high as the primary one, but its appearance is reliably consistent. What causes it?

Consider the earth's distinctive geography, which can easily restrict the supply (see figure).

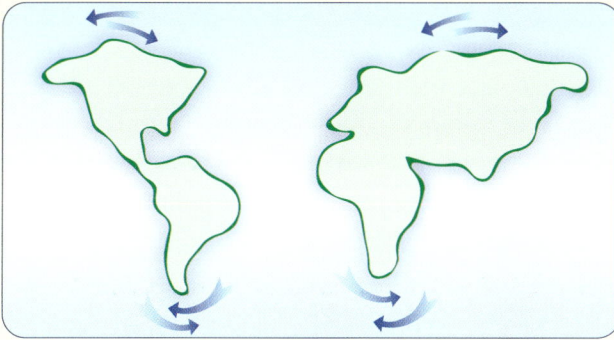

The presence of two daily tidal fluctuations instead of one may be the result of restricted passageways from one ocean to the next.

The earth contains two major land masses: North/South America, and Europe/Asia/Africa. Those land masses are separated by vast oceans. Flows between the two oceans can occur only through regions where the oceans connect. Unless you count the Panama and Suez Canals, those connections are present only at the narrow regions near the north and south poles. Hence, any feed from one ocean to the other will be limited, just as it is in the Gibraltar setting. Nevertheless, those connections imply that inter-ocean flows are possible.

Now, think of the situation when the overlying moon raises the water level on the proximate side of the earth. On that side, the water level should substantially exceed that on the moonless side, where there should be no pull at all. Depending on the hydrostatic pressure (height) difference between one side and the other, however, the high water from the illuminated side should tend to flow downhill toward the moonless side — but only as much as those narrow polar channels will allow. The consequence? Flow downhill

should act to build the secondary high tide on the distal side, lower than the primary high because of restricted supply, but inevitable, nonetheless.

Such ocean-to-ocean flow may help explain why, in the northern hemisphere, the largest tidal swings are found in regions far north.[w10] Lying closest to the source of the flowing waters, those regions can enjoy the bounty of flow more than regions more remote.

A final point with regard to tides: If the lunar gravitational pull arises from charge forces, then we'd expect pulling action from any source of charge. Consider foul-weather conditions. During serious storms, low-altitude clouds bring plenty of charge close to the earth (Chapter 9). That proximity should draw the water upward, creating higher than normal tides. And that's indeed what foul weather does — sometimes creating much higher tides.[w11] Thus, the charge paradigm helps explain why it's not just the moon that influences the tides, but also the weather. Beware those unexpectedly high tides!

case, the red-shifted wavelength might signify less about the universe's expansion than about the presence of cosmic electric fields.

Hence, the universe might *not* be expanding after all, as some prominent scientists have long argued.[4] Additional insights on this issue come from the Electric Universe leaders[5] as well as that group's experimental voice, the Thunderbolts Project.[w5] Indeed, the universe might turn out to be rather stable — surely a reassuring concept.

Summary

Mass-centered phenomena have long served as steadfast predictors of gravitational force. Why, then, do we bother to consider alternative hypotheses?

Conventional gravitational theory assuredly works, albeit inconsistently. Certain orbits are reasonably predicted (though not without error). Yet, a growing list of "anomalies" raise questions about the underlying theory's adequacy.

Anomalies are euphemistic terms for failures to fit theoretical expectations. Enough such failures inevitably provoke skepticism. Yet, in the case of gravitation, we hesitate. So familiar is the mass-based formulation that we seem hesitant to step back and ask how many inconsistencies may be necessary before the theory's adequacy begins to come into serious question. I believe that point has been reached. We need to acknowledge the large grey elephant standing in the room.

The three chapters in this section advance an alternative paradigm for gravitational force, based on the simple attraction of opposite charges. Those charges lie *within* the masses. Because they are part and parcel of the masses themselves, the deviation from convention is not as extreme as it might, at first, seem.

Charges also pervade the sky and the cosmos, and as we have seen (Chapter 1), charge-based forces can have magnitudes colossal enough to stir our collective imaginations. Charges of that scale could well have functional significance.

In considering gravitational phenomena throughout the solar system, we use Earth gravitation as a guide. Positive charges filling the atmosphere above pull the negative earth toward the sun. So should the positive charges residing in the solar wind. Reasons exist to suggest that similar attractive forces could act on the other planets, drawing them correspondingly sunward. In such a way, a charge-based gravitational pull toward the sun could prevail uniformly throughout the solar system.

Charges might also explain the planets' revolution around the sun. For the earth, two charge-based forces arguably drive that motion. The first comes from atmospheric positive charge, an orbitally directed component of which may help drive the planets' revolutionary motion. The second, "inertia-like" force arising from the charged environment may augment that drive: As the negatively charged planetary body moves through a sea of positive charges, those positive charges lying in the pathway ahead should compress. The compressed charges ought to create a forward-attractive force that perpetuates the orbital motion. Those two charge-based forces may help drive the earth, and perhaps likewise its sister planets, in their orbital directions.

The inertia-like forces outlined just above could apply not only in the cosmos but also to phenomena that we experience in everyday life. Which ones, we do not yet know. However, our eyes remain open.

For perpetuating all of these electrically based forces, the underlying energy arguably comes from the sun. So long as the sun continues to shine, the heavenly bodies ought to persist in moving as they do, and we should continue to feel the relentless pull of gravitation.

Prayers may be in order for continuing sunshine.

SECTION V

Learning to Fly

Gravitation keeps us grounded. Yet objects of every kind, from drones to dust, manage to linger above the earth, sometimes with minimum effort.

Lift entails a fight against gravitation. Given gravitation's proposed electrostatic basis, it would seem natural to consider whether lift may likewise depend on electrical charges. The next three chapters do exactly that: They explore whether achieving lift has a basis in electrostatics.

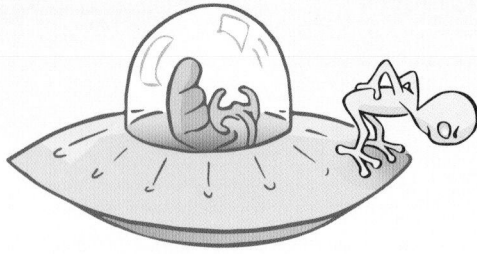

CHAPTER 14

Flying Objects: The Challenge of Staying Aloft

One Sunday morning in the spring of 1980, Seattle residents awoke to the blast of Mount St. Helens. We knew it was coming. For weeks, seismologists had recorded the nearby mountain's rumblings and filmed its belching steam. Onlookers were warned to stay clear. Then, suddenly, the mountain quite literally blew its top.

The Mount St. Helens eruption was timid compared to the 1883 blast of Krakatoa, an island set between Java and Sumatra, in Indonesia. During the days prior to that explosion, sailors in the area reported unusual activity. Then, in one huge paroxysm, the island exploded. Thousands died, 40-meter tsunamis inundated nearby islands, and waves rocked ships as far as South Africa. Most of the volcanic island of Krakatoa literally disappeared from the map.

Both of these volcanoes spewed boulders, gases, ash, and sundry materials high into the sky. The boulders returned quickly to earth. But the finer particles stayed aloft for some time — in the case of the Krakatoa eruption, circling the earth multiple times before finally settling to the land below. The airborne particles produced lingering optical effects — sunsets lasted longer, and they became redder. The intense red glow caused by the Mount St. Helens eruption reportedly triggered frantic calls to fire engines to quell the supposed fires. Those optical effects persisted for several *years*.

The lingering airborne particles also affected the weather. Since suspended debris scatters incoming sunlight, one might anticipate a temperature drop in the areas beneath. Indeed, the 1815 eruption of Mt. Tambora in Indonesia resulted in an essentially summer-free year for New England, while the particulate matter from Krakatoa reflected enough sunlight to reduce the global temperature for a period of, again, several years. The entire earth experienced a global cooling.

Figure 14.1. Ecuador's Tungurahua volcanic eruption spews a plume of ash thousands of meters into the sky.

Volcanic exotica tend to fascinate (**Fig. 14.1**). But at the risk of disappointing readers, I must acknowledge that the material that follows concerns volcanoes only indirectly. Our focus lies with a single volcanic feature: why those tiny erupting particles stay aloft for as long as they do. How can we understand their suspension in the atmosphere for several *years*?

We will then advance to the broader question: Can the principles underlying particle suspension explain the suspension of larger objects as well? Can those same principles tell us why Frisbees and paper airplanes can stay aloft for as long as they do?

Remaining Aloft from Charge?

We begin with those volcano-enhanced sunsets that we enjoy because of the particles' persisting suspension. This persistence is commonly ascribed to the particles' diminutive size: Finer particles experience relatively more air friction per unit mass, thereby restricting their descent speed. Hence, those minute particles take relatively longer to settle to the earth. They remain long suspended.

This explanation may seem plausible — but can it really account for *years* of continuous suspension?

By now, it will not surprise readers of this book to hear the suggestion that another explanation might hold sway — that particles may resist settling to earth because of repulsive charge forces. Charges commonly arise from air friction. (For details, please see the section, below, on *Triboelectricity*.) Small airborne dust particles and blown sand acquire negative charge. Wind-tunnel studies confirm increased negative charge with higher wind speed.[1,2] Hence, keeping those charged particles afloat could well arise out of their repulsion from the negative earth.

Evidence for particle charge comes directly from the volcanic eruptions, themselves. Strategically positioned cameras have recorded intense lightning discharges projecting from the erupting material to the ground (**Fig. 14.2**). This tells us that, much like lightning-generating thunderclouds, volcanic discharges must bear substantial electrical charge.

Figure 14.2. Lightning in Japan's Sakurajima volcanic eruption.

But how can that charge persist for so long? Long-term charge retention could occur in at least two ways. The first is through air friction (triboelectricity), a well-recognized phenomenon whose nature I will soon detail. Continued movement through the air could keep those particles charged over the long term. A second way is through hydration. If EZs from atmospheric moisture envelop those volcanic particles in the same way that they envelop water droplets (Chapter 7)[3], then those particles should bear negative charge. Infrared energy from the sun could then help sustain that charge.[3] By those two means, particles could remain charged over the long term, keeping them aloft for extended periods.

Similar reasoning applies to dust. Ordinary house dust comprises mainly flakes of skin and hair, as well as pollen and soot from outside.[w1] While the floating dynamics of tiny skin flakes might seem suspiciously similar to that of miniature kites, possibly kept aloft by their natural swirling in air currents, it's necessary to bear in mind a salient fact: all such particles are denser than air. According to conventional thinking, they should progressively sink. Yet, house dust doesn't readily settle to the ground: A beam of sunshine coming through your bedroom window reveals numerous dust particles dancing continuously in the air, with little tendency to sink.

Why so? Like most proteins, these protein-containing particles should bear negative charge, as should the enveloping EZ water. Air friction should augment that negativity. One wonders: Could that negative charge be sufficient to keep those denser-than-air dust particles repulsively suspended above the negatively charged earth?

Finally, on the subject of dust, consider dust storms. An Arizona dust storm in 2011 stretched for some 50 miles, the dust rising to an impressive height close to 8,000 feet (~2,500 meters).[w2] Electrostatic repulsion may once again bear relevance: The negative charge of those airborne particles repels the earth's negative charge, keeping the particles floating. Those particles also repel one another, augmenting the loft and sustaining those monster storms.

Thus, *staying aloft could involve nothing more than developing enough negative charge to repel the earth's negative charge, countering the effects of any gravitational pull*. No, this does not imply that negative people will eventually float away, but it does imply that objects with sufficient negative charge ought to levitate above the earth.

We will consider multiple potential manifestations of this simple principle in the remainder of this chapter and the next. The principle could hold relevance for practically everything that flies, from paper planes and Frisbees, all the way to gliders and eagles.

But aren't we going too far? A man with nothing more than a hammer at his disposal may think that everything looks like a nail. Could charge really explain practically *everything*? Surely, aerodynamic lift is one of those well-researched fields of science. Why concern ourselves with alternative points of view in a field whose fundamental principles would seem reinforced each day by the dearth of accidental plane crashes?

One reason is that the science currently ignores the immense forces that can be generated by electrical charge. We prefer to think of charge forces as trivial. Yet, I began this book with several examples showing exactly the opposite. In Chapter 1, I described the repulsive force experienced between two clusters of charge separated by 1 meter (3.3 feet), each cluster obtained by collecting one second's worth of electrons flowing through the filament of an incandescent 120-watt lightbulb. For convenience, I reproduce **Figure 1.1** here. If you recall, that repulsive force equaled the weight of roughly 50,000 stacked garbage trucks, or perhaps more fitting for this air-centered chapter, about 5,000 jumbo jets. Imagine!

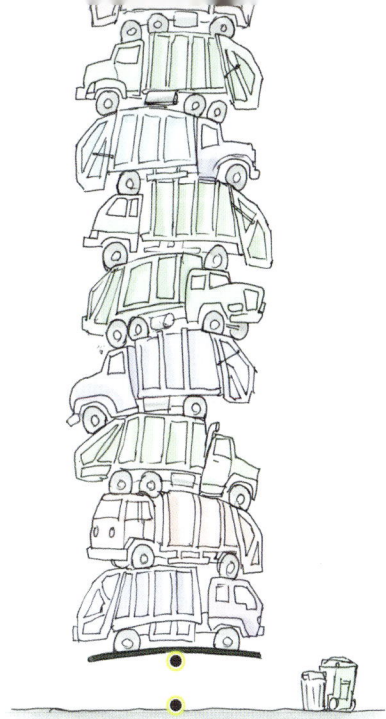

Figure 1.1. One-coulomb blobs of charge, spaced one meter apart, exert enough repulsive force to lift 50,000 garbage trucks (or 5,000 jumbo jets).

Rather than jumbo-jet lifting, consider the lifting of a smaller, garden-variety object. Imagine how *few* charges would be needed for that. I hope you'll bear this in mind as we proceed. I appreciate that this proposed line of thinking may seem foreign, but it comes from nothing more than a chapter out of a physics-101 text. Modest charges can develop enormous forces. If we fail to consider a potential role of these huge forces in producing lift, we would indeed be remiss.

Lifters and Charge

Concrete examples of charge-based levitation come from the "lifters," mentioned briefly in Chapter 10. When energized by high voltage, these devices rise mysteriously off the ground. Yet, they contain no engine of any sort. Their "engine" consists of two integrated ring-like electrodes, each lying in the horizontal plane, vertically separated from one another. A high-voltage power supply energizes the pair of conductors through flexible wires, allowing the lifter to move without much mechanical constraint. Somehow that electrical energy alone creates demonstrable lift.

Hobbyist demonstrations of these lifters are the sad remnants of a once-active field called "electrogravitics." A 2008 book by the physicist Paul LaViolette[4] details this field's fascinating history, which began almost a century ago and peaked during World War II. The author's account details developments not only by Americans but also by Germans and Russians. It cites multiple reports and patents.

But progress abruptly terminated. With the increasing recognition of the multiple patents secured by the American engineer T. Townsend Brown and the ensuing developments by the US Navy (which at one time reputedly spent as much as 5 percent of its budget on electrostatic lifting machines), the field suddenly went underground. That happened in the late 1950s, apparently for security reasons. Since then, academic researchers have largely ceased pursuing the subject, and accessible developments have come mainly from curious hobbyists.

Where those underground developments have led remains unknown to the public, although it's speculated that the B2 "Stealth" bomber gains lift in part from electrogravitic forces.[4] Meanwhile, Internet sources decry the entire field of electrogravitics as some kind of hoax or conspiracy theory. No doubt the lifters work: We've successfully explored two versions in our laboratory (**Fig. 14.3**). And, if you still harbor any doubts, you can easily build a working model yourself.[w3]

How lifters achieve their lift remains uncertain. The standard explanation posits that the lift arises from the so-called Biefeld-Brown effect,

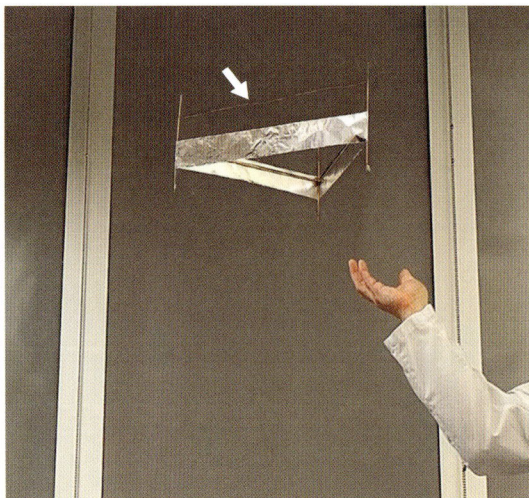

Figure 14.3. Lifter rising off the laboratory bench. Thin horizontal wire running around the top of the lifter (arrow) is the positive electrode; aluminum foil, the negative electrode. Energizing wires not easily visible.

which involves an ion current (sometimes called an "ion wind") running vertically through the air from one electrode to the other. Whether this explanation suffices seems unclear: Reports conflict on whether vacuum environments, theoretically devoid of ions, can sustain lift.

One lifter built in our laboratory taught us something useful (**Fig. 14.3**). The lifter could rise whether the negative electrode was the lower or upper one. That observation dashed our initial speculation that lift might arise because the negative pole was closer to the negative earth than the positive pole; so, repulsion might prevail. Drat!

Following that disappointment came a more promising explanation: a field of electrons repelling the earth. Driven by high voltage, electrons get emitted from the electrode into the air, drawing naturally toward the positive terminal and thereby creating a splay of electrons lodged between the two electrodes. If you could photograph electrons, you'd detect that stream of electrons, much as you could detect a stream of flowing water. At the same time, you might think the positive electrode would similarly emit protons, canceling any such electron-based field. But, protons remain securely embedded as integral parts of the electrodes' atoms; therefore, effectively, the positive electrode can't emit anything.

Because of that imbalance, electrons emitted from the negative terminal should dominate the scene (**Fig. 14.4**). The endless supply of electrons should ensure that the entire region between electrodes bears abundant negative charge. That negative charge should repel the negative earth, thereby lifting the lifter. Lifting should happen irrespective of whether the negative electrode lies above or below the positive electrode. Essentially,

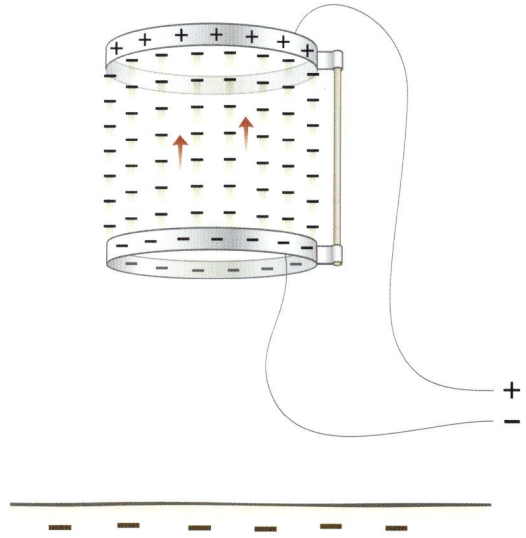

Figure 14.4. Possible lifter principle. Electrons emitted from the lower electrode travel to the upper (positive) electrode, creating a sheath of electrons. That negative sheath repels the earth's negative charge, creating lift.

the lifter acts as a blob of negative charge, rising from the negatively charged earth.

Hence, obtaining lift from charge might not be all that arcane. A simple device can easily manage it — possibly through simple repulsion from the earth.

No surprise, then, that MIT engineers have exploited a similar configuration of separated high-voltage electrodes to demonstrate eventual practical use in flight.[w4] Shaped like an airplane, the device is kept aloft through charge forces alone. Demonstrably, charge forces can lift "airplanes," or reasonable facsimiles thereof — a provocative demonstration for sure, and a feature we will consider in the subsequent chapter. Evidently, charges alone can create lift.

Triboelectricity: Acquiring Charge through Air Friction

If charge can create lift, then where might those charges come from? For the lifter, the charges originate from an electrical power supply. Is there a more natural source of charge for the other situations?

An important source of charge is friction, especially air friction. To illustrate, think of the ordinary hair dryer. By blowing streams of warm dry air past your just-showered locks, your hair dries. But there is more: From air friction, each strand also acquires charge; hence, the hairs repel one another and your hair fluffs. Air friction creates the charge that does the job.

While you may flatter yourself by looking good for a time, hair dryers are not the only creator of frictional charge. Rubbing any two substances past one another will accomplish the same. One surface becomes positively charged, while the other acquires negative charge. A good example is a thin sheet of paper rubbed on the wall of your room: Following a few vigorous strokes, the paper sticks to the wall because the two rubbed entities have acquired opposite charges. The effect lasts for a few seconds, after which the charges neutralize, and the paper drops to the floor. But friction nevertheless does separate charge.

Another example is shown in **Figure 14.5**. The cat's fur acquires charge by moving through the surrounding air. That charge induces opposite charge on nearby pieces of lightweight Styrofoam, creating a Styrofoam coat. Whether the cat is as amused as the rest of us is difficult to say.

Figure 14.5. Styrofoam peanuts clinging to a cat's fur. As a result of the cat's motions in air, an electrostatic charge builds on its fur. The resulting electric field draws the lightweight pieces, which cling.

The study of frictionally acquired charge is an established scientific discipline, called triboelectricity. "Tribo-," from the Greek, "rubbing," refers to friction. "Electricity," you already know. Triboelectricity deals with the charge transfers that attend the rubbing of different substances past one another. One substance pulls superficial electrons from the other, setting up a charge imbalance.[5]

Those transfers are summarized in the so-called triboelectric series (**Figure 14.6**).[w5] The series reveals what happens when any one substance rubs against another: Substances higher on the list become positively charged, while lower ones become negatively charged. Fur lies above vinyl in the series; so, when rubbed on vinyl, fur acquires positive charge, leaving the vinyl negative.

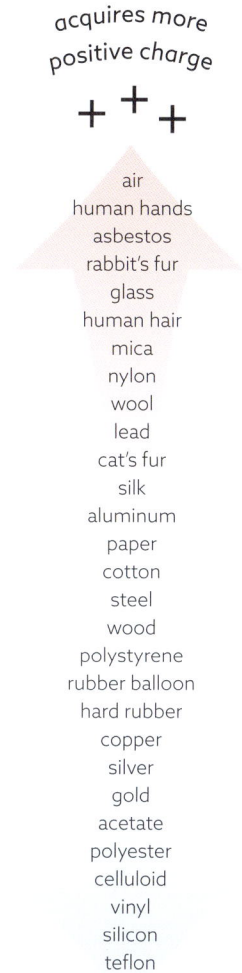

acquires more
positive charge

+ + +

air
human hands
asbestos
rabbit's fur
glass
human hair
mica
nylon
wool
lead
cat's fur
silk
aluminum
paper
cotton
steel
wood
polystyrene
rubber balloon
hard rubber
copper
silver
gold
acetate
polyester
celluloid
vinyl
silicon
teflon

− − −

acquires more
negative charge

Figure 14.6. Triboelectric series.[w5] Substances higher on the chart become positive when rubbed on substances lower, which acquire negativity.

The highest substance on the chart is air — implying that *air blown on anything acquires positive charge, while conferring negativity on its recipient*. The recipient may include you. Walking in a brisk wind confers negative charge, which builds EZ water and helps satisfy your need for a full complement of that negative charge for optimum function. The air, meanwhile, becomes positive. That's a powerful insight. It implies that the wind's impact can extend beyond just blowing stuff around; it can build negative charge on anything in its path.

Objects with acquired negativity, as we've surmised, can keep aloft through repulsion from the earth. Thus, triboelectric charging might explain long-term particle suspension, as from volcanoes or dust. It might also explain why wind gusts can blow dried autumn leaves not just sideways, but also high into the air. It might even help explain dust devils: At their tops, the particles become so intensely charged that they produce electric fields on the order of 100,000 volts per meter.[6] Those high fields could keep constituent particles suspended high in the air, thereby sustaining the devil. Even wind-blown sand, denser than dust, may acquire enough charge to rise high into the atmosphere. Indeed, the leading edge of the sandstorm is said to resemble a 1,500-meter-high charged wall[w6] (~5,000 feet).

To summarize, in all of these diverse examples, lift may arise from electrical charge, which, in turn, comes from triboelectric friction. Anything passing through the air should acquire negative charge. That charge repels the earth's charge, creating lift. Only modest amounts of charge are required.

Charge-Based Amusements

With some understanding of the role of air friction, we may now ask about various familiar flying objects. Why do Frisbees, boomerangs, and even kites float for as long as they do?

You might not know, but the modern Frisbee dates back as far as 1871, when William Russell Frisbie opened a small bakery known as the Frisbie Pie Company in Bridgeport, Connecticut. Frisbie's pies gained popularity among nearby Yale University students, who began tossing around the empty pie tins. They called those tins, "Frisbies." The first commercial production of these flying discs came in 1957. Later, they became popular when the Wham-O Company released their trademarked, "Frisbee."

Aficionados presume that the Frisbee's iconic shape must play a critical role in its persisting flotation. That is, the lift comes from the Frisbee's rounded edge, which creates a lower pressure above, rather than below, the disc, so the Frisbee floats. The spin confers angular momentum, which stabilizes the floating disc from wobble.

This interpretation acquired broad acceptance until someone invented the almost flat Frisbee (**Fig. 14.7**). Lacking the iconic rim, the washer-like Frisbee has set distance records. The absence of the iconic bowl-like, rounded edge purportedly necessary for creating that pressure difference implies that something else must be responsible for the Frisbee's lift.

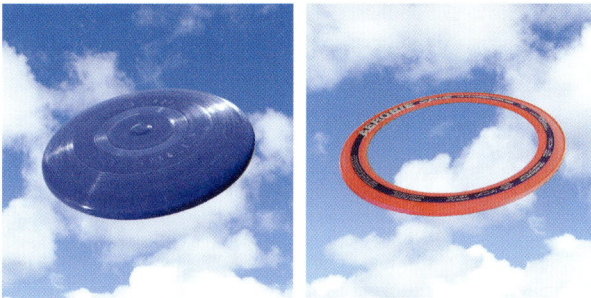

Figure 14.7. Two types of Frisbee — the classic one with rounded edge, and the flat donut.

I suggest charge. If the Frisbee acquires enough negative charge as it passes through the air, it should rise. According to the triboelectric series (**Figure 14.6**), the acquired charge polarity should be negative. The highest negative charge should reside at the Frisbee's edge because the rim's high rotational speed creates the highest friction. Rotation is key: Once the rotational motion slows, the Frisbee should begin losing its charge and quickly sink — as common experience confirms.

But wait! If the rim is charged, then how come you don't get shocked each time you catch the Frisbee? One option is that the charge theory is simply inadequate. Another option is that the charge is not sufficiently strong to create that shock. Feeling an electrical discharge requires substantial current flow from Frisbee to fingers. Because the Frisbee material is non-conducting, however, it may be only the charges you touch that contribute, not the full complement of charges lodged on the Frisbee's edge.

Flying Spiders and Other Terrestrial Bugs

As terrestrial creatures, spiders don't ordinarily fly. Yet, under certain circumstances they may levitate, and one wonders whether the responsible agent might reside in acquired charge.

It was Charles Darwin, then a young naturalist aboard the HMS Beagle, who reported something noteworthy: Red spiders, native to Argentina, suddenly appeared in large numbers aboard his ship — situated 100 km from the nearest land. To arrive, Darwin surmised, those terrestrial spiders must somehow have levitated across that sizable expanse of water. Indeed, spiders are known to "balloon" — *i.e.*, to take off following the release of a strand of silk.[w7] Spiders have been detected high in the atmosphere, above 10,000 km, and more than 1,000 km (~600 miles) out to sea.[w8] How could they have gotten there?

In 2013, the scientist Peter Gorham demonstrated the physical plausibility of taking flight by electrostatic means; the numbers fit.[7] Subsequent investigations provided experimental confirmation.[8,w9] Electric fields comparable to those just above the earth were imposed on spiders. When the field was turned on, the spiders levitated; when off, the spiders returned to the ground. The authors argued that the earth's electric field may be critical for flight, implying that spiders may levitate by repelling the earth with their negative charge.

Plausibly, then, it might have been electrostatic forces that kept those red spiders levitating long enough to get blown by the wind toward the HMS Beagle — a charge-based principle similar to that surmised to keep sand and dust floating in the air.

More mundane, perhaps, than flying hundreds of kilometers is the spider's capacity to build webs. Imagine yourself, instead of one of those agile creatures, spitting out silk and using that silk to begin building a web. Descending from above poses no problem – gravitation pulls you downward from a fixed point as you release an initial strand of silk. But, consider the next step, building upward. Without some levitational force to propel your rise, how do you advance upward— or even sideways — to build the subsequent strand?

To see how the spider actually builds her web, you can watch a video.[w10] At 28 seconds into this video, the spider latches onto the end of a dangling strand. It then releases additional silk, drawing that latter strand sideways and upward. Since no ladder is present to facilitate, the spider must achieve that vertical climb through some kind of levitational force. Later (2.50 minutes into the video), you can witness

Viktor Grebennikov rising from the ground (left) on his flying machine.[w12]

similar upward movement involved in constructing the web's circumferential strands. Here, some nearby latch points appear to facilitate; but upward movement takes place also in regions where those convenient latch points are absent. Without levitational force, which Darwin deemed necessary, web construction might be impossible.

Another example of levitation is that seen in certain beetles. Most beetles don't fly. However, some species have wing-like structures tucked under covers. The wings themselves have been shown to levitate, presumably facilitating the beetles' capacity to leap.[w11]

The entomologist Viktor Grebennikov (1927-2001) pioneered the study of these beetles in the steppes of Russia. He first noticed that the wings had the capacity to levitate. He describes these observations in a book.[9, see also w11]

Grebennikov became controversial when, in the same book, he describes something seemingly impossible (see figure). He lined the inside of a box-like enclosure with beetle wings. Then, he stood atop the box, and, with the assistance of a control mechanism mounted on a front post with handlebar, he demonstrated that he could levitate. (Nobody seems to know the nature of the control mechanism, but speculation is that it might induce wing vibration.) Grebennikov reports traveling hundreds of meters on his flying platform.

Considering that levitation must be impossible, observers have dismissed these observations as fantasy. One can't be sure. If Grebennikov's experiments could be reproduced, on the other hand, then they would seem to offer potentially fertile ground for investigating the role of electric charge in achieving lift.

The repulsion principle may likewise apply to the boomerang, which also rotates through the air. Like Frisbees, boomerangs stay aloft longer when thrown with their characteristic twirl. The resulting air friction should create negative charge, which could bear responsibility for the boomerang's persisting flight.

Figure 14.8. Helicopter blades display visible electric discharge.[w13]

Sometimes the effects of triboelectric charge can become visible (**Fig. 14.8**). Under the right atmospheric circumstances, helicopter rotors twirling through the air can show visible electrical discharge. Astonishing videos reveal this phenomenon clearly, and bloggers offer possible explanations.[w14] A simple basis may be triboelectric: Slicing through the air, the high-speed propeller builds abundant charge; when the voltage exceeds a threshold and atmospheric conditions provide enough conductivity, the result is visible electrical discharge. Ground personnel are urged to take precautions.

Spinning-wheel discs launched from gun-like hobby devices may work similarly: Start those lightweight discs spinning using a hand-held launcher and they will fly like Frisbees, falling only when rotation ceases. Such devices can be purchased on the Internet. Kids love them. Few think to inquire why those spinning disks fly, and for as long as they do.

Indeed, charge-based educational toys have recently hit the market — and they are labeled as such. An example is an electrical wand. Powered by an internal moving-belt Van der Graaf generator, the wand acquires charge. When it touches or even approaches certain objects, it confers charge on those objects, which then exhibit seemingly "magical" properties, such as levitation.[w15]

Charge-based levitation brings to mind the flying of kites. Wind certainly bears responsibility for keeping kites suspended in the air. But is it the wind *per se* pressing on the kite that keeps it aloft, or the air friction creating negative charge?

Thinking about kites and charge conjures an image of Benjamin Franklin's audacious kite-flying experiments. Franklin was canny — he did not actually perform his experiment during a period of lightning discharge, which, in all likelihood, would have brought quick electrocution. Instead, Franklin's experiments took place in the charged atmosphere just prior to the encroaching storm, at which time the presence of electricity in kite flying became abundantly clear.

Modern experiments amplify Franklin's conclusions: Carried out mostly in uncharged environments, they nevertheless show that electrical discharge (corona) can occur around the kite's edges.[w16] Clearly, kites bear charge — presumably triboelectric — and one wonders about the extent to which those charges may keep the kite suspended in the air.

Finally, with regard to lift mechanisms, we turn our attention to paper airplanes (**Fig. 14.9**). Paper planes first appeared in ancient China around 500 BCE, with the advent of widespread paper manufacture. Then they appeared in Japan in conjunction with the newly developed art of paper folding (origami). And they appear today in classrooms full of bored students. You might be surprised to learn that even the Wright brothers flew paper planes. Credited with building the first successful powered aircraft, the Wrights used paper planes extensively to test their designs in wind-tunnel experiments.

Figure 14.9. Common paper airplane.

Paper-plane flight begins when someone imparts forward thrust. The plane then soars. Most paper planes fly limited distances, others surprisingly far — the current distance record being just shy of 70 meters

Eric Laithwaite, Gyroscopes, and the Stability of Your Bicycle

The triboelectric-charging mechanism implies that wheels ought to be easier to lift when spinning than when not: The anticipated assist comes from the repulsion between the spinning wheel's acquired negative charge and the earth's negative charge.

By demonstrating exactly that, the prominent British electrical engineering professor Eric Laithwaite once stunned an otherwise dour Faraday Society audience. A heavy wheel mounted on a pole was a struggle to lift; however, when the wheel was made to rotate, anyone could lift it. A video captures that feat.[w17] Faraday Society members deemed it so radical that they abstained from their otherwise routine practice of publishing such demonstrations. Levitation of that kind was simply too extreme.

A modern demonstration of same can be seen online[w18] along with a somewhat different take on what's happening.[w19]

One wonders whether any such levitational behavior could play a role in the workings of the gyroscope. Comprising little more than a wheel spinning in the horizontal plane around a vertical post, the gyroscope somehow manages to maintain a relatively stable orientation.

So long as the wheel rotates, the post tends to remain nearly vertical, and that feature makes gyroscopes useful for knowing which way is up. Understanding the working principle of the gyroscope can be formidable: A visit to the Internet or perusal of any handy physics text will provide abundant information on angular momentum and rotational inertia, mostly non-intuitive, and for some, rather murky. Check it out!

A potentially simpler explanation (but for one thorny objection that I'll mention in a moment) could lie in triboelectric forces. Spinning rapidly through the air, the rotating wheel ought to acquire substantial negative charge. That negative charge should repel the earth's negative charge. Should the wheel slightly tilt, its downward edge will repel the earth more strongly than its upper edge because of the diminished distance from the earth's negative charge; and that marginally stronger repulsion could help right the wheel, maintaining the gyroscope's stability.

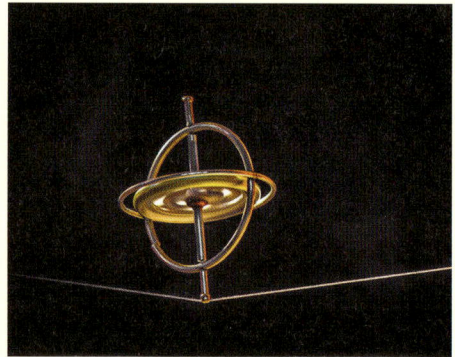

The potentially thorny objection is that gyroscopes can work in space, where there's no air, and hence no air friction. But consider the physical conditions. Satellite-based gyroscopes are packed into chambers that are either filled with nitrogen or argon or evacuated. Vacuum-based triboelectricity is a subject not well understood: In one study conducted in a 99.997% vacuum, triboelectric charging was not lost as expected; it was reduced by only about 50%.[w20] So, triboelectric charging could well persist. And likewise, in instances in which nitrogen or argon replace air, again, there is no reason to think that triboelectric charging should necessarily vanish.

So, in these seemingly inopportune circumstances, gyroscopes could still employ the triboelectric principle. As long as the satellite remains within reasonable distance from our planet, meaningful repulsion could still exist. Charge-based repulsive forces could provide a simple explanation for the gyroscope's vaunted stability.

OUT-ON-A-LIMB METER

Given that interpretation, one wonders whether that same principle applies to the stability of bicycles. Oddly, nobody's sure why fast-moving bicycles remain stable, while slowly moving or stalled ones will commonly falter to one side or the other. Probably, you've experienced that phenomenon. What creates that speed-based stability?

Opinions vary. Some say that the mechanism remains unknown. Others opine that the tire behaves like a gyroscope — which leads directly to the following speculation: A bicycle's fast-rotating tire should acquire substantial negative charge as it passes rapidly through the air. Charge acquisition in rotating tires is not only confirmed (although attributed to causes other than triboelectricity), but also ample enough to be considered for practical use in the recharging of electric-car batteries.[w21]

Which brings us back to bicycles. Could tire charge lend stability in the same way as proposed for the gyroscope? If so, then the higher the bike's speed, the more plentiful the tire's charge and the higher the stability—as commonly observed.

Whether any such charge may be substantial enough to bring stability to the moving bicycle remains an open question. On the other hand, if that's not the explanation, then the question lingers: What is responsible for maintaining the bike's stability?

(230 feet), roughly the length of a football field. The obvious question: What keeps paper planes aloft for so long? Rising air would seem an unlikely candidate, especially indoors, where vertical air currents ought to be minimal.

Nor does the "Bernoulli mechanism" help. Considered in detail in the next chapter, that standard explanation for lift relies on typical shape difference between the wing's upper and lower surfaces — curved above and flat on bottom. The curved top supposedly leads to lower pressure above than below; and that confers lift.[w22]

It is tempting to invoke Bernoulli's principle to explain paper-plane dynamics except for one inconvenient fact: Paper-plane wings typically have flat, or almost flat, tops. They lack the classic curvature required for creating Bernoulli's lift. Yet, they fly just fine, even in wind tunnels. Some other mechanism must keep paper planes aloft, and if neither Bernoulli's principle nor rising air makes sense to explain this phenomenon, we must search elsewhere.

We come once again to electrical charge. According to our triboelectric chart (**Fig. 14.6**), any paper passing through air will acquire negative charge. That charge should confer lift by electrostatic repulsion from the earth. The charge (and therefore the lift) should diminish as the plane slows and slowing is typically when the paper plane finally crashes to the ground.

Hence, negative charge could easily explain why paper planes remain aloft for as long as they do. So long as they keep moving and sustain that charge, the plane should keep flying. The principle is nothing new or radical; it's the same as suggested to explain the lift of all the varied objects considered in this chapter. In all cases, electrostatics could well prevail.

Summary

This chapter considered many "flying" objects. We examined whether those objects could stay aloft through electrical repulsion from the negatively charged earth.

A potentially key feature turned out to be the so-called triboelectric mechanism — the transfer of charge that occurs when one material rubs against another. One material becomes positive, the other negative. The mechanism is entirely conventional; physicists study it routinely.

An eye-catching feature of that mechanism is the fact that the most positive entry in the triboelectric table is air. Thus, air passing over any material will acquire positive charge as it loses electrons to the bypassed material. Thereby the bypassed material acquires negative charge. Indeed, by this principle, all objects passing through the air should become negatively charged, at least transiently.

Applying that principle can lead to a simple understanding of why objects heavier than air can remain aloft. Those objects range from spiders all the way to paper planes. While nothing is proven beyond reasonable doubt, we could find good reason to support the negative-charge hypothesis — *i.e.,* by passing through the air, these various objects acquire negative charge, repelling the negative charge of the earth and thereby keeping afloat — much like clouds keep afloat.

Many of the points in this chapter should be experimentally testable. Placing ultra-light weight charge sensors on the various airborne objects, for example, could test for the expected presence of negative charge. All that's required are motivation and resources. Such pursuits could help determine the extent to which charges may indeed play a role in achieving loft.

Omitted from consideration were the most obvious of flying machines: airplanes. The flights of both non-powered glider planes and common powered planes will be considered in the next chapter. Do those planes stay aloft by employing conventional flight principles? Or might electrostatic charge find unexpected relevance there as well?

CHAPTER 15

Winged Flight

Of all objects capable of taking flight, the most obvious must surely be the airplanes buzzing noisily overhead. I'm reminded incessantly of their presence. A major flight path to Seattle's airport runs nearly above my home, and, during peak hours, the noise can be distracting. But flight frequency declined, temporarily, some time ago. I recall during the pandemic/response that threatened our lives for several years, one of the few benefits was the dearth of airplane flights. With few people daring to expose themselves to the viruses infesting those flight cabins during the peak of the pandemic, people stayed home. Airlines were not pleased, but the relative silence brought with it a measure of peace.

Airplanes have been with us since 1903. It was then that Wilbur and Orville Wright, experimenting in Kittyhawk, North Carolina, achieved what was arguably the first successful flight: The plane didn't crash land. It's been more than a century since then. With abundant technological advances under our collective belts, surely the principles of flight must be fully revealed. The underlying mechanisms must be perfectly clear.

But is that really so?

I can practically *feel* the readers' reaction to merely raising such a question. Any hint that we might not fully understand the essential principles of flight must surely bespeak some kind of author madness. Impossible! The buzzing of those planes overhead must have infested this author's brain with creeping neurological tangles. If conventionally accepted flying mechanisms were truly questionable, then how could the designs of airplanes have progressed to their current advanced state?

So, let me commend you to a 2020 article in *Scientific American*[1] — a journal not known for radical stances. The article's innocuous title reads: "The Enigma of Aerodynamic Lift," while its more provocative subtitle opines, "No one can completely explain why planes stay in the air." Imagine! — the field's acknowledged experts have yet to figure out how your plane manages to get you from New York to Chicago.

That conclusion may come across as jarring. Yet, the field's experts are quite clear: To put it bluntly, *"we simply don't yet understand how planes fly."*

Bernoulli and Friends

According to conventional thinking, lift should come from Bernoulli's principle. As briefly mentioned in the previous chapter, Bernoulli-based lift derives from the asymmetry of the wing cross-section — curved on top and flat on bottom. That classical airfoil shape is presumed to create a pressure differential — lower on top than on bottom, which pushes the wing up — and voila! The plane lifts from the tarmac.

While that mechanism may seem neat and tidy, challenges arise. One obvious paradox is the fact that planes can fly upside down (**Fig, 15.1**). If airfoil shape bears responsibility for creating upward lift, then you can imagine the calamitous impact expected from flying upside down. Yet, certain pilots may do that routinely. One friend, a former test pilot, claims that they can fly upside-down just as easily as right side up, but who would choose to see the world upside down?

Figure 15.1. F-16s in death-defying feat. Thunder Over the Rock Airshow, 2010.

Those intimate with flying may claim that the upside-down feat does not necessarily compromise the validity of Bernoulli's principle. Upside-down flight could take place simply because of a judiciously altered angle-of-attack, *i.e.*, the angle between wing and direction of movement. It's like sticking your hand out of the window of a fast-moving car: At the proper angle, your hand lifts. Upside down flying could theoretically occur in a similar manner, without the need to invoke any Bernoulli-related concerns.

While that expedient may have hypothetical merit, it does not always suffice. Unlike the tilt-of-hand scenario, upside-down flight can occur with the wings remaining perfectly horizontal. To confirm, check out the example shown in this web reference (which contains some spine-tingling drama).[w1]

Even helicopters can fly upside down. Videos nicely demonstrate that feat,[w2,w3] which you can also see in **Figure 15.2**. These demonstrations don't necessarily address the Bernoulli issue, but they do raise a more general question of what's responsible for keeping aircraft suspended in the air. (**Figure 14.8** may provide a hint.)

But I digress. Returning to the Bernoulli question, I raise another relevant point: bird flight. Like planes, birds supposedly use the Bernoulli principle to keep afloat as they glide through the air. However, birds, too, can fly upside down. In certain regions of Iceland, observers report seeing ravens executing routine upside-down maneuvers over considerable distances.[2] The reasons for such seemingly anomalous behavior remain unclear. What is clear is the conflict with the expectations of the Bernoulli principle. If the Bernoulli mechanism were to keep birds aloft, then upside down gliding should be impossible.

Figure 15.2. Sikorsky S-70 Blackhawk helicopter flying upside down. Flown by Turkish Pilot Major Yusuf Keles, 1998.

Further to the Bernoulli issue, consider the progressive disappearance of the classical airfoil wing shape, the latter supposedly *required* for the Bernoulli mechanism to operate. Beginning in the 1950s, wing designs increasingly diverged from the classic airfoil cross-section to so-called supercritical designs that are flatter on top, sometimes genuinely flat (**Fig. 15.3**).

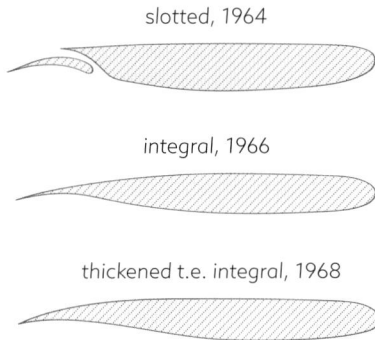

slotted, 1964

integral, 1966

thickened t.e. integral, 1968

Figure 15.3. Cross sections of three supercritical wing designs. Redrawn from a company flyer.[w4]

Modern examples of flattened tops can be seen just by peering at airplane wings at your nearest airport. Classic airfoil shapes with curved top and flat bottom are increasingly difficult to spot. Yet, planes with those modern supercritical designs fly perfectly well — perhaps even better than those with the classical airfoil designs. If Bernoulli-based top curvature were necessary for flight, then how could planes with flat wings fly?

Looking at hobby-plane designs reinforces concern about the Bernoulli mechanism. Those planes are commonly made of Styrofoam, covered with fabric. They can be flown as gliders or fitted with small motors for powered flight. The wings of those foam models come in a variety of styles ranging from those with classical cross-sections to symmetrical ones — the latter having *identical* wing curvatures on top and bottom.[w5] If planes, even toy planes, with symmetrical wing cross-sections can fly perfectly well, then what does that say about the centrality of Bernoulli-based lift?

I thought I might have been alone in questioning Mr. Bernoulli but discovered otherwise. I found people doubting not only the applicability of the Bernoulli lift principle, but also the validity of the principle itself. Two examples come to mind. The first comes from a book with the colorful title, *Stop Abusing Bernoulli: How Airplanes Really Fly (Craig, 1997)*.[3] The book contains food for thought that will challenge your thinking.

The second is a study by the Cambridge University Engineering Professor Holger Babinsky. Babinsky demonstrates experimentally that the foundational assumption underlying the formulation of the Bernoulli principle — air passing above and below the wing reaches the rear at the same time — is, after all, invalid.[w6] With that observation, he challenges the relevance of the classical Bernoulli mechanism.

In toto, these multiple challenges raise serious questions about the adequacy of the long-standing Bernoulli mechanism for explaining airplane lift. Despite the principle's dominant familiarity, good reason exists to surmise that, at the very least, it cannot tell the whole story.

If Bernoulli's principle fails to account for lift, then something else must. Planes do fly. Recognizing the limitations of the Bernoulli principle, some have argued for an alternative lift mechanism involving wing tilt — the so-called "angle-of-attack" mechanism mentioned earlier. But does that argument suffice? Returning to the hand-outside-the-window analogy, recall that, with proper tilt, your hand gets pushed upward; but, it also drags backward. So, while planes could theoretically use this wing-angle mechanism for gaining lift, the huge drag incurred by that maneuver is problematic. For engineless gliders particularly, where no fuel is available for countering the drag, gliders would quickly stall and fall, much like paper airplanes. So, angle-of-attack expedients don't necessarily rescue the Bernoulli challenge.

Given these multiple concerns, it seems difficult to reflexively accept the Bernoulli mechanism as the unequivocal basis for lift. At the very least, something else would need to supplement that mechanism, and at the extreme, perhaps even replace it. Indeed, the practically instinctive adherence to the Bernoulli mechanism may bring some understanding to the issue of why *"we simply don't yet understand how planes fly."*

Flying Machines

To probe possible mechanisms of flight, we first consider gliders, the engineless man-made birds of the sky (**Fig. 15.4,** *see next page*). Commonly towed by powered planes or fast-moving cars and then released to fly independently, these quiet aircraft create endless joy for enthusiasts, as well as a notable challenge for those trying to understand the underlying principles. How might sailplanes remain aloft without the benefit of engine power?

Figure 15.4. A modern glider.[w7]

I think you can guess the suggestion I'd like to make. But first, some background. According to the prevailing view, gliders employ the standard Bernoulli principle to stay aloft. However, drag forces slow them down. Slowing reduces the lift force, so gliders would ordinarily descend. To maintain altitude, they depend on the presence of rising air.

Experienced pilots know four likely sources of rising air:

(i) Rows of cumulus clouds, beneath which warmer air appears to rise ("thermals");

(ii) Areas where air masses converge, forcing air upward;

(iii) Sharply rising cliffs, against which strong winds have no choice but to rise; and,

(iv) Regions where strong winds blow over mountains, forming so-called mountain waves.

No doubt air can rise (Chapter 9). However, the criteria that pilots use to infer rising air can mislead. If the glider rises, then it is *presumed* that the air around it must be rising, for what else, it's thought, could lift the glider? Rising air has seemed a no-brainer — but can we be sure?

The rising-air scenario faces a subtle, but potentially daunting problem: Upward airflow at one location implies an equal downward airflow somewhere else. Otherwise, our atmosphere would quickly dissipate into space, leaving us gasping for air. Downflows would wreak havoc on gliders — if the upward airflows bear responsibility for lifting the plane, then the downward flows would drive the plane crashing into the earth. Imagine the fate of the hapless glider pilot inadvertently meandering into sinking air. Such downflow-encounter opportunities must surely exist in abundance, especially on glider flights as long as 2,000 km (1250 miles).[w8] But gliders rarely crash. So. updrafts would seem unlikely explanations for the full story of lift. Some other force must be at work, helping to keep those gliders aloft.

In this vein, scant consideration has been given to the possible role of electrical charge forces. I argued in the previous chapter that the lift of diverse objects could arise from the development of negative charge. Through repulsion from the negative earth, that charge would produce lift. With respect to airplanes, gliding through the air should build plenty of triboelectrically induced negative charge, particularly on the wings and nose. If gliders gain lift from that mechanism, then their pilots need not face the harrowing prospect of a sinking-air disaster. Simple repulsion would keep them aloft — just as it appears to keep clouds aloft.

At present, any such mechanism remains theoretical. Experimental tests are needed. Should the charge mechanism suffice, on the other hand, it could easily provide new strategies for improved aircraft design. Since the triboelectric accumulation of electric charge depends on the surface area exposed to the air, increasing that surface area should create more charge and hence more lift. Grid-like fenestrations on the front of wings could accomplish this; and, as exotic as it sounds, I'm told that Boeing aircraft engineers have tried exactly that (with some gains in performance — although too expensive to routinely implement). Perhaps fenestrations may one day appear on gliders as well.

Forward Thrust

While negative charge may help solve the problem of lift, it skirts the issue of forward motion. Gliders don't simply hover like helicopters; even as they rise, they advance continuously. What drives that forward motion?

I posit that either of two forces may propel those engineless planes forward. The first is the "inertial" force that I outlined earlier (Chapter 13). Recall the proposed mechanism: A moving body compresses the positive atmospheric charges ahead. Those compressed positive charges attract the moving body's negative charge, drawing the body forward;

meanwhile, the void in the wake of the moving body limits any backward restraining force. The unbalanced forces ahead *vs.* behind counter the inevitable drag and keep the object going in the same direction as before. That same principle should apply to moving objects of any kind, including aircraft. Thus, forward-moving gliders should keep flying forward.

A second forward-directed force, also based on charge, requires a brief explanation. Consider a classic example from physics: a dielectric (insulating) rod suspended in an electric field. For simplicity, we ignore any gravitational effects.

Figure 15.5 illustrates the situation. It shows a series of capacitors, each one consisting of parallel plates. Inserted between those plates is a dielectric rod. One capacitor envelops the tip of the rod; the others envelop the shaft. (Additional capacitors could be added along the shaft, with no change of the essential arguments.) Each capacitor generates an electric field between its separated plates. On the left-hand capacitors, the fields induce equal and opposite charges along the enveloped rod segments; that results in equal upward and downward forces. So, the rod remains suspended midway between the plates. It's as though the capacitors were not present at all.

The right-hand capacitor does much the same, but with a twist. The *partial* insertion of the shaft between that capacitor's plates creates an asymmetric effect: Since much of the capacitor's charge lies to the right of the rod, those charges exert a rightward pull. Hence, the tip of the rod gets drawn rightward. The rod gets sucked into the capacitor, thereby undergoing "forward" motion.

All of this is textbook material — nothing exotic or radical. The argument implicitly presumes no counter-balancing asymmetry at the rod's tail end, and that presumption seems reasonably satisfied.

An equivalent description of this drawing force is that the electric field bends toward the rod's tip, as shown in **Figure 15.5**. The field (and therefore the force acting on the rod) then has both vertical and horizontal components. The vertical components cancel one another. The horizontal component creates a lateral force pulling the rod further into the capacitor — the same as deduced in the paragraphs above. So long as asymmetry exists, the rod will inevitably advance; it will always move "forward."

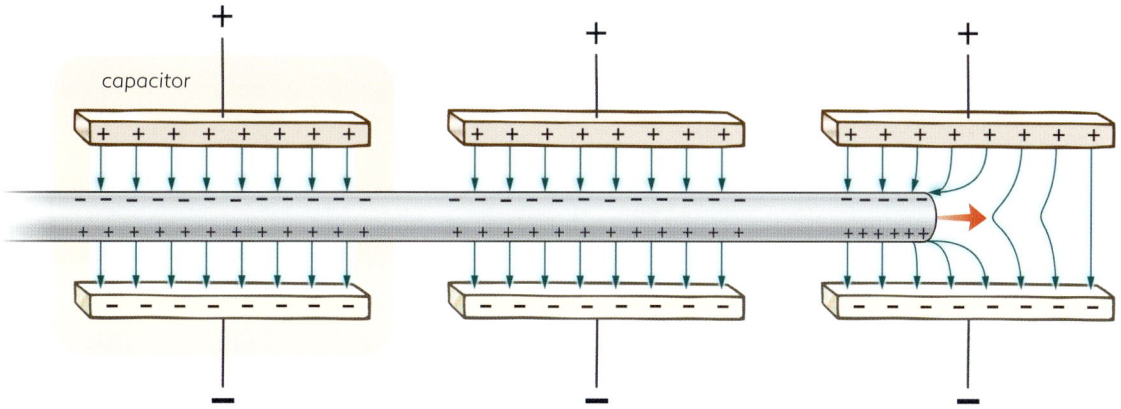

Figure 15.5. Dielectric rod suspended within the plates of multiple capacitors. Right-hand capacitor draws the rod's charges toward its tip, creating an attraction that pulls the rod into the capacitor's field. Other capacitors induce charges on the rod, resulting in vertical forces that balance one another.

Next, consider the same scenario but suppose that the rod bears a net negative charge, as might a wing (**Fig. 15.6**). The result is similar to that depicted in the previous figure except for one additional feature: a net upward force. That happens because the positive top plates pull the negatively charged rod upward, while the negative plates beneath exert a repulsive upward push. Thus, both forces generate lift. The right-hand capacitor, meanwhile, acts similarly to the one in **Figure 15.5**, producing a drawing force, possibly intensified because of the higher concentration of negativity at the rod's tip. Hence, so long as some asymmetry exists, negatively charged bodies in electric fields can experience both vertical and lateral forces.

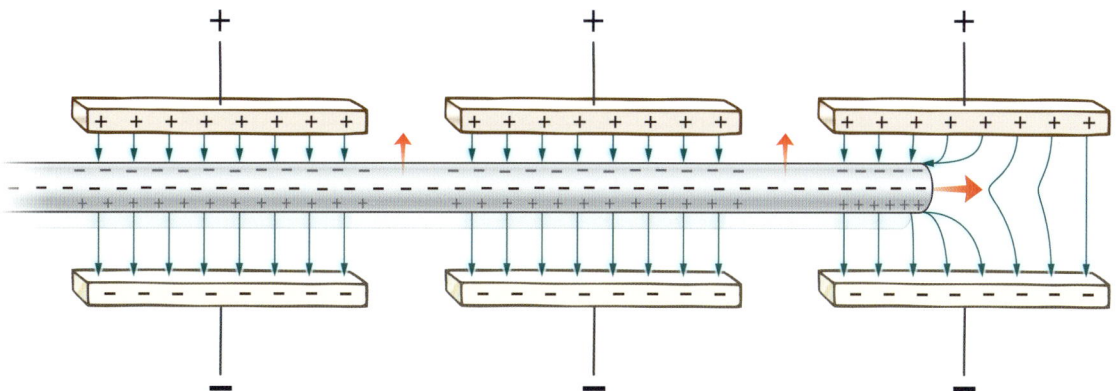

Figure 15.6. Similar to *Figure 15.5*, except that the rod contains net negative charge.

Those forces, as we shall see in a moment, can create both lift and forward motion in airplanes.

The required asymmetry may arise naturally in wing-shaped devices such as those found on airplanes. **Figure 15.7** shows why. Much of the negative charge will develop where the wind hits the wing most directly — at its front edge. As a result, the passing air will acquire equal and opposite positive charges, with the expected distribution shown in the figure. That backward-streaming, positively charged air should act to offset any negative charge buildup on the rear flanks of the wing. Hence, the wing's front edge should wind up with the highest level of net negativity.

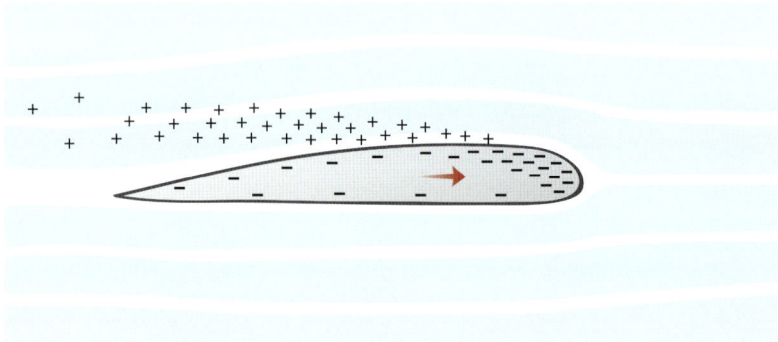

Figure 15.7. High-velocity air passing over the wing's front edge should build negative charge. Meanwhile, the air acquires positive charge, partially neutralizing the wing's rear-edge charge.

The wing depicted in **Figure 15.7,** therefore, resembles the rod in **Figure 15.6**. The earth's electric field replaces the capacitor's electric field; and, the field lines should bend toward the wing in the same way as they bend toward the rod. Hence, the wing should advance, in the same way as the rod advances.

A surprising expectation of both this mechanism, and the quasi-inertial mechanism, is that forward motion can occur even *against* the wind. Wind ordinarily generates frictional drag, which ought to produce a backward-directed force; the plane should move rearward. However, the two mechanisms just described generate forward-directed motion. Provided that the wind generates enough negative wing charge, the plane should stand ready to move *into* the wind.

In sum, a straightforward rationale exists for understanding how simple flying machines could fly, in ways that do not require invoking the Bernoulli mechanism. At least theoretically, charge-based forces can account not only for the lift but also for the forward motion.

I can appreciate the reader's likely reluctance to entertain the posited mechanism — that electrical charge could play a role in flight. It's a radical departure from conventional thinking. Yet, if the experts themselves admit to the inadequacy of the prevailing mechanism, then we should feel obliged to look elsewhere. The previous chapter argued that charge-based forces could account for the levitation of many diverse objects. If those forces apply to airplanes as well, then, possibly, one general mechanism might explain the dynamics of all objects that fly. Sir William of Ockham might have been pleased.

Powered Flight

While hobbyists may revel in the excitement of engineless glider flight, common airplanes mostly run on fuel. Engines create thrust. Thrust drives the plane forward, helping to overwhelm the drag forces that slow the plane down and thereby compromise lift.

The earlier-outlined charge-based mechanism may apply not only for gliders, but also for powered flight. The charge mechanism requires neither the classic airfoil shape nor special wing tilt. Achieving lift merely requires the accumulation of charge, and that should happen naturally as the aircraft's nose and wings pass through the air. So, the charge mechanism can apply equally for powered and non-powered aircraft.

Charge also develops on rotating propellers (see **Fig. 14.8**). To witness the possible impact of that charge, check out the following competitions,[w9, w10] where you can see hobby planes performing impressive stunts. Noteworthy are situations in which the planes point directly upward, either rising steadily, or — jaw-dropping — standing suspended in the air. How Bernoulli's principle could play in stationary objects boggles the mind. Propeller-based charge, on the other hand, could easily serve to keep the plane suspended, just as it might for the upside-down helicopter (**Fig. 15.2**).

Charge development on airplanes is no mere conjecture. A six-year study by the US Army and Navy showed that aircraft generated fields of 2,000 volts per meter while flying through ordinary city haze and up to 45,000 volts per meter — practically enough for coronal discharge — while flying through dry crystalline snow.[4] No wonder that airlines take special measures to dissipate excess charge to prevent fuel-tank explosion.

In fact, aircraft charge often does produce coronal discharge. My pilot friend reports that such electrical discharges commonly occur around a plane's forward regions. That includes the windshield as well as the wing-fuselage joints. Suffering the highest triboelectric wind shear, those frontal regions should develop the highest electrical charge, explaining the observed discharges occurring in those regions.

If airplanes and the earth both bear sufficient negative charge, then lift should be inevitable. In this context, I was not surprised to find a patent application dealing with aerodynamic effects arising from wing triboelectricity.[p1] Thus, others have started to recognize at least some aerodynamic consequences of electrical charge. This author is evidently not alone.

A critical question remaining: How much lift can charges generate? Is the lift force sufficient to keep a plane aloft? Theoretical models can address that question; however, multiple assumptions will necessarily invest any such model, leaving any theoretical attempt to test the adequacy of the proposed lift mechanism open to question. We don't exactly know the answer.

Nevertheless, one may assert at least that charges can exert unexpectedly powerful forces. Recall again from Chapter 1 that a collection of one second's worth of electrons flowing through a lightbulb filament can lift the equivalent of 5,000 jumbo jets (**Fig. 1.1**). With that gauge as reference, you'd think that charge-based lifting of a single jet, rather than 5,000, ought to be as easy as falling off a log.

With charge-based forces in play, we can understand what happens as a plane accelerates for takeoff. Once the plane reaches a critical speed, it begins to rise. It's not necessarily the pressure differential below and above the wing that produces the upward force, but arguably the charge buildup on the plane's wings and nose. Once that buildup grows sufficient to overcome the downward gravitational pull, the plane's nose lifts, and off it goes. Averting any subsequent mishaps, you may then make it all the way to Chicago.

So, we conclude that charge forces could matter for flight. They might not tell the entire story, but the arguments above imply that charges may play at least some role in the magic of airplane flight.

Summary

While many substances may bear negative charge, the laws of triboelectricity imply that all substances can acquire substantially more of that charge by merely passing through the air. Thus, *any object moving with enough speed should become negatively charged*. Toys such as Frisbees and boomerangs ought to acquire substantial negativity as they twirl rapidly through the air, repelling the negatively charged earth and potentially explaining those objects' longer-than-expected loft.

The same principle may apply to gliders and powered planes. As they pass rapidly through the air, those flying machines ought to acquire appreciable negative charge. Exploiting that charge for achieving lift may be achieved by optimizing those machines' characteristic shapes.

While simple charge repulsion may help explain lift, forward motion could derive from two different charge-based forces: the quasi-inertial force mechanism, which compresses charges ahead and thereby pulls strongly forward; and the classical forward-pulling force arising from electric-field asymmetry. Design considerations may maximize these forces, explaining why some planes perform better than others.

Many of the arguments offered in this chapter, and the one previous, should be testable. Placing charge sensors on the front and rear of an airplane wing, for example, could test for the expected spatial distribution of negative charge. Especially revealing could be observations drawn from ultra-lightweight sensors mounted on paper-airplane wings. Tests of this nature could confirm or deny the hypotheses set forth here. Again, all that's needed are motivation and resources.

While the proposed hypotheses may seem radical, we need to bear in mind the context: Flight mechanisms centered around Bernoulli's principle are suspect; they could well be as misleading as the computation of planetary orbits based on an earth-centered solar system. Multiple arguments raise concern.

If those arguments bear any validity at all, then building on the Bernoulli principle will likely fail to bring us to the promised land. Rather, such restricted focus may help explain why today's aerodynamicists must reluctantly admit that years of study notwithstanding, we still don't yet know how airplanes manage to fly. As unsettling as this revelation must be, it implies that the need to look elsewhere for full mechanistic understanding may qualify as a veritable no-brainer.

That's what I've attempted to do in this chapter. Whether the proffered charge-based mechanism will ultimately suffice remains to be determined. Nevertheless, for readers hungry for fresh sustenance, the proposed mechanism may offer some meat to chew on.

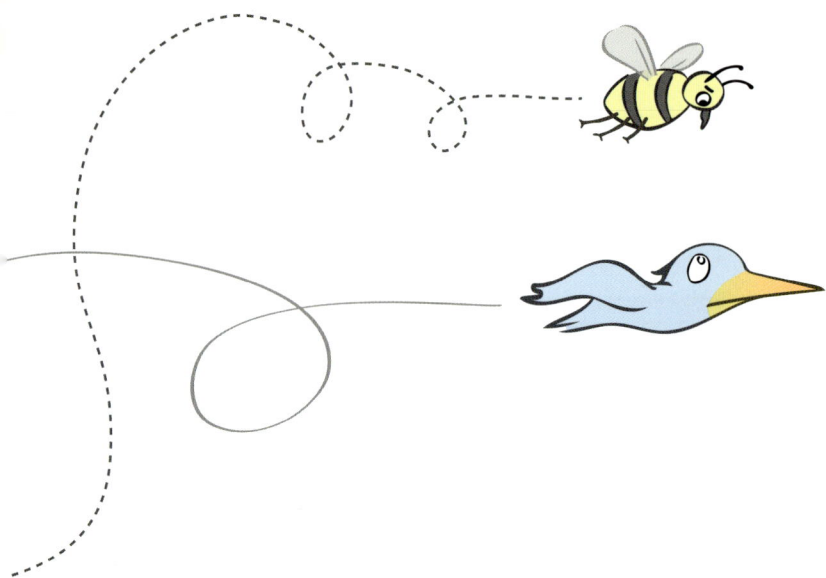

CHAPTER 16

The Birds and the Bees: Uncensored Secrets of Natural Flight

We in the Pacific Northwest love ferries. Puget Sound dominates our regional landscape, making seaborne transportation expedient and popular — not only among humans, but also among seagulls. Gulls understand that flying behind the ferries may net them free food, as passengers offer tasty scraps in appreciation of the spectacle of their effortless glide. Seagulls seem to have forgotten what most everyone knows: Flying requires wing flapping (**Fig. 16.1**).

For those non-flapping seagulls, keeping pace with the fast-moving ferry would seem a challenge. The ferry moves ahead at, say, 30 km per hour (18 mph), sometimes into a 20 km per hour (12 mph) headwind. That puts the gulls' effective forward velocity at 50 km per hour (~30 mph). Headwinds notwithstanding, the gulls proceed effortlessly into the wind, seemingly as relaxed as sunbathers on a Hawaiian beach. All of this with hardly a flap of their wings.

Figure 16.1. Seagulls gliding alongside the Seattle-to-Winslow ferry.

Such a sublime ability must have some explanation.

One possible explanation lies in the low-pressure air vortex following the ferry. The bird gets behind the ferry and uses the pull of the vortex in the same way as a car gains a forward assist by positioning itself just behind a fast-moving truck.

While this explanation may seem plausible, please think about it. On the rear of the ferry, a child offers a gull some popcorn. The gull nabs a kernel and even seems to thank the child with a subtle nod. Cool! But some deviation in that offering reveals the underlying enigma: For fear that the gulls might nip their fingers, children may release the treat prematurely. And when they do, the popcorn immediately flies *backward*, vanishing off the rear of the ferry. Everything flies backward — the ship's flag, your hat, and perhaps even your empty lunch bag. If most everything flies backward, why, then should the non-flapping bird pull itself forward, as it does?

A plausible rescue: shape matters. Lunch bags may fly backwards, but the birds' airfoil-shaped wings save them from that ignominious fate; therefore, birds can more easily get sucked forward. Before you spend a lot of time thinking through this option, let me assure you that those gulls glide not only at the ferry's rear, but also at its side, occasionally in front — and even at the ferry landing. So, any theory of gliding based on ferries' rear vortices can amount to no more than a partial explanation.

Gliding is the seagulls' habit. They do it daily, anywhere they dwell, and without obvious effort. Sometimes they flap their wings, but just as often they will glide effortlessly for long distances without any loss of altitude, even into a headwind. It's not just seagulls that glide — eagles, vultures, albatrosses, and countless other airborne species glide or soar by default over vast stretches of land or sea, and for extended periods of time. For them, soaring is a part of everyday life.

Instead of dismissing flapless flight as one of those anomalies reserved for future consideration, this chapter will deal squarely with the possible underlying mechanism. Once again, I will argue that electrical charge may play a key role — not only in flapless flight, but also in more "conventional" flight with ordinary wing flapping.

In so doing, we will move beyond the mere mechanics of flight. We will see how such a charge-centered mechanism may help account for various paradoxical features of natural flight, like flying nonstop for a week, or flying over Earth's tallest mountain peaks instead of through the more hospitable valleys.

So, if you'll pardon my chutzpah, I ask the direct question: How indeed do birds fly?

RISING AIR

The Fall of the Rising-Air Hypothesis

We learn from childhood that birds fly by flapping their wings. Just ask any middle-school student. Nevertheless, gliding and soaring occur commonly even among birds that ordinarily flap. Once we appreciate the frequency with which those birds employ those "alternative" modes, it becomes difficult to dismiss them as aberrant; they stand as ordinary features of flying, nothing to do with ferryboat scenarios. Any understanding of the mechanism of flight needs to explain both modes — flapping and non-flapping.

We may think we already understand the non-flapping mode. With no visible hint of flapping, airplanes manage to fly perfectly well; so, we presume that by now the underlying principles must be fully understood — but the previous chapter argued otherwise. Even if the mechanisms advanced in that chapter bear some measure of validity, that doesn't mean that we fully understand non-flapping mechanisms. "How" and "why" questions remain.

If those lingering questions can be adequately answered, then perhaps we can better understand how and why birds can manage the feat of flying without even a hint of flapping, and, perhaps at the same time, permit a more considered approach to the engineering of (bird-like) aircraft. We begin with a *how* question.

Since the default mechanism of bird flying is widely considered to be based on wing flapping, any non-flapping flight is reflexively presumed to arise from some extraneous feature. Usually, that's much the same as for the glider: rising air. Such rise presumably keeps the bird aloft, while inertia ostensibly maintains its forward motion. Problem solved!

Or is it? Some common observations bear on that supposition.

The first observation: Ask yourself how often you've felt that upward-moving air. If you've not bothered to notice, then try the following experiment. Proceed to an area where birds commonly soar at low altitudes. Extend your hand, and check for any sense of vertical airflow. If you notice what I've noticed, then you'll be hard pressed to detect even a hint of rising air.

But perhaps soaring requires extremely modest upward flow, undetectable in that hand experiment. To test this, hang a lightweight model bird from several thin strings so that it dangles with wings extended horizontally. Then, place an upward-directed fan beneath the bird and crank up the speed until, finally, the bird rises. Now place your hand above the fan to sense how much updraft was required. Can you feel that much updraft in areas where birds soar? If not, then, can you be certain that it's updrafts that explain flapless flight?

Finally, there's the balloon test. Updrafts strong enough to lift birds should likely lift air-filled balloons. I tried the balloon test myself. One day, just a few meters above the deck of my Seattle home, birds glided by with no evident loss of altitude. Quickly, I grabbed an air-filled balloon left from my daughter's birthday party. Holding the balloon high, I let go. Immediately, it floated downward, to the ground. Try it yourself. Not a definitive experiment for sure, but perhaps another indication that the rising-air hypothesis offers limited promise in explaining flapless flight.

In any case, for birds (as for gliders), depending entirely on rising air for transportation poses quite the risk, for that kind of "fuel" could be hard to find. Should the fortunate bird happen upon such an opportune circumstance, it would then face the same hazard noted earlier for gliders (Chapter 15): an abrupt plummet.

The reason: Unless replaced, those rising air molecules would leave a void beneath. But nature abhors a vacuum; so, replacement air must arrive from lateral sources. The laterally shifted molecules must themselves be replaced by molecules that can only come from falling air. But falling air can be treacherous: A bird gliding into a torrent of falling air could easily suffer a hopeless plummet (**Fig. 16.2**).

Of course, the bird could avert any such mortifying fate if it had the capacity to detect downward flowing air. That's certainly possible; but

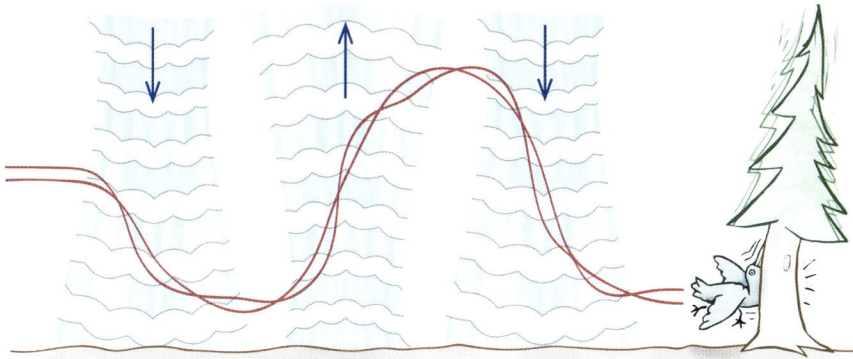

Figure 16.2. Birds lifted by rising air may fall victim to falling air.

have you ever looked to the clear blue sky to discern whether the air was moving upward *vs.* downward? How a bird could achieve any such competency is certainly not beyond reason, but also less than obvious. Downflows could wreak havoc.

Besides this confounding *how* issue, other contradictions plague the rising-air hypothesis. Filling the void beneath the rising column with air drawn from elsewhere requires horizontal airflow — *i.e.,* wind (**Fig. 16.3**). Yet, in the midday's near-perfect calm, vultures regularly soar over vast, sunny regions. I've witnessed that myself in Africa — birds gliding with no detectable wind at all, either on my face or on the tree leaves situated high above.

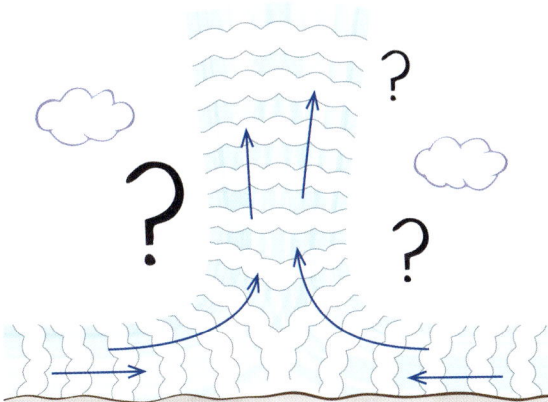

Figure 16.3. Rising air molecules leave a void behind, which nearby air must then fill; that airflow amounts to surface wind.

And much the same is accomplished by seagulls: As I draft this chapter on the Island of Ischia, just offshore from Naples, seagulls happily glide over the nearby docks just outside my hotel room, while the air hardly stirs. Wind is absent — but seagulls nevertheless soar. I'm told that the gulls do this day after day, year after year, seemingly carefree despite their supposed dependence on the vagaries of the rising air that one cannot consistently detect. How do they manage?

And beyond the question of *how*, we ask the question *why*. Why would those gulls engage in such feats if success depends on features beyond their control?

By no means am I suggesting that birds (and gliders) *never* exploit rising air; at times they surely do. However, a force so ephemeral and unreliable hardly begins to explain the commonplace acts of gliding and soaring. Something beyond rising air must bear at least some measure of responsibility. The question: What might that something be?

A ROLE OF ELECTRICAL CHARGE?

Birds, Bees, and Electrostatics

Figure 16.4.
Ulrich Warnke. 1945 –

Since gliders, I have argued, might remain aloft by acquiring negative charge (Chapter 15), an obvious question arises: Might birds and flying insects exploit that same principle?

Flying insects and birds certainly bear charge. An entire literature exists on the subject, with prolific contributions from the German scientist Ulrich Warnke (**Fig. 16.4**). Warnke describes how electrometers (charge-measuring devices), placed beneath flocks of birds flying 40 meters (~130 feet) above, register up to 6,000 volts.[1] That reading implies substantial charge.

Additional detail comes from measurements made on insects. When a flying insect passes near an electrometer, an oscillating pattern is recorded. **Figure 16.5** shows a representative pattern obtained as a bee approaches such an instrument. The electrical potential's magnitude increases as the bee comes nearer. More strikingly, the potential oscillates

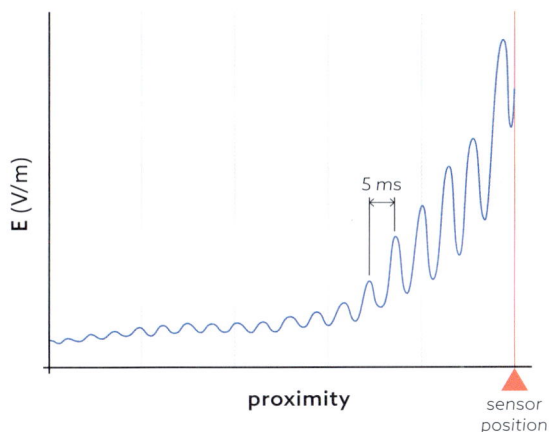

Figure 16.5. Recording of the electric field (arbitrary units) as a bee approaches an electrometer, positioned at the panel's right. The abscissa's "proximity" label could be replaced by "time," since the bee is continuously approaching the sensor. After Warnke.[2]

with each beat of the wings. This latter feature provides a mechanistic clue, which we will consider in the next section.

While the presence of charge in flying creatures seems clear, less clear are the charges' magnitudes and polarities. Polarity often goes unreported. Some studies, in fact, report positive polarity. To play a functional role in flight, however, the polarity should be negative so as to repel the negative earth; and that repulsive force must go a long way toward lifting the bird from the earth's surface. Absent these two features, invoking charge is akin to tilting at windmills.

The confounding issue: Charge-polarity measurements can be ambiguous. To appreciate the ambiguity and how to resolve it, consider the following thought experiment. Imagine an object with negative charge suspended in the air. That object will attract positive charges from the atmosphere. Some may penetrate, but soon a cloud of positive charge will surround the negatively charged object.

An electrometer probe genuinely touching the object should report a negative polarity (**Fig. 16.6,** *see next page*). However, if the probe lies at some distance from the object, it will sample the local environment and may thus read positive — the *opposite* polarity. (A similar issue arose when we considered the atmosphere abutting the earth; **Figure 3.2**). Not only will the probe report the opposite charge polarity, but the detected magnitude will also underrepresent the object's charge: Since the cloud's positive charges are spread widely over space, the measured magnitude may grossly underestimate the object's actual charge.

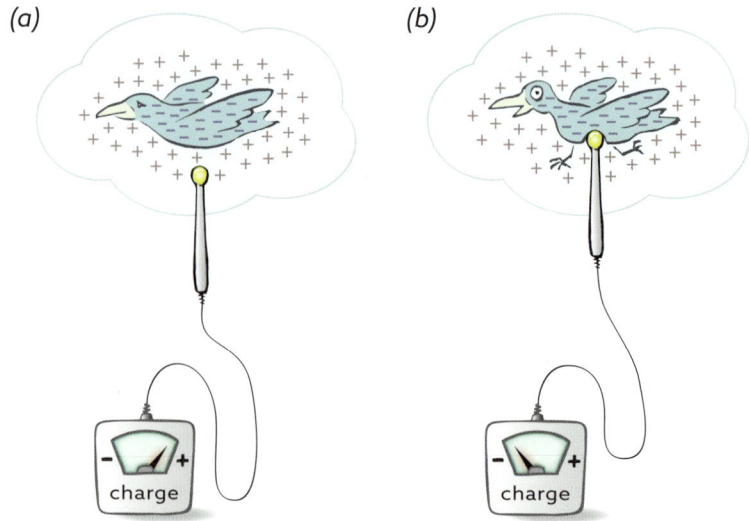

(a)   (b)

Figure. 16.6. An electrical probe positioned near the bird may report a positive charge *(a)*, while the bird itself bears a negative charge *(b)*.

charge

charge

With those considerations in mind, let us examine several representative studies that may help establish the actual charge on the flying creature.

First, consider an experiment carried out on grasshoppers flying in a wind tunnel.[3] The incoming air was enriched with negative ions. Those ions should have tended to neutralize any positive-ion clouds enveloping the negatively charged grasshopper, thereby increasing the insect's loft. That expectation was confirmed: When the negative ions were added, the flying insects rose up; and when those ions were withdrawn, the insects descended. Although the polarity and magnitude of insect charge were not measured directly in that experiment, the author concluded much the same as we've suggested: Electrostatic forces must contribute to vertical lift.

In a second experiment, investigators measured the electrical potential of foraging bees, 1.5 seconds after they landed.[4] An electrometer placed half a centimeter from the bee's 100-mg body measured the charge as 23 pico-Coulombs. This calculates to some five or six orders of magnitude less than that required to keep the bee suspended against the force of gravity.

However, a qualitatively higher-than-measured charge on the bee is implied by two factors. First, the measurement probes were placed in the counter-charge cloud surrounding the bees, where the broad distribution of positive charges should underrepresent the bee's more concentrated charge. Second, the bee had already landed. Since its wings had ceased beating, the in-flight charge magnitude would have diminished steeply.

A delay of 1.5 seconds constitutes a veritable eternity given the *millisecond* time scale of the bee's wingbeat (**Fig. 16.5**). Indeed, the authors confirm that the electrical potential values during flight were higher than after landing, although those in-flight values were not reported, as the authors' interest was in foraging, not flying.

Clearly, flying creatures bear charge. If we could take measurements on the animal itself rather than in the surrounding cloud (not a trivial task), then we might record much higher levels of charge than so far measured — perhaps equal to the amount actually needed to keep those creatures aloft. At this stage we cannot be certain. Where more certainty exists is in the actual bearing of charge: Flying birds and insects evidently do carry electrical charge.

Next question: How might flapless creatures acquire that charge, and why might it make a difference?

Air Friction Creates Charge

According to the laws of triboelectricity (Chapter 14), air flowing past any object confers negative charge on that object. For the insect or bird, the wing is the main object of focus. Each stroke drives the wing past the air — or equivalently, the air past the wing. The wing thereby acquires negative charge, which should help create loft as that negative charge repels the earth's negative charge.

On the other hand, those acquired charges should not necessarily persist for very long. Positive ions from the air will always rush to neutralize the deposited negative charges. Measurements of bird and bee charges illustrate such gain and loss. Each stroke builds charge, while the more relaxed return stroke permits some of that charge to dissipate. Repeated cycles show some mean value with superimposed oscillation corresponding to the wingbeat frequency (see **Fig. 16.5**).

Wing beating is not the sole means by which flying creatures can acquire charge. Gliding accomplishes the same. During a glide, air flows past the wings and head, conferring negative charge (as in the glider plane). This mode of charging can be particularly effective because it averts charge neutralization: The oppositely charged air falls behind (see **Fig. 15.7**), unable therefore to accumulate around the constantly advancing object and thus failing to neutralize it. The bird can thereby acquire hefty charge, merely by passing through the air.

So, it's not just the flapping of wings that keeps birds aloft. The flow of air past even non-flapping wings (and bodies) could well play a significant role in lifting. Both of those flight modes should build negative charge, driving the charge bearer upward. The decision whether to flap or not to flap is left for the bird to make, depending on situational demands.

Regarding the sufficiency of the lift force, we need to remind ourselves how light birds really are. A ruby-throated hummingbird weighs only 3.2 grams (about a tenth of an ounce); more common birds such as blue jays weigh in at about 88 grams (~3 ounces).[wl] Our housecats regularly dispatch inattentive birds, proudly returning their trophies home. As the designated agent of disposal, I can report that those poor birds seem practically weightless. Hollow bones and large internal air sacs minimize body density. Because of their extremely light weight, keeping birds aloft should require only relatively modest amounts of charge.

Calculating the needed amount of charge is straightforward. Of course, we can't know for sure whether mechanical or other forces might contribute to lift; but, if we assume for simplicity that charges bear full responsibility, then we can calculate the required amount — the theoretical minimum charge needed for staying aloft. The upward force is the product of charge and electric-field strength. With a known near-earth field strength of 100 volts per meter (Chapter 2), and, say, a 100-gram bird, the required excess negative charge computes to 10^{17} electrons. Considering that one mole (gram-molecular weight) of a substance contains 6×10^{23} molecules, the quantity 10^{17} implies that keeping the bird aloft should take *less than one negative charge per million* of the bird's molecules, a number that hardly seems unreasonable.

Besides maintaining altitude, birds have been known to move forward. They fly ahead. For driving that forward motion, recall the two glider-centered mechanisms discussed in Chapter 15: compressing positive charges ahead to create excess pulling force; and bending the electric field ahead, also producing a forward-directed force (**Fig. 15.6**). Why these same two mechanisms would not similarly apply to birds is unclear, for gliders and birds share many physical attributes — especially birds operating in the gliding mode.

Thus, triboelectric charge forces appear capable of driving both lift and forward motion in birds. The modesty of the charge requirement may be hard to fathom; but recall that something the volume of the earth can lift off the planet with only one extra negative charge per hundred (see Roadmap). Think of it! For the featherweight bird, the lifting requirement is evidently far more modest than for the earth — only one extra charge *per million*. That's not a whole lot of charge for keeping the bird in the sky above.

Still, one may wonder: From where comes the required energy? A bird moving forward above the earth does work to overcome air friction. Doing work requires energy. The source of that energy, I remind you, is by no means magical. It derives from the sun. Solar energy separates atmospheric and earth charges, setting the stage for all triboelectric action.

I don't mean to suggest that *all* energy is triboelectrically sourced. Hummingbirds, for example, eat prodigious meals before initiating the heroic bouts of wing flapping that facilitate hovering. Food, evidently, can supply fuel. Nevertheless, a goodly amount of the energy required for forward motion and lift may well come naturally from separated charge, especially when flying occurs in the gliding mode.

If charge separation does contribute meaningfully, then you might expect that the bird's design would exploit that separation maximally. Indeed, it does. We next consider three features that maximize the bird's negative charge advantage.

MAXIMIZING NEGATIVE CHARGE

Can Feathers Build Charge?

One prospective contributor to charge buildup is feathers. All birds have feathers, the saying goes, while everything that has feathers is a bird. Birds and feathers seem practically synonymous.

Why so?

One possibility is that birds require warmth, and, like the down comforters on old-fashioned featherbeds, those feathers help to trap the bird's natural body heat. But warmth can hardly be an issue in the tropics, where birds remain amply feathered despite the frequently blistering heat. Would you wear a fur coat in Saudi Arabia? Another oft-cited suggestion is that feathers make the birds lighter. But that argument lacks cogency: Plucked chickens certainly weigh less than unplucked chickens (**Fig. 16.7**).

Figure 16.7. Contrary to a common perception, feathers add weight.

A more plausible explanation lies in the potential utility of feathers for flying. Is there some way that feathers could facilitate flight?

Feathers are made chiefly of keratin, the main protein found in human hair. Hair, in turn, occupies a position well beneath that of air in the triboelectric table (**Figure 14.6**) — implying that a feather moving through air should acquire negative charge. (Recall the hair-dryer effect.) If negative charge matters for flying, then feathers ought to help; they should facilitate negative charge buildup.

Figure 16.8. Typical feather structure, highlighting the grid-like pattern.

In terms of their charge-carrying capacity, feathers seem judiciously designed. Resembling a bug screen, their grid-like architecture endows them with numerous surfaces that can acquire charge as the air passes by them, through each contiguous pore or fenestration (**Fig. 16.8**). You might opt for that motif if you were designing an efficient charge carrier. Hence, if negative charge matters for flying, feathers might help.

A side issue — the notion that flapping is mainly about pushing air — is worthy of mention. If the point of wing flapping were to push the air, then an advantage would surely go to wing feathers without pores (analogous to competitive rowing, where nobody would think to use paddles with lots of drilled holes). Surely, if wings were solely about pushing air, then solid, non-fenestrated feathers would rule the day. But they do not.

Feather-based charge acquisition is a concept understandable through common experience. Think of those old-fashioned bed pillows that may have lent comfort when sleeping at grandma's house. When beaten, those down-filled pillows fluff splendidly. Plausibly, the fluff comes from the feathers' easy acquisition of negative charge, gained as the air inside the pillow brushes past those feathers. It's pure triboelectricity. The forced air readily escapes the pillow; but the feathers remain, along with their freshly acquired negative charge. Negative-charge repulsion then creates the abundant fluff, which may not only help confer pleasant feelings, but also sweet dreams.

It's not just pillow feathers that fluff in this way, but bird feathers as well. Commonly suggested reasons for bird fluffing include not only providing insulation for keeping warm, but also creating the appearance of a larger and hence more formidable target to ward off potential predators. Either way, bird-feather fluffing implies the presence of internal repulsion, and the simplest option for creating that repulsion comes from the presence of charge. Negative charge on each feathery unit should produce bird-feather fluff, just as it produces pillow fluff.

Hence, feathers may offer the ideal structure for facilitating flight. Their seemingly robust charge-carrying capacity should help the bird build more of what's needed — not only for gliding, but also for the flapping mode, *i.e.*, for conferring more robust charge with each flap. On the other hand, feathers are by no means obligatory for flying. Extra charge can build in other ways: For the bat, it can arise from appreciable wing-surface area; and, for flying insects, it could come from high wing-beat frequency.

For birds, on the other hand, negative-charge enhancers would seem to reside largely in feathers. Feather-borne charge could be especially helpful during takeoff, when birds will often fluff their wing feathers prior to taking flight — an act that should create a surge of negativity. That surge could make the difference between a smooth takeoff and an ungainly plummet.

As useful as they may seem, bird feathers did not appear suddenly on the evolutionary scene. Although various theories have been put forth,[w2] feathers seem to have evolved from reptiles' scales. In some reptiles, in fact, vestigial feathers remain on the lower parts of legs and feet.

Given such vestigial feathery scales, no stretch of the imagination is required to envision how reptiles might have evolved into birds. We can suppose that over time, those scales became more like feathers in some species than in others. Creatures better endowed could presumably fly awkwardly over at least short distances. With that survival advantage, those species could have evolved into the large flying reptiles whose bones grace the halls of major museums — and ultimately into today's birds, which stand as nature's premier flying machines. In this way, we can make sense of the evolution from scaly reptiles to agile birds.

So, the practically synonymous linkage between feathers and birds does make sense within the context of charge-driven flight. Feathers possess an arguably prodigious capacity to acquire charge; and that feature could help facilitate charge-based flight.

Intrinsic Negativity?

Besides the capacity of feathered wings to acquire negativity, birds bear negative charge even without flying — as do all plants and animals. Living beings are composed mostly of cells, and electrodes stuck into those cells routinely report negative electrical potentials. This is common knowledge among physiologists and confirmed in many experiments, including those in my own laboratory. It implies that cells have excess negative charge. More recently, rudimentary experiments by students could confirm the presence of a net overall negative charge in whole plants, as well as in (whole) people — as expected.

Early on, I wondered whether such body negativity might suffice for staying aloft. This seemed plausible at first, given birds' extreme lightness. But birds also land. Hence, any such negative charge must be of limited quantity; otherwise, those birds might be constrained to eternal flight, something like the Flying Dutchman. They could never settle down on land. Despite its limited magnitude, any such permanent avian negativity cannot help but assist in creating at least some lift, in much the same way that negative charge may foster vertically directed growth in plants. Negative charge, after all, repels the earth.

Regarding the negativity inside cells, I must digress once again, as readers interested in biology might find the subject of some interest. Cells are filled mainly with structured ("EZ") water.[5] Since EZ water commonly bears negative charge, the abundance of that water may suffice for explaining the cell's negative electrical potential.[6] Textbooks argue for more complicated, membrane-centered mechanisms, but most texts were written before it became apparent that the lion's share of the water inside cells bears a negative charge.

Of course, animals consist not only of cells but also the material outside of cells, which could theoretically balance the cells' negativity. However, that seems not to be the case. Of non-cellular materials, the most abundant are negatively charged connective tissues such as collagen and elastin, which themselves nucleate the formation of negatively charged EZ water. Hence, most of the body's mass would appear to bear negative charge.

But then, what happened to the complementary positive charges? If the EZ's negative charge originates from the negatively charged component of water, then surely the positively charged component must lie somewhere in evidence. But where?

Positive charges seem to accumulate in compartments that expel. The urinary system contributes to this by jettisoning protons each time you pee. Sweat expels protons as well, with its generally low pH value. And so does expired air, as test exposures to litmus paper readily show: With each exhalation, positively charged protons exit. Through every conceivable orifice, our bodies seem designed to expel positivity. Meanwhile, the complementary negativity remains within the bulk of the body's tissues. Preserving as much of that negative charge (or alkalinity) as possible may be a fundamental feature of biological viability.[7] It assures bodily retention of ample potential energy, stored in the form of concentrated negative charges.

The cellular origin of body negativity may seem merely a peripheral issue. The main point here concerns birds. By retaining negative charge, tissues of the bird can lend yet another hand in facilitating their rise from the earth. Intrinsic negative charge can't perform the entire job, but it can certainly assist.

Why Can't You Levitate?

If you, yourself, bear intrinsic negative charge, and negatively charged masses rise from the earth, then how come you don't levitate? (Or perhaps you do?)

Rising from the earth depends on a balance between upward and downward forces. The downward gravitational pull is proposed to come from charge induction (Chapter 11) — the earth's net negative charge inducing positive charge on the object's underside and thereby drawing said object toward the earth. Meanwhile, a complementary upward push may arise from repulsion between the body's net negative charge and the earth's negative charge.

For you and me, gravitation seems to win out. Plenty of dipoles fill our bodies. With the positive ends of those dipoles oriented toward the negative earth, we experience a downward gravitational pull. The upward force may come from any excess of negative body charge;

but, if the dipoles dominate, as is commonly the case for dense creatures like ourselves, then we ought to remain incessantly stuck to mother earth. Gravitation should dominate.

On the other hand, levitation could theoretically arise from any process that builds enough negative charge in your body, or that reverses the polarity of your dipoles. With those two physical options in mind, along with evidence that charged objects can manifestly levitate (Chapter 14), it may be prudent to refrain from reflexively dismissing levitation as a hoax, although plenty of hoaxes have been identified. Possibly, the practitioners of various spiritually oriented belief systems ranging from Buddhism and Hinduism, all the way to Transcendental Meditation and Christianity, have figured out something about levitation that the rest of us have not.[8] Meanwhile, most of us remain firmly grounded.

Revised Respiration: Does Bird Anatomy Maximize Body Negativity?

Another feature that facilitates acquisition of negative charge may be a bird's unique respiratory system. If respiratory systems enhance negativity by expelling protons, then you might expect a bird's respiratory system design to optimize that process.

Compare a bird's respiratory system to your own. In the human respiratory system (**Fig. 16.9**), air enters through tracheal and bronchial conduits, making its way to the lungs; there it exchanges with the blood. Exhalation occurs through those very same conduits, but in reverse. Some of that conduit air never makes it as far as the lungs before getting exhaled. That volume of non-exchanged air is well recognized, and commonly referred to as "dead space." Dead space limits oxygen exchange. It diminishes the efficiency of the respiratory process, as energy is expended to pull in air without exploiting that energy for extracting maximum oxygen.

No such limitation exists in the bird's respiratory system. Air inhaled passes through a series of air-holding sacs, eventually exiting from the nostrils (**Fig. 16.10**). It's essentially a one-way path. Dead space is absent. All protons produced by a bird's respiratory system should get expelled, making this idiosyncratic construction well suited for maintaining the highest possible negative charge.

Respiration may also assist in takeoff. Immediately prior to flight, birds will sometimes pant. This behavior is widely thought

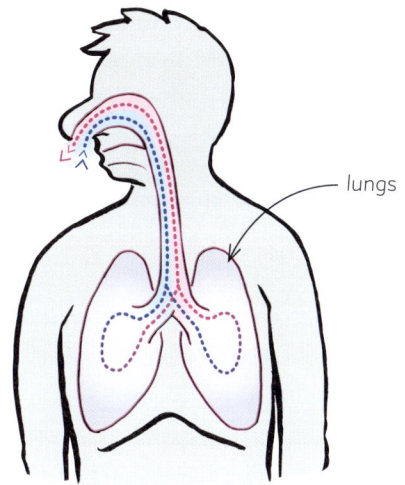

Figure 16.9. Human respiratory system. Exhaled air follows the same path as inhaled air, but in reverse.

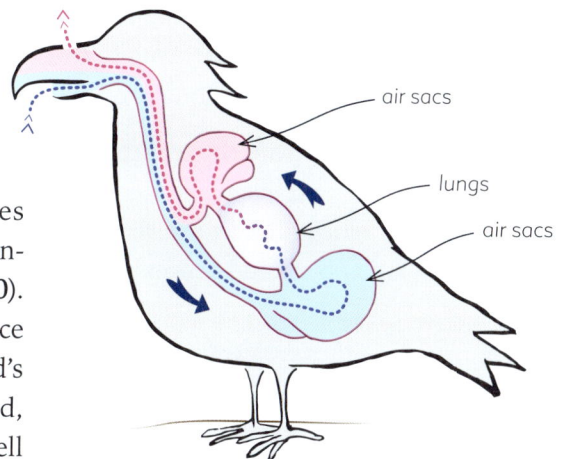

Figure 16.10. Bird respiratory system. Air follows a one-way path.

to signify some kind of physical weakness, the afflicted bird requiring a hefty dose of oxygen to prime its engines. Panting could rather be a more routine maneuver, a strategy for quickly blowing off positive charge to increase net negativity — functioning much like preflight fluffing. By enhancing negativity, both strategies could take the edge off takeoff.

Yet another charge-based takeoff strategy involves discarding the white stuff that birds often drop, sometimes at the time of takeoff. Urine and feces make up this excreta, commonly known as guano. Of course the bird loses weight; but much like your own excretions, a bird's excretions are proton rich, with low pH values.[w3] Perhaps that positive charge explains why those droppings show some propensity to land on negatively charged objects — like your head. It may not be your bad fortune, but merely your negativity.

Excreting that bolus of positive charge represents yet another strategy by which birds could enhance their net negativity. Birds will often defecate just prior to launch and also when frightened. While the conventional explanation involving tactical weight reduction may well hold validity, defecation also instantly augments the bird's net negativity, useful not only for liftoff but critical for the rapid elevation gain that an escaping bird may require.

Thus, in multiple ways, Mother Nature appears to have optimized flying creatures' anatomy and physiology so as to maximize negative charge, and arguably, therefore, to facilitate flight.

With flying creatures so packed with electrical charge, you might expect each landing to generate an electrical shock. That could happen if legs and feet were made of conductive materials. However, in the case of birds and bees (and probably other flying insects), non-conductive legs and feet should insulate the body from the ground. Thus, when landing, these creatures need not live in fear of electrocution.

On the other hand, these settled creatures need to take off again. If positive ions in the air soon bring neutralization, then those neutralized negative charges must be replenished for a subsequent launch. In the discussion above, we identified three mechanisms for achieving quick negative-charge replenishment: panting, wing-fluffing, and defecation. Absent those expedients, the bird might opt for a downward-directed takeoff, which builds negative charge through triboelectric air friction, preparing the bird for flight. More rarely, as in the case of the albatross, a running start with wings spread (or, in the case of crows, merely hopping ahead) represents yet another charge-building strategy. All of these mechanisms should quickly rebuild the requisite surplus of negative charge, enabling takeoff.

But the main point in this section is not those handy takeoff expedients; it's the buildup of negative charge based on the bird's idiosyncratic pulmonary anatomy. The respiratory system's linear design seems perfectly conceived to expel as many protons as possible, leaving the bird with plenty of excess negativity. So, we add this feature to the growing list of attributes seemingly designed to build avian negativity: triboelectric friction (of wings especially), feather charge, intrinsic body negativity, and now, streamlined respiratory anatomy.

If charge is indeed involved in bird flight, it would seem that Mother Nature has done an exemplary job of design. All conceivable expedients appear to have been brought into play to make flying as easy and natural as breathing.

PARADOXICAL BEHAVIORS

The Energetic Cost of Flying

Having dealt with the mechanics of flying, we move on to flight energetics. Surely, you've witnessed the following idyllic scene: birds meandering aimlessly through the air, seemingly free of any worldly concerns. So enticing that you almost wish you could take wing and join them.

Why do those birds meander? And what does this have to do with energetics?

Meandering cannot be a hunting strategy; birds often fly too high to spot ground food or low-flying insects. But surely some rationale must exist, for staying aloft would seem to require at least some expenditure of energy. Even if those birds mainly soar, intermittent flapping requires energy for powering the requisite flight muscles. If the energy costs of meandering are commensurate with the energy costs of commercial airplane flight, then making such an expenditure without purpose would seem unwise — it might even qualify as "bird-brained."

On the other hand, birds may have discovered something that Boeing engineers have not: Bird flight could come cheaper than expected. If so, then aimless flight might cost the bird no more than you might expend on a pleasant spring-day's stroll in the woods — both activities undertaken mainly to enrich the soul and requiring no excessive expenditure of energy.

Indeed, evidence indicates that birds may expend little of their own energy, at least in long-range flight. In a now-famous study published in the journal *Nature*, scientists studied energy utilization in songbirds migrating from Panama all the way to Canada.[9] The birds flew spans on the order of 700 km (435 miles) without stopping for snacks. According to convention, the only energy available to those birds should have been metabolic: *i.e.*, the birds should have lost weight from flying, in the same way that you'd lose weight from running the Boston marathon. The migrating birds did lose weight; however, the loss was no more than a trivial fraction of the calculated expectation.

Flying those distances may therefore require little of each bird's own energy. On the other hand, the laws of physics tell us that the required energy must come from somewhere.

If staying aloft depends on charge-based repulsion, then maintaining those separated like-charges is where we expect much of the energy to be expended. Here, solar energy should be central. Energy from the sun separates charge on the earth's surface (Chapter 4), leaving a negatively charged earth that arguably helps keep the negatively charged bird aloft. If so, then the main source of energy may lie *outside* the bird; reliance on the bird's own metabolic reserves may be unnecessary. The bird could meander aimlessly for extended periods over land or sea, or migrate over long distances, without experiencing very much weight loss — as confirmed. Hence, it's solar energy that endows those birds with the capacity to migrate over those long distances without starving to death.

An impressive exemplar of such free flight is the bar-tailed godwit. That bird routinely migrates all the way from New Zealand to Alaska — a span equal to 30 percent of the earth's circumference — making only a single stop for food. On its even more remarkable return flight, it flies for eight days nonstop.[w4] Attaining this seemingly miraculous feat would appear much more feasible if the lion's share of the energy came ultimately from the sun, not the bird. And that appears to be the case.

Low energy costs aside, the question remains: *why* do birds bother to meander?

My Ecuadorian friend Arturo Cepeda offered the best explanation I've heard so far: "Because it feels good." Before dismissing his explanation as superficial, think of the experience of the breeze upon your face, or of walking barefoot on the beach — pleasant experiences for most. Both activities confer negative charge. Adding that negative charge to cells that may be incompletely filled with their requisite negativity ought to nourish those cells back toward full functionality. That should include brain cells, whose default state should include a sense of well-being. So, birds may meander for no reason other than to experience bliss. Why not? We do much the same.

Thus, while ordinary flying may require lots of energy for powering wingbeats (supplied through bounteous food consumption), longer term migration that includes copious gliding comes practically free. The bird gets away with expending precious little of its own energy. Arguably, whatever energy may be required comes mainly from the separated charges around the earth, which in turn derives from the rays of the sun. Birds seem to make good use of solar energy.

By massively exploiting the sun's energy, birds might be thought of as clever — the "bird-brained" moniker perhaps an unfortunate misnomer. On the other hand, it's not the bird itself, but Mother Nature, who bears responsibility for designing the birds' energetic system. The birds merely follow the rules. Nevertheless, those birds seem to have learned to power flight, especially long-range flight, through rather direct use of solar energy. Kudos to them!

Resolving Migratory Paradoxes: Do Birds Snore?

Beyond the modesty of its energetic cost, long-range flying brings several additional paradoxical features that merit consideration. What are those paradoxes, and how might the proposed charge paradigm offer satisfying resolutions?

The first paradox is sleep — or the lack thereof. Ordinarily, birds sleep daily, with patterns similar to those of humans. If you skipped sleeping for a week, you'd probably exhibit psychiatric symptoms: Even several days of sleep deprivation, according to numerous studies, can cause long-lasting psychological harm. In theory, birds might be less sensitive to sleep deprivation, but for preserving sanity during a weeklong migration, an occasional snooze would nevertheless seem obligatory. Yet some migrating birds do not stop even for naps.

Might birds sleep *during* their migratory flights? And if so, then how could they possibly navigate? While we don't know the answers, we can certainly theorize.

The electrostatic flying mechanism could run on automatic. Neither the forward thrust mechanism nor the lift mechanism requires any of the bird's attention: Once the bird sets itself in motion it could continue without thinking. If so, then as the birds migrate, they could indeed snooze. This idea is not original; it has been the subject of considerable speculation.[w5] If true, then migrating birds would behave much like orbiting satellites, circumnavigating the globe on automatic, while at the same time enjoying needed sleep on the wing. Bird sanity could thus be preserved.

A second migratory paradox is altitude. Migrating birds frequently cross towering mountain ranges. Surely, any bird with a modicum of sense would prefer valleys over those cold, inhospitable peaks. Yet, migrating birds commonly opt for the high-altitude routes, whether preferentially or otherwise. The bar-headed goose, for example, after wintering on the Indian subcontinent, migrates high across the Himalayas. A flock has been spotted crossing directly over the summit of Mount Everest.[10]

Flying at such frigid altitudes happens routinely. Ruppel's Griffon, a large vulture native to central Africa, has been recorded soaring at 11,300 meters, or 37,000 feet.[11] At such altitudes, birds may well collide with jetliners, and several reports detail feathers caught in engines.

Flying at such high altitudes presents additional perils. Besides the frigid temperatures, the partial pressure of oxygen is a paltry 20 percent of that at sea level, making breathing a struggle. Even a birdbrain would have enough sense to choose more hospitable altitudes. But snoozing birds may lack the capacity to choose. Half asleep and behaving like negatively charged projectiles, birds would go wherever natural forces take them, evidently including routes crossing high above the Himalayan peaks (**Fig. 16.11**).

Figure 16.11. Birds may sleep during long-range flights.

Such high-altitude routes may owe their allure to the intense electric fields present above those peaks. Mountains are unique. Their upward-projecting peaks lie closest to the upper atmosphere's ample positive charge. That charge can induce plenty of negative charge at those sharp mountaintop points. Compared to an electric field of a hundred volts per meter at ordinary ground level, mountaintop electric fields can rise to thousands of volts per meter. Those intense fields create problems for mountaintop antennae, which experience frequent electrical discharges that cause persistent buzzes. Touching a car parked high on a mountain can bring massive shock.

Those intense high-altitude electric fields may attract migrating birds in the same way that fields draw other charged objects (**Fig. 15.6**). The greater the field strength, the stronger the attraction. Moreover, the "inertial" pull from the compressed charges ahead should be strongest where atmospheric positivity is highest — near mountaintops (**Fig. 16.12**). For both those reasons, the regions above high mountain peaks could preferentially draw half-asleep birds. They can't help it. Summiting Mount Everest may challenge humans, but for dozing birds the intense field attracts like moths to light.

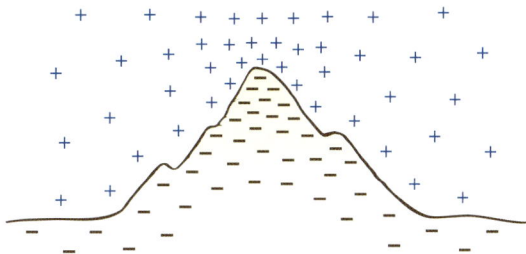

Figure 16.12. Charges accumulate at sharp points, like mountain peaks.

So, migrating birds might not always choose the most "sensible" routes. If they're asleep during their migration, then their path may include altitudes that they would surely sense as inhospitable. It's like undergoing major surgery: painful for sure, but not if you're fast asleep.

Finally, a third paradox: electrical power lines. Why do birds commonly sit on those wires (**Fig. 16.13**)? Might they find comfort in the lines' design? Or, is something more physical going on?

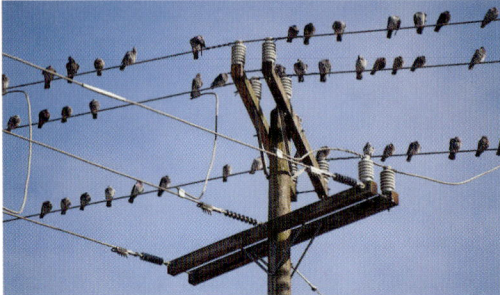

Figure 16.13. Birds frequently park on electrical power lines.

Again, the birds may be drawn there. Let me explain.

Power lines produce strong electric fields, oscillating at 60 times per second. During the positive phase, the line should attract the bird's excess negative charge, drawing the bird closer. During the negative phase, you'd anticipate repulsion of equal magnitude driving the bird away — except for one subtle issue: The relevant distance is slightly greater. The line's negative phase pushes the bird's negative charge toward the bird's far edge. Because of the distance difference, the attraction should be slightly stronger than the repulsion. Hence, the bird ought to be drawn toward the power line, however weakly.

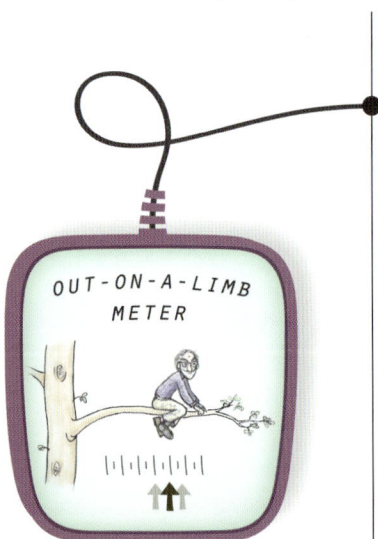

OUT-ON-A-LIMB METER

Once the bird gets closer to the power line, however, that attractive force should grow larger. The reason lies in simple arithmetic: As the bird nears the wire, the oscillatory shift of the bird's negative charge becomes a larger fraction of the bird-wire separation. Hence, the net attraction should grow increasingly stronger than the repulsion. Once the bird sits on the wire, that attraction should tend to keep it in place.

Until, of course, the bird opts to push off. According to the argument just presented, you'd think that the bird might stick forever. But escaping the clutches of those wires merely requires the bird to develop a pushing force that exceeds its attraction to the wire. That act should come naturally: Having sat on the wire for a period, the bird will have lost most of its triboelectric charge, diminishing its attraction to the wire. So, flying off the wire should be no more difficult than flying off a tree limb.

In any event, the phenomenon of powerline sitting could well arise not because birds are fond of electrical wires, but plausibly because they are attracted to them.

In theory, those power lines should attract virtually any object with excess charges of either polarity. The same principle as above ought to prevail. That would include positively charged bird droppings (guano), which also tend also to accumulate on power lines, often causing disruption.[w6] Surely, those droppings don't decide to take aim; rather, like the birds, they may experience an inexorable attraction, with no choice but to stick.

I'm not suggesting that the multiple anomalies introduced in this section are fully resolved by these electrical considerations, but only that the central paradigm of charge-based flight offers a possible pathway toward their resolution.

ACROBATICS

Flight of the Bumblebee

The principles presented in this chapter, if valid, should apply not only to birds but also to flying insects, although the latter pose special complications. In 1934, the prominent French entomologist August Magnan calculated that bee flight should be aerodynamically impossible. The flapping of the bees' flimsy wings, according to Magnan, should be inadequate for keeping those bugs aloft — something akin to tissue-paper wings lifting a bus.

Considering that bees do seem to fly, and by no agreed-upon mechanism other than wing flapping, the entomologist Michael Dickinson and colleagues posited a rescue: that an unconventional combination of short choppy wing strokes, rapid rotation of the wings as they flop over and reverse direction, and a very fast wing-beat frequency, combine to create complicated vortices that may succeed in propelling the bee. Perhaps the esteemed Magnan was wrong after all?

On the other hand, even if such intricate mechanisms were at play, questions about flights of other insects might still remain unresolved. Moths, for example, can move practically straight up with only minor flapping. Some other insects can hover for extended periods. Dragonflies can fly backwards. Such acrobatic feats have confounded scientists, whose focus tends to lie rather exclusively on mechanical wing-beating mechanisms.

An advantage of the electrostatic mechanism is that it renders the specifics of wing dynamics less critical; so long as the wings beat and negative charge builds, the insect can fly. Additional maneuvers can help create acrobatics.

Among those acrobatic feats, the simplest is hovering. Maintaining that steady position has two requirements within the proposed framework: a stable negative charge for sustaining altitude; and the absence of any forward-directed force. Retaining steady charge should be attainable by flapping the wings at a proper frequency. Achieving stable lateral positioning should be realizable by angling the body and wings appropriately: Hummingbirds, for example, hover by adopting a fairly upright orientation near 30 degrees from the vertical.[w7] Also, wings twist during their wing stroke, plausibly helping to maintain that lateral stability.

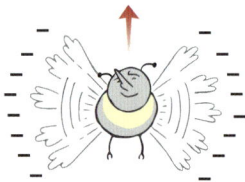

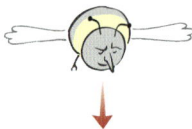

More formidable, perhaps, is explaining the straight vertical flight that many insects can achieve. Here, Bernoulli once again fails (Chapter 15), for the insect's flight is strictly vertical. Hence, no possible lift can come from Bernoulli's principle.

Figure 16.14. Lift can be achieved through more vigorous wing flapping, which builds negative charge.

Within the charge framework, however, the rise enjoys a possible explanation (**Fig. 16.14**): Vertical flight could be accomplished through progressive gain of negativity, which, in turn, could be achieved triboelectrically: through a progressively increasing wing-beat frequency. Once that upward course begins, it could continue naturally: Air flowing more swiftly past the insect would confer increasing negative charge, facilitating upward acceleration. Away you go!

Then come the insects' turns (**Fig. 16.15**). Those directional changes could be achieved through mechanically induced charge imbalance. A slightly stronger left-wing stroke should confer higher left-side charge. The additional left charge leads to more intense left-side lift. More significantly, it also results in more left-side forward pull, from the principle illustrated in **Figure 15.5**: higher rod-tip charge yields more forceful pull into the capacitor. That action would provoke a right turn. The opposite action should drive a left turn (**Fig. 16.15**).

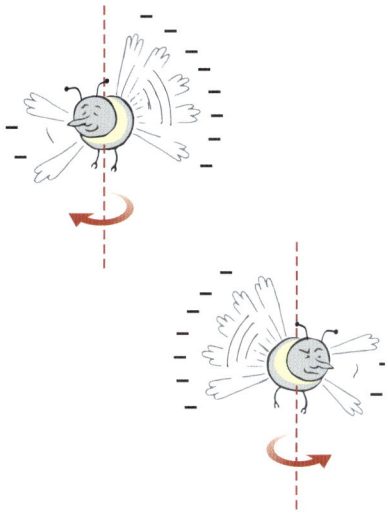

Figure 16.15. Turning mechanism. More charge on one wing will tilt the insect and increase the forward pull on that more highly charged wing, producing a turn.

Maintaining strictly straight flight would demand the absence of any charge inequality, lest the insect fly like a drunk. Nature takes care to prevent any such awkwardness: The wings of some insects touch one another at the end of each upstroke, effectively equalizing wing charge and maintaining directional consistency. So, insects can mostly fly dead ahead.

On the other hand, an insect's ability to acquire more charge on one or the other wing could produce dazzling acrobatic feats, like flies abruptly flipping upside down so they can suspend themselves from ceilings. Just watch those insects and envy their amazingly athletic agility — maybe even more impressive than that of birds.

Perspectives

Here we draw this discussion of the flight of birds and bees to a close. Most everyone surmises that wing flapping is practically synonymous with flying; *i.e.*, that flying arises as the mechanical consequence of flapping. In so thinking, we ignore the fact that, like gliders, birds can fly perfectly well without any flapping at all. Hence, there must be more to the story of flying than current understanding offers.

Current thinking also tends to ignore a bird's forward motion. To achieve that motion by orthodox means, the bird's wings must somehow push the air backwards; and the wings of some birds do exhibit a modicum of backward motion as they stroke.

But not all wings. To check, watch the seagulls from some high perch near the sea. Focus your attention on those birds flying directly toward you. When I watch from the same height as the oncoming, wing-flapping, birds, I see no hint of any wing-angle change that would denote backward push; from my front view, the wings look consistently thin as they flap, *i.e.*, rather pure up-down motion. Please have a look and judge for yourself. In the absence of any serious rear-directed motion, understanding how wing flapping could create the robust forward motion that keeps birds moving ahead is not so easy.

Hence, the popularly held view that wing flapping must be central to flying ahead seems to fail on two counts: (i) flapping is optional, not required; and (ii) flapping fails to account for forward motion. Habit may nevertheless push us to continue thinking of flight propulsion as purely mechanical in origin; but something beyond mechanical seems inevitably at play — not necessarily as the sole driver of flying but at least in some important role. And that feature, I argue, may well be electrical charge.

Summary

As children, we learn that birds fly by flapping their wings. Flapping, however, is not the only mode of flying. Eagles, hawks, seagulls, and other birds can stay aloft practically indefinitely without any flapping at all. For those birds, some other mechanism must be at play.

That mechanism, I posit, centers around electrical charge. Charge builds triboelectrically. Negative charge gets extracted from the air by anything moving through it. That acquired negative charge repels the earth's negative charge, prompting the object, or bird, to rise up.

Within that mechanistic framework, wing flapping merely serves to increase that negative charge. It does so by augmenting the airflow past the wing material, thereby enhancing triboelectric charge buildup. Feathers may amplify that buildup: Their screen-like architecture provides a multiplicity of surfaces, which should build ample charge as air passes rapidly through those multiple pores.

After gaining enough speed, airborne creatures may fly without any wing flapping at all. They may keep flying in the same way as do gliders (Chapter 15) — arguably through ample buildup of triboelectric charge as they pass through the air.

Charge-based flying may require scant energy expenditure from the creature, especially during long-range flight. Much of the energy comes indirectly from the sun, whose energetic output gets used to separate the earthly charges that make electrostatic lift possible. This feature may help explain why birds often feel free to meander aimlessly in the sky, as if for fun, and why even weeklong migrations seem to require only meager consumption of the bird's metabolic energy.

Charge-based mechanisms may also explain seemingly awesome acrobatics of flying insects. Having achieved liftoff through the expedient of wing flapping, an insect may then go on to use its body parts to redirect its flight path largely as desired — even straight up. Vertical flight could be readily achieved by progressively increasing negative charge, the latter attainable by nothing more than increasingly vigorous wing flapping.

Think of it. How remarkable it would be if flying creatures had long ago managed to find a way of exploiting ambient electrical charges to their advantage. Obtaining energy would then merely require tapping into naturally existing energetic resources. Nothing exotic. Eons back in their evolution, those creatures might have succeeded in harnessing energies that, until today, seem practically magical. But perhaps not magical at all to those flying creatures.

SECTION VI

Moving Ahead

The previous three chapters mainly addressed the challenge of staying aloft, *i.e.*, of vertical lift. Here, we lend additional consideration to a second aspect of kinetics: moving forward.

The forward-motion issue has already been considered in the context of birds and planes. Here we deal with entities in which forward motion is primary: sailboats and fish. In both cases, I raise relevant questions about the generally accepted mechanisms, and go on to suggest alternatives that build on themes considered in earlier chapters.

I argue that the alternative explanations fit the evidence in a more natural and consistent way. Hence, they may be worthy of your consideration.

CHAPTER 17

Forward-Thrust Wind Machines: Sailing

Sailors love the Pacific Northwest. The city of Seattle, where I dwell, once boasted more sailboats *per capita* than any major city in the US. This surfeit results in part from the area's dominant population of transplanted Scandinavians, whose genes seem to code for boating; and, also, from the Northwest's plentiful winds.

Arriving some years ago in Seattle's water-bound geography, I promptly enrolled in my university's sailing course. I was immediately captivated. The art of downwind sailing seemed immediately obvious — the wind simply pushes on the sail, dragging the boat along with it. No problem. Upwind sailing, *i.e.*, against the wind, made less sense: The instructor taught us that sailboats could indeed move upwind — not straight into the wind, but heading up to about 45° off the wind. This seemed counterintuitive. Having had ample exposure to the discipline of static and dynamic force balance, I could not fathom how a sailboat could move even partially *against* the wind, although it clearly did happen. If you were pushed toward the south, would you respond by heading north?

The sailing textbook explained it using vector diagrams, but still I couldn't make sense of it. Physics 101 taught me that pushing in one direction moves the pushed object in that same direction. Driven golf balls experience this all the time. Could you imagine the ball flying in reverse? Yet, sailboats seem to do exactly that: Even if the wind pushes in one direction, the boat may move in a rather opposite direction — *into* the wind. Crazy!

How could pushing one way create motion in an almost *opposite* direction?

That paradox lingered in the dark recesses of my mind for years, as a bumbling sailor who won no races and even suffered some indignities. In one embarrassing incident, while sailing toward the north with my family one day, we found ourselves drifting southward — caught in a powerful tidal flow that was rapidly taking us out to sea. A nearby Coast Guard patrol boat megaphoned an offer of help. Proudly, I refused. My family overruled, and I suffered the indignity of getting towed home. That incident reinforced how very little I understood about sailing.

Over the years, those textbook principles (which you can check on the Web) continued to elude me. Intuition resisted. Support for my lingering doubts came when John Kimball, a professor of physics at the State University of New York, handed me his new book on sailing.[1] The book dealt with the physics of sailing. According to Kimball, downwind sailing was a snap, but sailing into the wind required multiple chapters to explain — implying far more complexity than even my basic sailing textbook had suggested. Sailing upwind was not so simple after all.

But full vindication only came when I learned about a certain class of sailing vehicle whose motion is simple enough to require no vector diagrams to understand. I'm referring to wind-powered contraptions designed to sail on land, or even on ice. "Land yachts" course on flat land using low-friction wheels. Iceboats employ thin rails secured beneath their hulls, which glide smoothly over bodies of frozen water.

Astonishing feats can be performed by these wind-powered vehicles. As I'll describe in a moment, they can head practically dead into the wind. No vector diagrams required. The mind-blowing ability of these vehicles to buck the wind practically head on makes clear that something beyond the mechanical force of wind pressure must be at play; otherwise, Newton must have gotten it terribly wrong. Could a force pushing on an object in one direction cause that object to move in the *opposite* direction? Something beyond that pushing force has got to be involved.

By now, you have likely guessed what I will be suggesting in this chapter: that the explanation for such apparently incomprehensible behavior might lie in electrical charge. Before I attempt to detail how, let us first see what those exotic sailing vehicles can actually accomplish, and why those seemingly remarkable feats cannot be explained by wind thrust alone. Following that, we will transition to the explanation mode: Could such seemingly magical capacity reside in uncomplicated forces brought by electrical charge? Could the explanation lie in simple, unforeseen electrical forces? And if so, then how?

Running on Thin Ice

If you happen to live in Scandinavia, or in North Central US, or in the frigid zones of Canada, or almost anywhere in Russia, then you are probably familiar with iceboats. They sail on ice. Iceboats first gained popularity 150 years ago among fishermen for wintertime transport to locales with abundant fish. For that, those vehicles served admirably. Eventually redesigned for more general transport, some iceboats became versatile enough to traverse either ice or water, depending on what terrain their long-distance journey might encounter.

Modern iceboats are designed mainly for adventure. They travel on runners resembling elongated ice skate blades (**Fig. 17.1**). Many design variants exist, some rigged with multiple sails for faster movement. Steering draws on a movable blade at front or rear.

Figure 17.1. Racing iceboat.

Iceboats enjoy a singular advantage over conventional sailing boats: no need to plow through that viscous water. Hence, iceboats experience minimal drag to slow them down. Absent that restraining force, those boats run as fast as the wind-borne forces dictate, restrained principally by the sailor's fear of instant high-speed death.

A typical start runs downwind. The pace of the boat picks up, soon reaching the speed of the wind. Because both the boat and wind velocity are then equal, the sailor (also the sail) does not sense any *apparent* wind — nothing, ostensibly, to drive the boat. If the boat were to slow down, it would feel an apparent wind from behind, pushing it to speed up; conversely, if the boat were to gain speed, it would feel an apparent wind from ahead, presumably slowing it down. Yet, it does not slow down. The mysterious feature is this: The vehicle *continues* to pick up speed even after it outpaces the wind, and after the sailor begins to feel wind on his or her face. Effectively, the boat is now heading *into* the wind, yet still accelerating. Iceboats do this regularly, on a course almost dead into the apparent wind. How can this be possible?

The concept of "apparent" wind sometimes confuses, so let me pause for a moment to consider it. A helpful analogy is driving in a car, with no wind outside. You crank down the window and then stick your hand out, feeling an "apparent wind" on your hand. In terms of the force felt by your hand, what counts is that apparent wind, not the real wind.

Likewise for the sailboat, what counts is the immediate environment around the boat — the wind experienced by sailor and boat at any given time. We call that the "apparent" wind. And, that's what matters to the boat. The terminology is unfortunate, because the word "apparent" conveys the sense of being unreal, or ephemeral; yet, that wind is quite real — it's exactly what the boat senses, and precisely what the sail responds to, much like your projecting hand.

Given the irrelevance of *real* wind velocity for steady state dynamics, it might not surprise you to learn that iceboat velocities can substantially exceed the real wind velocities. Even in modest winds, those boats commonly reach speeds of up to 150 kilometers per hour (93 miles per hour). Imagine! There are many accounts of steady sailing practically into the wind (just 7° off the apparent wind, to keep the sails filled) at speeds up to three times the real wind velocity.[w1] Some websites claim speeds even higher, up to five to ten times the speed of the wind. Those boats move in the same direction as the true wind — but faster. They *outrun* the real wind. That's why sailors feel wind on their faces.

When dealing with iceboats, therefore, the application of standard Newtonian physics can be simplified: Heading almost directly into the wind instead of at some angle obviates the need to deal with vector diagrams and, moving on low-friction ice rails averts the need to deal with uncertain drag forces. The situation is now simpler. If the real wind comes, say, from the south, the force should theoretically drive the boat almost directly to the north. It does — at least at first. But the boat then goes on to pick up so much speed that it soon experiences an apparent wind coming from the north. That's the wind that ultimately counts — the "apparent" wind that the boat *feels* as it moves on ahead. The resulting enigma: The iceboat now heads almost directly into that apparent wind, exactly opposite to naive expectations.

No physicist has been able to explain to me how this could happen using standard Newtonian force-balance principles. How can you move spontaneously into a wind that's pushing you in the opposite direction? Some term seems missing from the equation of motion, or more precisely, from the balance of forces. The enigma persists. And it demands explanation.

Meandering on Land: Land-Sail Vehicles

The second type of wind-defying vehicle is the land-sailboat, or "land yacht" (**Fig. 17.2**). First appearing in China more than 1,500 years ago, these "wind-driven carriages" could transport dozens of people from one place to another. Intrigued by this novel invention, visiting Europeans brought home detailed reports and drawings, and by the year 1600, European royalty were entertaining guests on these exotic contraptions.

Recognizing the broad utility of these sailing vehicles, the Europeans began using them for regular transport by 1900, and eventually for racing, as the advent of lightweight fiberglass hulls and stiff battens could increase their speed (**Fig. 17.3**).

Figure 17.2.
An early 20th century sail wagon.

Figure 17.3.
Modern land yacht.

Today, in flat areas like the Sahara Desert or Utah's salt flats, such vehicles appear in concentrations high enough to create traffic jams. They frequently race. The land-speed record, achieved in 2009, is 203 km/h (126 mph), in this case plowing *almost directly* into the apparent wind at roughly three times the real wind speed.[w2] Hence, the same question facing the iceboats also face those land yachts: How can they sail *into* the apparent wind?

Figure 17.4. The land yacht, "Blackbird."

A more recent incarnation is the propeller-driven land yacht (**Fig. 17.4**). That enigmatic vehicle was expressly designed to reinforce the notion that vehicles can run at steady speeds much higher than the real wind. While not a sailboat, the vehicle nevertheless moves solely on wind power, just like the sailboats. No engine drives the propeller — just the wind alone. Hence, including this species in our consideration seems appropriate.

This vehicle mechanically couples the propeller to its wheels. As the apparent wind drives the propeller, the wheels turn, and the vehicle moves ahead. That's it. Like its sail-bearing cousin, this vehicle begins running downwind, gradually picking up speed. Eventually, it reaches speeds up to 2.8 times the real wind speed.[w3] Again, the pilot feels a stiff facial wind — coming from ahead. So, the vehicle runs *into* the apparent wind. Almost two years after setting the above-mentioned record, it set another record — moving at 2.1 times the true wind speed, *dead upwind,* with no internal power source, yet, once again, directly into the apparent wind.

It seems the designers of these various ingenious vehicles have proven their point. No vector complications to deal with — just vehicles that run essentially dead into the apparent wind, with no internal power source to buck that wind.

How can that be?

Physics Undone?

Conventional physics has offered no explanation that I'm aware of. These vehicles run into the apparent wind, some of them *directly* into that wind. The operator feels wind on his or her face; but the vehicle nevertheless advances into that wind, sometimes at impressive speeds. Wind pressure should accomplish the opposite. It seems clear that some force beyond just the pressure of the wind must drive the vehicle forward, *into* that wind.

Traditional attempts to understand upwind sailing deal with headings well off the wind, *i.e*, angled relative to the direction from which the wind comes. Sailboats can reach close to 45°, while racing yachts can make it to roughly 30° and catamarans sometimes even closer to the wind. But the complexities of angled sailing, including water drag, keel vortices, and other complications preclude simple analyses; to get some grasp on the paradox of upwind sailing, we become mired in the intricacies of those details.

The vehicles described in the previous sections circumvent all those complexities. No hull or keel plows through the water to create friction; instead, the vehicles travel on wheels or ice, so frictional complexities practically vanish. Further, the vehicles run practically dead into the apparent wind, hence no angled vectors to contend with. Yet, the vehicles head in directions exactly opposite to the directions of the applied force. If Newton's laws of motion are to remain applicable, then the observed motion must derive from some force beyond the simple push of the wind. The logic seems inescapable.

From whence might that other force appear?

Consider once again the phenomenon of triboelectric charge. Standard principles of triboelectricity (Chapter 14) tell us that wind-blown sails must develop negative charge, the higher the relative speed, the higher that charge.

Evidence for the existence of electrical charge on sails comes from the phenomenon of St. Elmo's fire. Sailors have long reported blue-violet discharges — the "fire" — emerging from masts and other sharp-pointed objects under certain weather conditions. Caesar, Columbus, and Magellan all reported witnessing them, just as aircraft pilots report similar discharges off airplane wings and cockpit windshields.

Visible discharges of this nature typically require potential differences amounting to tens of thousands of volts. That's not trivial. The high negative charge of the masts and sails would appear to discharge into the surrounding, positively charged atmosphere. Even at dockside, some people report getting shocked from touching sails.[w4] Evidently, sails do carry charge, as the triboelectric principle implies they should.

The most natural site for maximum charge buildup is probably the luff of the sail, corresponding to the front of the airplane wing. That front sail region should naturally experience the highest wind shear and ought, therefore, to develop the most triboelectric charge.

So, forward motion against the wind needn't imply an undoing of conventional physics. If sail charge could somehow drive forward motion, then the paradox of sailing directly into the wind might resolve. We next consider such a possibility.

Charge-Based Forward Drive?

Given such sail charge, let me suggest two ways in which said charge could propel forward motion (**Fig. 17.5**).

The first operates much like the suggested forward-propulsion mechanism of an airplane wing. Wind-filled sails closely resemble classical airfoil shapes, although here, the "wing" is oriented more nearly upright. The lines of the atmospheric electric field, ordinarily running vertically, should bend toward the sail's leading edge, just as they bend toward the leading edge of the charged wing (Chapter 15, **Figs 15.6, 15.7**). This creates a forward-directed force. It matters little whether the airfoil is horizontal (airplane) or angled (sail). Under the right conditions, the field should draw the sail forward, just as it draws the glider forward.

The second mechanism involves the charge-compaction ("inertial") force (Chapter 13). Through triboelectricity, a sail passing through the air should bear negative charge. That negative charge should draw the boat forward because the sail's forward motion compacts the atmosphere ahead, concentrating its positive charges, which then pull the negatively charged sail forward. Meanwhile, the void left behind the

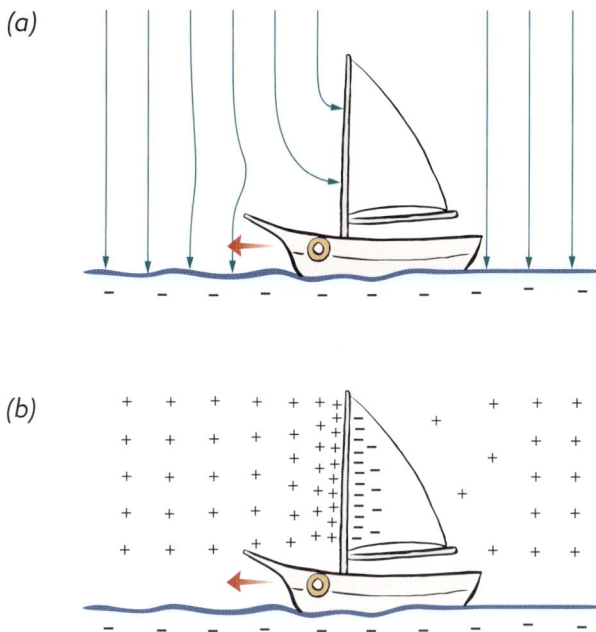

Figure 17.5. Forces drawing sailboat forward. *(a)* electric field bends toward negatively charged sail, pulling boat leftward. *(b)* Leftward-moving sail compresses atmospheric positive charges ahead, intensifying leftward pull on sail.

sail provides little or no compensating backward attraction. So, the net force is decidedly forward. The quicker the boat moves through the air, the more powerful is this forward-driving force.

Now the critical question: Could these two mechanisms generate forward force of sufficient magnitude to do the job? Some number crunching implies they can.

The relevant variables in the first mechanism are the sail's charge and the electric field's horizontal component (**Fig. 17.5a**). While not simple to calculate precisely, the magnitude of that horizontal component should be comparable to that of the vertical component, something under 100 volts per meter. The sail's triboelectric charge, the other variable needed to compute the force, should concentrate at the sail's luff, just behind the mast. Suppose each atom of that surface equally contributes a single electron charge. For a representative atom dimension of 300 picometers,[w5] and hence a density of ~10 atoms per square nanometer of surface, the number of charges computes to 10^{19} elementary charges, or 1.6 Coulombs, per square meter. For a sail-luff having one square meter of surface area, the total charge then comes to 1.6 Coulombs. The product of 1.6 Coulombs and 100 volts per meter field strength gives a pulling force of 160 Newtons.

The charge-compaction force, the second propelling mechanism, should augment that 160-Newton force. Once the boat gets fully underway and compaction increases, that second component could dominate. However, its magnitude is difficult to calculate with any degree of certainty; so, for the moment we ignore its contribution and press on with consideration of the first mechanism.

Is a pulling force on the order of 160 Newtons sufficient to produce the unexplained forward motion (**Fig. 17.6**)? To put this number in perspective, 160 Newtons is roughly equivalent to 35 pounds of force. Could that quantity of force exceed the drag forces on those exotic vehicles? Given the minimal drag of wheels, or of runners on ice, perhaps it could.

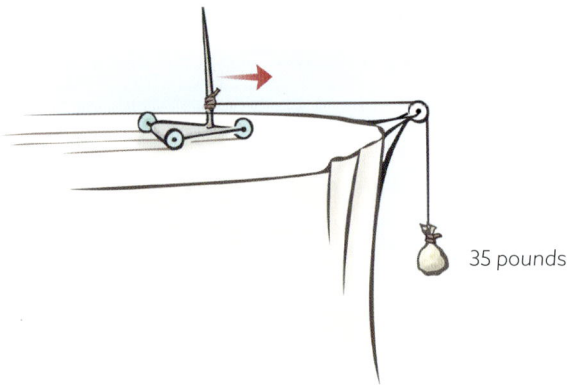

35 pounds

Figure 17.6. Could a weight equivalent to a bushel of apples move a low-friction vehicle forward?

If such a mechanism were to suffice, then charge forces could resolve the otherwise enigmatic forward-motion paradox. The sail's charge could be central.

Here is a plausible sequence of events. Initially, the vehicle would advance in the conventional manner, propelled from behind by pressure force of wind against the sail. As the vehicle accelerates and the sail charge builds, the aforementioned charge-based force mechanisms would begin dominating. Those forces would accelerate the vehicle. Speed should continue to increase until the vehicle-drag forces (that generally increase with the square of the velocity) grow high enough to balance the charge-based forward-propulsive force. At that point, the vehicle would move along steadily.

But wait. If sailboats acquire enough negative charge, shouldn't they eventually fly? Shouldn't their negative charge repel the earth's negative charge, ultimately lifting the boat out of the water?

Well, apparently, sailboats *can* lift off. This increasingly popular sport is called "foiling." The fun draws hobbyists, as well as racers;[w6] in fact, foiling boats now dominate the America's Cup sailing competition.[w7] The lift is commonly ascribed to submerged ski-like appendages. How much of that lift actually derives from those appendages *vs.* the upward force from acquired charge remains unclear — nobody has yet investigated. However, no question exists that sailboats can lift off, just as speeding cars can take off.[w8] Does some reason exist to reject the charge mechanism?

In bringing forth this charge-based hypothesis, which I hope someone has motivation enough to test, I'm not suggesting that sailing is unrelated to the force exerted by the wind. I'm suggesting that the pressure of the wind plays a limited role. Beyond wind pressure, triboelectric charge may contribute to sailing dynamics in a meaningful way. Indeed, charge may constitute the missing piece needed for explaining the seemingly unexplainable: What propels a boat to advance directly into the opposing wind?

In the proposed framework, no physics is violated. We simply acknowledge the contribution of triboelectricity in helping to drive wind-driven vehicles forward. That triboelectric contribution should apply generally — not just to the exotic vehicles considered here, but also to conventional sailing vehicles. That includes the clunker that eventually led to the personal ignominy of having to be towed to shore by the Coast Guard.

What had seemed physically questionable when I first started sailing finally reaches *possible* resolution through an understanding of the role of charge. Newton (**Fig. 17.7**) is not violated. He can rest easy.

Figure 17.7. Sir Isaac Newton (1643-1727).

Summary

Modern sailors assure themselves that sailing upwind is perfectly explainable in terms of conventional force vectors. Such confidence may be unfounded: With the advent of lesser-known wind-based vehicles that can advance practically dead into the wind, conventional force analysis breaks down, for it implies that an applied force can generate motion in the direction *opposite* to the applied force. Such behavior would surely raise an inquisitive eyebrow from Newton.

Recognizing the role of charge-based forces may help resolve this paradox. Negative charge should develop triboelectrically on the leading edge of the sail, just as it arguably develops on the leading edge of an airplane wing or a bird wing. That charge matters. It should create a horizontal component of the nominally vertical atmospheric electric field, which then draws the vehicle forward. Moreover, the advancing sail (along with the mast and the boat's prow) should compress the atmospheric positive charges ahead, intensifying their attraction to the negatively charged sail behind and thereby adding to the forward-directed force. With these charge-based forces, the boat could theoretically sail into the apparent wind.

Simple calculation shows that the anticipated force magnitudes seem provisionally adequate to account for that forward propulsion. Hence, no magic may be needed to explain sailing against the wind if it derives from electrical charge forces. And, if that explanation suffices for the more exotic boats, conceivably, it helps explain the dynamics of ordinary sailboats as well.

So, next time you find yourself tacking into the refreshing breeze, do consider whether charge may be playing some unsuspected role.

In the next chapter, we wrap up our consideration of forward-directed forces by considering the swimming of fish. Fish may dart forward at unexpected speeds. Could charges play a role?

CHAPTER 18

Against the Tide: Swimming Upstream

Having dealt with the forward motions of both flying and terrestrial objects, we turn our attention to undersea creatures. Fish can thrust forward at remarkable speeds. Attaining those speeds in the face of water's appreciable viscosity is no small matter. How do these creatures manage it?

Standard explanations maintain that fish propel themselves forward by two visually obvious means: flipping their tail fins and/or undulating their bodies.[w1] Such maneuvers arguably create forward thrust — though of uncertain magnitude. Here we explore an additional option: that fish might exploit a set of forces based on acquired charge. Any such charge-based proposal should not come as a surprise to the reader by now. On the other hand, good reasons do exist to consider charge-centered mechanisms, at least as adjuncts to conventionally accepted mechanisms, if not more.

In this chapter, we begin with some paradoxical features of fish locomotion, which I hope may raise some eyebrows as they did my own.

Can Fish Climb Ladders?

The vision of a hapless fish awkwardly negotiating the rungs of a ladder must surely evoke a scene from a children's fairy tale. But fish do climb ladders — of a kind.

Following their Pacific migration, Northwest salmon instinctively return to their spawning sites to perpetuate the birth cycle. Those sites inevitably lie upstream in freshwater streams and rivers. Despite being capable swimmers, even the most athletic of salmon will find the 30-meter-high vertical dam walls (98 feet) placed across rivers to generate hydroelectric power too formidable to climb; so, engineers responsible for building those dams thought to provide stepped walls nearby to assist the salmon. Emboldened with natural resolve, the salmon find it within themselves to climb these ladders — one step at a time.

The fish ladder at the Hiram Chittenden Locks in Seattle, just one among many similar marvels of engineering, contains an abundance of steps. To mimic the downstream rush of swiftly flowing streams, engineers designed waterfalls that allow the water to cascade down from pool to pool. In season, crowds watch the salmon heroically leap upward, ascending from one pool to the next, against that daunting flow. Salmon will typically linger some minutes at each level, summoning the energy to spring to the next. The power of these jumps is astonishing to witness.

To facilitate public engagement, engineers of the Hiram Chittenden Locks installed a viewing window that allows visitors to look from the side into one of those larger pools. The window resembles an extra-wide movie screen. Visitors gawk at the salmon, hovering stoically against the fast-moving downstream current as they muster the energy required for their next jump.

The scene reveals something paradoxical (**Fig. 18.1**). The salmon in the figure all point upstream (rightward). Occasionally, one fish will slacken and fall back, or dart slightly forward, or perhaps even meander a bit; but, mostly the salmon remain more or less in place like a column

Figure 18.1. Salmon swimming upstream in viewing tank at Seattle's Hiram Chittenden Locks. Lines depict leftward flowing bubble tracks.

of right-facing soldiers standing at ease. The water, meanwhile, flows toward the left, past the fish. You can detect the swift downstream flow from water-borne debris and bubbles, which create visible horizontal tracks as they flow past the largely stationary fish (**Fig. 18.1**).

The fish may appear stationary to us, but the water flows continuously past them, from head to tail. Even with their notably low-friction bodies, the fish ought to get dragged backward; but they do not. Effectively, the salmon are swimming steadily upstream, against the current.

How do they manage such a feat? The prevailing explanation of how fish swim invokes two mechanisms: whole-body undulatory motions, and flapping of tailfins.[w1] While both types of motion can be observed in the viewing tank, those movements are relatively sparse, and generally feebler than one might expect for propelling the fish steadily forward against the rapidly flowing stream. Viewers with enough patience will note periods lasting up to several seconds during which a fish can be detected exhibiting no body motion at all; yet the motionless fish does not fall behind; it remains in position, effectively swimming against the current.

I am not the first to note this paradox. Over the course of his life, the legendary naturalist Viktor Schauberger (1885–1958) watched trout swim in their native Alpine habitats. His fascination exceeded mine. He noted again and again how, except for the occasional flick of their tailfins, large mountain trout could lie *indefinitely* motionless in the strongest of currents.[1,2] Schauberger's trout and Seattle's salmon pose the same conundrum: Swimming supposedly requires bodily motions, yet the fish can effectively swim against the stream with barely any detectable movements. The fish *seem* to get something for nothing.

A related enigma can be observed in stationary water. A huge saltwater tank at the Seattle Aquarium allows viewing of all types of sea creatures swimming to their heart's content. I watched how a dolphin flicked its tail once or twice, and then progressively accelerated to an impressive speed. That action seemed perfectly normal — until I thought about it more carefully.

According to Newton, acceleration comes from force. So long as force is being applied, a body can accelerate. For the dolphin, that accelerative force would appear to come from swishing its tail. If its tail were to *keep* swishing, then the dolphin could plausibly continue to gain speed, just as if you wished to accelerate while swimming in the water, you'd need to keep kicking your feet. However, the dolphin I observed managed to gain speed well *after* those initial tail flips terminated — as though the dolphin's acceleration might have relied on some other, yet unspecified force.

Plenty of videos depict dolphins' swimming patterns, presumably because of those dolphins' never-ending charm. While no videos I could find captured the aquarium scene I just described, plenty of them show natural swimming. Representative examples appear in this video[w2] where you can observe plenty of vigorous tail flapping. On the other hand, you can also see periods of steady swimming at hefty speeds through viscous water with tails remaining steady, *i.e.,* with neither any observable swishing nor any detectable body undulation. Like the stationary salmon and trout, those fast-swimming dolphins hint that something seems missing from the equation of underwater motion. If not the conventional mechanisms, then what propels those creatures through that viscous medium?

As someone who has spent much of his academic career studying muscles, I have familiarized myself with a multiplicity of propulsive mechanisms. While undulating bodies and flapping tails can certainly produce forward motion, the extent of their effectiveness has always seemed limited. My sailing experience reinforces that concern: Finding myself too often stranded without wind, I learned to employ the standard rescue procedure — stick an oar out of the boat's stern and swing it laterally to and fro in an arc-like motion, mimicking the swing of the fish's tailfin. That action certainly moved the boat forward, but at speeds that might only slightly outpace that of a crawling snail. Forward propulsion was feeble.

It would seem that engineers harbor similar doubts about efficacy, at least implicitly. To my knowledge, no marine engineer has yet proposed a swinging-paddle mechanism to drive the Queen Mary II, or any other ship for that matter (**Fig. 18.2**). If Mother Nature found such a mechanism useful for propelling fish forward at high speed, you'd think that engineers might have thought to exploit that same mechanism in ships. And yet, swinging stern paddles continue to be something of a rarity in marine engineering, as do undulating submarine bodies. Ships generally move forward by pushing water backward — raising the question of whether fish might somehow do the same.

Figure 18.2. An engineering feat reserved for the future? Or not?

In the following sections I will offer evidence that they may do exactly that. I will suggest two complementary backward-push mechanisms, along with a third variant involving forward pull — all three originating from electrical charge. One mechanism works well for local maneuvering, the other two for producing vigorous forward thrust.

It helps to begin with the obvious.

Gill Slits and Backward Flow

In attempting to deduce plausible mechanisms, we consider the most notable of piscine features, which must surely include those gently curved structures situated just behind the head on either side of the fish, commonly known as gill slits (**Fig. 18.3**). All fish have them.

Figure 18.3. Gill slits on either side of the fish, used for expelling imbibed water.

Gill slits provide outlets for wastewater. Fish take in water (and food) through their mouths. That incoming water passes through the gills, which extract oxygen. The residual wastewater then gets expelled backward through the gill slits. One could imagine the backward thrust of water pushing the fish forward in the same way as a propeller's backward thrust pushes a ship forward. However, the serpent of magnitude rears its threatening head: How could fish expel any more than just a passive trickle of water?

The answer to the magnitude question had long remained unclear and any such mechanism consequently stood as doubtful. Nevertheless, the attractiveness of such a simple source of forward thrust sustained my interest and that of my students, and, as I describe below, we could eventually confirm experimentally that a rapid, rearward expulsion did occur. This allowed us to move on and try to deduce the inner workings of that mechanism.

Our start along the path to understanding began during a casual discussion with a colleague, Rainer Stahlberg. As an East German, Rainer spent his youth studying in the Soviet Union. He recalled a visiting professor showing a film illustrating how fish feed. The fish actively sucked the food before them — together with the water in which that food was suspended.

Active sucking was the film's main feature; and, by now, such aquatic behavior has become well known. A subsidiary feature of the film was the backward jetting of water exiting from the gill slits. Such jetting should be a no-brainer: If a fish forcefully sucks the water, it must expel that water with equal rapidity, otherwise, the fish would swell and burst. In theory, such backward jetting could drive the fish forward, just as underwater propellers drive ships forward.

With the contents of that old film in mind, my students and I began our own experiments using a simple protocol. Putting a small tropical fish into a clear, narrow, cylindrical chamber filled with water, we injected a drop of dye into the cylinder just ahead of the fish's head. The fish immediately sucked the dyed water ahead, expelling the colored exhaust water rapidly from its gill slits. That result confirmed exactly what Rainer had recalled — the backward jetting flow was forceful — although for technical reasons we couldn't tell whether that jet was indeed strong enough to propel the fish rapidly forward: The fish did advance toward the food, but the small size of the experimental chamber constrained the fish from moving freely. So, speed quantification was impractical. Nevertheless, the students carrying out those experiments did notice an additional feature of interest.

Usually, the expelled water shot directly backward. But sometimes the ejection path lay slightly upward or slightly downward. When ejecting upward, the reactive force tilted the fish head downward; the opposite with downward-directed ejection. It seemed that the fish could use the backward-flow mechanism to generate not only forward thrust but also up-down tilt. And by inference, something more: Any difference of left-right thrust could easily alter the fish's lateral direction. The gill-expulsion mechanism seemed to demonstrate unexpected versatility.

Still, the question of mechanism lingered: How could the fish forcefully suck and then squirt water backward? Without some sort of internal drawing/pumping system, you'd expect nothing more than a slow trickle. Instead of opening the fish for clues, we proceeded to identify the mechanism using experiments that had nothing to do with fish, but everything to do with generation of flow.[3,4]

Water-Propulsion Surprises

As mentioned earlier in this book, I almost fell out of my chair when an undergraduate researcher came rushing into my office to tell me what he'd found. Arthur Yu had been working with tubes made of Nafion. That is one of the polymers that we'd been using to study the EZ water that forms when water makes contact with certain materials. EZ builds splendidly next to Nafion. The polymer ordinarily comes in sheets, but Arthur had been playing with the tubular variant, 1 to 2 millimeters in diameter.

Arthur's experiment involved suspending tiny particles in a water-filled bath. Into that microsphere suspension, he immersed one of his Nafion tubes, which promptly settled to the chamber floor. The microspheres allowed him to identify EZs, which exclude those microspheres. EZs formed as expected, as annuli just inside and also outside the tubular wall. That was not surprising. But Arthur noticed something additional that was completely unexpected: Water began flowing *through* the tube, just as it might flow through a straw. It flowed practically endlessly. Only after many hours did it finally cease.

By the time of Arthur's discovery, I already had suspicions about what might drive such flow. We knew that water absorbs radiant energy from the environment. We also knew that the absorbed energy could split water molecules into separated charged fragments — leaving the EZ water negatively charged, and the contiguous bulk water positively charged. I suspected that this battery-like separation of charges might somehow bear responsibility for driving the flow, although it took some time to figure out just how.

Numerous experiments on those tubular flows helped narrow the possible mechanisms.[3-5] Ruling out various trivial explanations, we first replaced Nafion with other hydrophilic (water-loving) materials. All half a dozen materials we used produced similar results — clearly, the flow was not Nafion specific. We then confirmed the anticipated role of incident radiant energy: Adding more light beyond the ambient level increased the speed by up to five times.[4,5]

We also began to understand specifics of that flow generator (**Fig. 18.4**). The annular ring of EZ water lining the inside of the tube bears negative charge (*panel a*). Complementary positive charges (protons, or hydronium ions) fill the core. Those positive charges repel one another, building a pressure that forces the core water out at one end or the other (*panel b*). That initiates the flow.

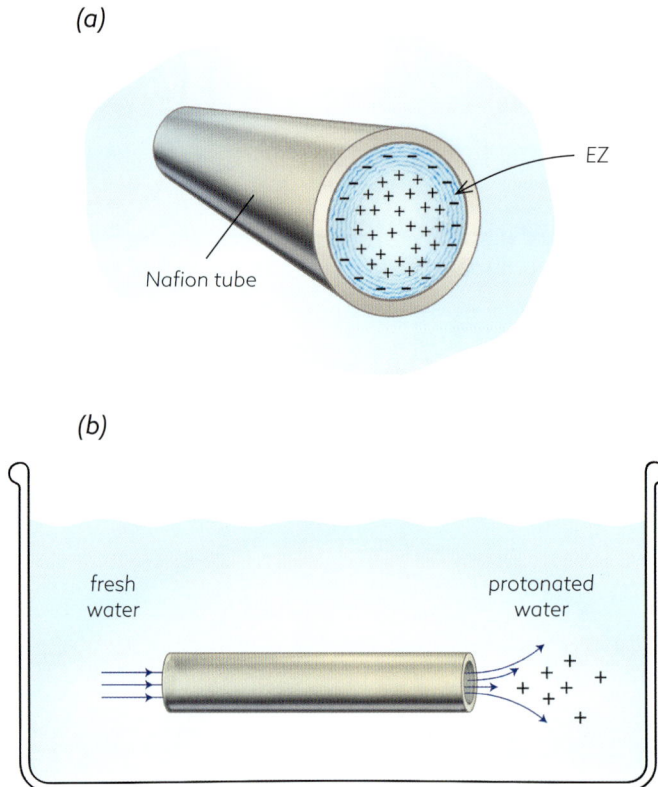

(a)

(b)

Figure 18.4. Mechanism of intra-tubular flow.
For more, see *Pollack, 2013*.[6]

Meanwhile, the EZ's innermost layers "feel" that shearing flow. Charged oppositely to the flowing hydronium ions, those inner layers attract the flowing fluid. That attraction, combined with inner-layer fragility,[7] predispose those inner layers to shearing off and following the flow. Hence, the exiting species should contain not only abundant core water but also a modicum of EZ water.

Meanwhile, that exiting mass of water leaves a void, which must be filled by fresh water sucked in from the tube's opposite end. Part of that incoming water becomes the raw material for replenishing the lost EZ fraction. That fresh EZ buildup releases protons to the core, perpetuating the flow and completing the cycle. In that way, flow continues on and on.

By adding pH-sensitive dye to the chamber water, we could confirm that protons flowed out of the tubular core and into the chamber proper. Prior to the flow's onset, the dye color indicated a near-neutral value of chamber-water pH. As the flow began, that changed. The chamber water turned progressively redder, indicating more and more protons.[4,8] Once the proton concentration rose to substantial levels, the intra-tubular flow began to subside. That made sense: The proton gradient between core water and chamber water grew smaller, so the driving force for flow out of the tube grew weaker. Eventually, the flow ceased. Stoppage always correlated with a significant rise in the chamber's proton concentration, corresponding to a pH change of up to several units.

The underlying mechanism is described in more detail in my book on water's fourth phase,[6] which includes potential real-world applications such as the internal flow of water to the tops of tall trees. Clearly, the mechanism produces flow, together with protonated water as a byproduct.

Although the proton outflow from a single tube was modest, we found that placing many tubes side by side could produce more robust flows, generating protons at an increased rate, effectively constituting an efficient proton generator. In other words, those tubes produced two products: flow and protons.

The emerging question: Could such a flow-generation mechanism operate in the gills of fish to generate backward flow? Precedent for backflow propulsion exists: Octopi and jellyfish propel themselves forward by squirting water backward.[w3] Might gills do something similar, employing the above-described flow-generating mechanism?

Flow Driver in Gills

To determine whether the aforementioned flow-propulsion mechanism might operate in gills, let us first consider gill structure (**Fig. 18.5**). Although typically curved, the simplified gill may be thought of as a flat, disc-like filter, processing all water entering through the fish's mouth. The disc "filters" that incoming water. Between the disc's front and rear faces lie numerous closely packed, parallel filaments. The spaces surrounding those filaments contain the incoming water. Hence, in the gill (and also tubule, above) all water lies near some protein surface; and since protein surfaces create EZs around them,[6] filaments and tubules stand as cousins: If protein tubules can generate water flow, then protein filaments might do so as well.

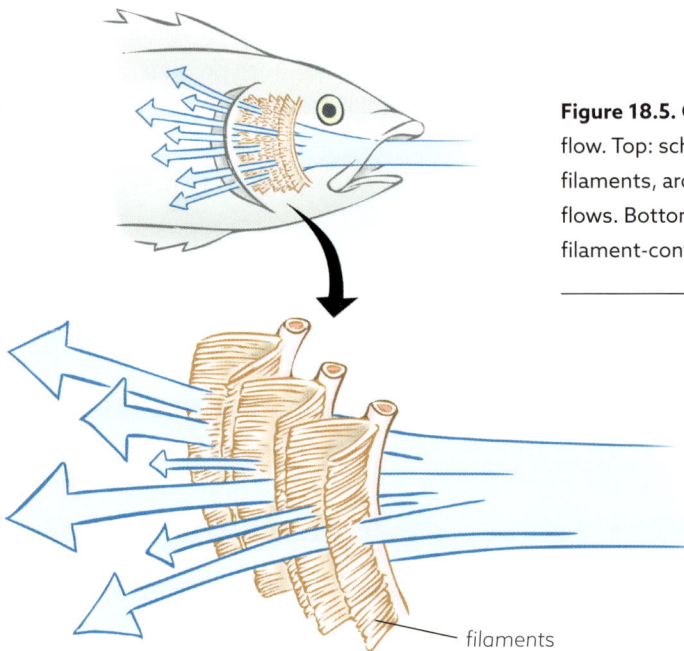

Figure 18.5. Gill structure and flow. Top: schematic view of filaments, around which water flows. Bottom: expanded view of filament-containing structures.

filaments

Evidence for filament-driven flow comes from a common biological structure: the microvillus (**Fig. 18.6**). Microvilli comprise bundles of long, parallel protein filaments, cross-linked along their length. Those bundles project from cell borders like bristles of a brush; hence, the nomenclature, "brush border." Brush-border microvilli, prominent in epithelial cells

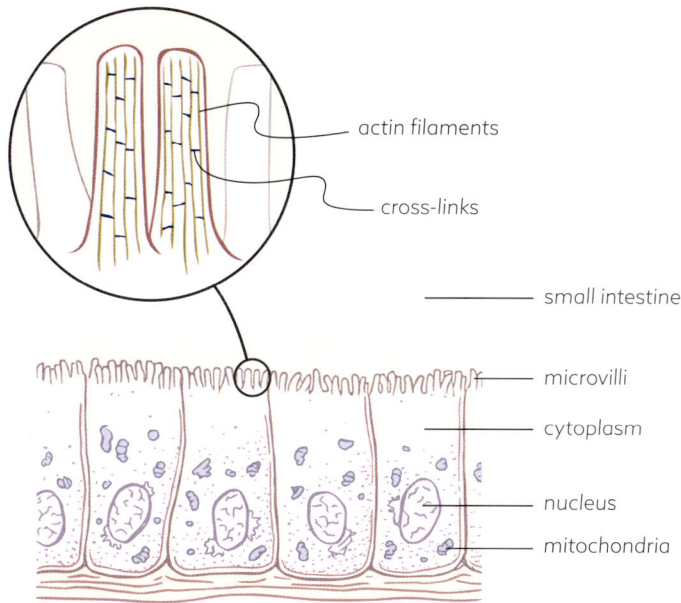

Figure 18.6. Microvilli form a brush-like border of the epithelial cells that line the small intestine. Detailed in the expanded view, the microvilli transport water from the gut into those cells.

of the gut, transport water and nutrients from the lumen of the small intestine into surrounding epithelial cells. The nutrient-containing water streams along the protein filaments. I am not the first to suggest that such streaming derives naturally from protein-filament bundles; others have argued much the same.[9,10]

Thus, precedent exists for streaming along protein filaments. To propose that the same may happen along filaments of the gill would seem like no major stretch. Indeed, the water-streaming mechanism may be rather ubiquitous, with confirmed examples (**Fig. 18.5**), along both filaments and tubules.

If so, then we can understand how propulsive rearward flow might emerge naturally from the fish's gill slits: The gill filaments could drive that flow, as it does in the filaments of microvilli. Propulsion of water may thus stand as an inevitable byproduct of gill structure, the filaments drawing water (with food) rapidly into the fish, and simultaneously expelling water out of the fish. In theory, such rapid, rear-oriented expulsion could propel the fish forward.

But there is more to the story. In the case of the Nafion-tube flow, recall that the exiting water contains many protons.[4] If the gill-flow mechanism resembles the tube-flow mechanism, then we might anticipate much the same: protons exiting from the gills. Beyond the action of the rapidly exiting water, could any such exiting charge play a role in driving fish forward?

I stress the role of exiting protons (actually, hydronium ions) because like-charges repel. Repulsion creates pressure. When the gill slits remain closed (the default situation), pressure should thus build in the tubules' exit compartments. When those slits open, that amassed pressure should thrust the water backward — at the same time pushing the fish forward. The exiting charge would augment that push, functioning in much the same way as a rocket's ion thruster. Hence, forward propulsion in the fish could occur not merely from the thrust of backward-directed water expulsion, but also from an additional force produced by the repulsion of concentrated hydronium ions pressing on the fish from behind. Whoosh! Repulsive charges at work, pushing the fish forward.

The critical question: Do fish gills actually release protons in that way? It appears so. Studies show that the water exiting the gills is more acidic than the water that enters. A paper on rainbow trout reported a pH drop on the order of 0.7 to 0.9 pH units.[11] (Note: One pH unit corresponds to a factor of 10 in proton concentration.) The suggested mechanism was the standard one involving carbon-dioxide hydration; nevertheless, the anticipated lowering of pH was confirmed. It would appear, then, that the gills eject not only water, but apparently also positively charged hydronium ions — exactly as we found in the water exiting the Nafion tubes.

For some time, I felt satisfied that we had possibly nailed the mechanism of fish propulsion. It seemed natural, evidence based, potentially effective, and in principle, similar to common ship-propulsion mechanisms. Those features lent confidence. Then, one day, somebody dashed all my hopes by reminding me that certain large sea creatures swim brilliantly without benefit of gills or gill slits. Dolphins, whales, and other aquatic mammals can even perform acrobatic feats; yet they boast no gills. If forward push from gill-slit efflux were a critical feature of swimming, then how might mammalian sea creatures, lacking those slits, propel themselves forward?

Bummer!

Panicked into fast thinking, I immediately remembered blowholes. Whales and dolphins empty their lungs through dorsal blowholes (**Fig. 18.7**). The blowholes function much like nostrils, discharging a mix of air, water vapor and carbon dioxide, the latter two species combining to create hydronium ions. Thus, water and hydronium ions blast out of the blowholes, sometimes at fire-hose speeds that can exceed 100 miles per hour. I quickly thought: Could those blowholes possibly function as proposed for gills, driving water and charge forcefully backward, and therefore, the creature responsively forward? Relief in sight? Could blowhole action rescue me from the horns of the mammalian dilemma?

Figure 18.7. Whale exhaling through blowhole.

Any such potential relief turned out to be short-lived. For the above-mentioned mechanism to work, blowholes would need to discharge not only into the air as commonly seen, but also underwater. And, for effectiveness, that underwater discharge would need to be directed at least slightly backward. Since I could find no evidence for either one, even after contacting some of the field's experts, I found myself inclined to dismiss that mechanism as an option. Score one loss.

On the other hand, a promising alternative solution for mammalian swimming did arise, and I hope you will oblige me by waiting until later in the chapter, where I consider that option in a more suitable context. The mechanism involves charge forces, acting in a straightforward way.

Meanwhile, we continue to explore the propulsion of fish. The gill mechanism stands as a natural candidate for driving the fish forward. But, is it the only candidate?

Beyond Gill Slits: Complementary Propulsion Mechanisms?

In searching for potential mechanisms, we continue along the original route — identifying physical features shared commonly among fish and identifying how those features might hold functional significance. Beyond the ubiquitous gill slits, what additional attributes do fish share in common?

Slipperiness is certainly one of them. All fish are slippery. Slipperiness obviously diminishes frictional drag, but I'll argue in a moment that the gel slime responsible for slipperiness could also work to propel fish forward.

Other widely shared features include body texture and shape. Small- to medium-sized fish are covered in scales, whose idiosyncratic profiles could matter for function. So could body taper. Among medium and larger fish that swim rapidly, bodies taper gently from just behind the head all the way to the tail. Think of sharks and tuna.

Because of their ubiquity, all three of those features — slimy surface gels, scales, and body taper — constitute potential candidates for function. Why else would they manifest as ubiquitously as they do? Could one, or some combination of those features, play a role in generating forward thrust?

We consider those features next. In each case, we will explore whether the particular attribute leads to a forward-directed force. We begin by considering those low-friction surfaces.

Slime and Slipperiness

Grabbing a live fish can bring utter frustration — the slimy fish flails maniacally, slithers forward, and quickly escapes your grasp. It's practically unavoidable. The fish's mucus coating evidently reduces friction, allowing the creature to slip readily through your hands. In a similar way, that mucus coating ought to help fish slip readily through the water. That seems clear.

Beyond slipperiness, however, one wonders whether those gel coatings could help produce the forward thrust that drives the fish out of your clutches. While no such mechanism seems immediately obvious, we'll see in a moment a plausible mechanism that could indeed help propel the fish forward, whether from your hands or within the water.

Slimy coats are a relatively ancient natural feature. As such, they imply a likely functional significance. A good example of that ancient sliminess is seen in the hagfish, *Myxine* — almost unchanged from species living 300 million years ago and therefore considered a living fossil. *Myxine* releases mucus from some 100 glands or invaginations running along its flanks. The abundance of this released mucus can clog a predator's gills, causing suffocation.

Myxine's mucus coat is, in fact, so slippery that the long, eel-like hagfish can easily tie itself into a knot. Knot-tying behavior constitutes an escape tactic: predator bites hagfish, hagfish releases copious mucus, hagfish sloughs off the mucus by tying itself in a knot that works its way from head to tail, scraping off the slime as it does, and presumably leaving the predator holding only the mucus, not the hagfish. Handy tactic!

All fish have slimy mucus coatings, just as all birds have feathers. The mucus comes from glandular cells dispersed throughout the epidermis. Those glands produce a glycoprotein called mucin, which gels as it mixes with water to produce the slimy coat. The ubiquity of the slime coat, from primitive fish upward, implies functional significance, one manifestation of which is the obvious slipperiness.

To probe further, we ask: Why are gels slippery?

In our laboratory, we have studied many "hydrogels" — so called because they contain mostly water. The water contained within those gels consists mainly of fourth-phase (EZ) water. Not only does EZ water invest the gel itself, but also it builds outward from the gel-surface, constituting an extended zone adjacent to the gel.[6] That projecting EZ commonly bears negative charge: As water molecules split, their negative fractions build the EZ, while complementary positive protons accumulate as hydronium ions in the water beyond. Numerous experiments confirm this charge-splitting process as well as the associated proton (hydronium-ion) buildup.[6] As any gel builds, the nearby water's proton concentration should build at the same time.

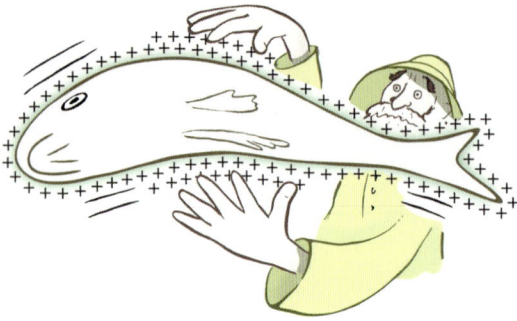

Figure 18.8. Gels lining the fish body release protons as they build. Repulsion among protons pushes away surrounding bodies, ensuring low friction. Thus, fish can easily slither through your hands.

Those released protons could be critical for explaining the fish's slipperiness (**Fig. 18.8**). Imagine many protons (beyond the modest number that would create the like-likes-like attraction) sandwiched between the fish and your hand. Protons repel one another, ensuring that your hand and the fish surface don't connect. That distancing eliminates, or at least greatly reduces, shear friction. A good analogy is sandpaper: When you rub two pieces together, you experience enormous friction; but if you keep those surfaces even slightly separated, the friction practically vanishes. The surface-generated protons act similarly: By repelling one another, those protons keep any gel-coated object from truly being touched. That, I surmise, is why fish exhibit low friction: It's the repelling protons.

Released in copious amounts as the gel's EZ builds to replace sloughed-off EZ, those protons do the job. Proton-proton repulsion keeps the respective surfaces from rubbing against one another, ensuring abundant slipperiness.

Slime-Based Propulsion?

Now, let us go a step further: Could those released protons also produce forward thrust? And still further, might that forward thrust benefit from fish scales and/or body taper?

Protons could create forward thrust in several ways. In the hand-grabbing scenario, the protons should keep your hand separated from the fish. Squeezing harder merely concentrates the protons, thereby increasing the repulsive pressure. That pressure pushes hard in all directions, forcing the fish rapidly forward or backward and out of your grip. By pushing on each scale's exposed rear edge, however, the protons' principal thrust would be forward. The fish would slip ahead, as it most often does.

This mildly embarrassing scenario provides a clue as to what might occur in the water. Fish expel mucin molecules from sub-surface glands onto their outer surface. Exposed to the aqueous environment, those released mucin molecules combine with nearby water molecules to create surface mucus, the slimy stuff. To achieve mucus formation, those water molecules must split: The negatively charged OH^- fraction builds the EZ of the mucus gel, while the complementary H^+ charges eject into the region beyond the gel and its associated EZ. Some of those positive protons (now hydronium ions as they combine with water molecules) may cling electrostatically to the negatively charged mucus gel on the fish's surface; but the more-distant hydronium ions, lying beyond the grip of the gel's negativity, should primarily repel one another, acting to increase their separation.

How could those latter hydronium ions help create propulsion? Begin by recognizing that repulsive forces push in all directions (**Fig. 18.9**). They push the more distant hydronium ions even farther from the gel surface, while at the same time pushing the closer hydronium ions toward the fish's gel surface. That latter force counts: It pushes against the fish. If the fish tapers toward the rear, then a component of that pushing force should act to drive the fish forward, in the same way that your grasping hands may drive the fish forward.

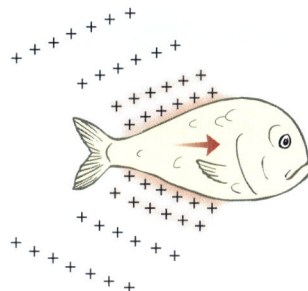

Figure 18.9. Hydronium ion repulsive forces push on fish.

That pushing force might seem rather weak at first blush, as the underlying hydronium ions should be free to disperse. However, any such dispersal should be slowed by water's notable viscosity. Further, hydronium ions should continuously replenish: As the gel continues to build, it should continue to release protons. Hence, the hydronium ion concentration near the fish ought to remain high enough to facilitate a forward-directed force of reasonable magnitude.

Forward motion arising from gel interactions has been experimentally demonstrated in other contexts. **Figure 18.10** shows the setup: A gel is enclosed within a boat-like shell made of aluminum-foil, with an opening at the rear. Solvent enters through that rear transom, reacts with the gel, and causes the boat to advance at speeds up to 80 mm per second.[12]

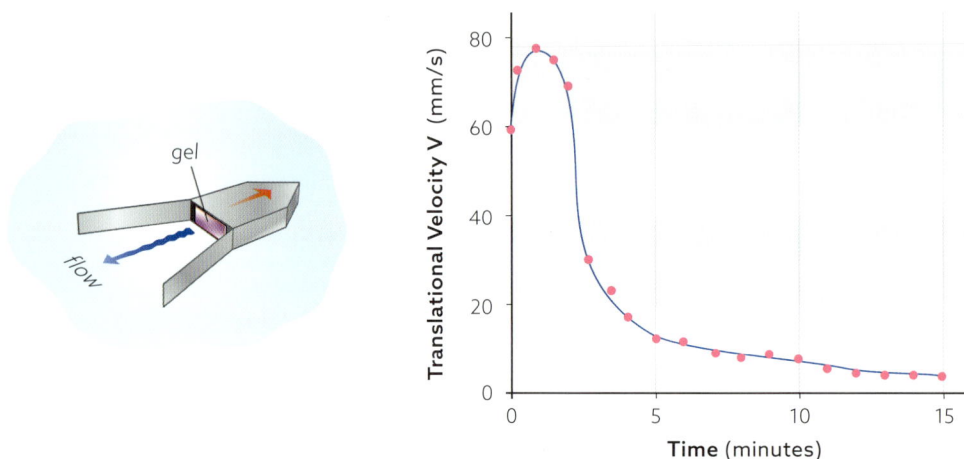

Figure 18.10. Gel boat moves forward as solvent interacts with the enclosed gel, replacing the original solvent used to create the gel. In reacting to the push backward, the boat moves forward. After Mitsumata *et al.*, 2000.[12]

We confirmed similar behavior in our own laboratory using a slightly revised boat model, filled with chia seed. Chia hydrates to create a sticky gel containing abundant EZ water, releasing protons as it does. The chia boat moved resolutely from one side of the tank to the other. With pH-sensitive dye, we could detect hydronium ions getting ejected from the rear, presumably bearing responsibility for driving the craft forward. Thus, gel hydration can demonstrably produce forward motion — as suggested for the fish.

In the case of the fish, it's the mucin-water interaction that creates the protons. Released protons should push against the angled body, driving the fish forward. Because charge forces can be unexpectedly high

(Chapter 1), any fish exploiting this mechanism could theoretically achieve substantial speeds.

Thus, a natural consequence of the gel-hydration mechanism is an accelerative force. In the diagram of **Figure 18.9**, the forward-force component may seem modest because most of the force pushes laterally, not forward. Once the fish begins to pick up speed, however, the freshly released protons will fall farther behind. The trailing position of those released protons should alter the direction of the pushing force to a more axial one, increasing the forward component and accelerating the fish. With that mechanism in play, we can understand how a fish could quickly gain momentum enough to dash through the water.

This forward-drive mechanism also helps clarify the role of scales (**Fig. 18.11**). Scales have a texture like shingles on a roof — one piece overlapping another. In fish, the exposed edges always face the rear. Hence, any hydronium ions exerting forward-directed force ought to push squarely on those exposed rear edges, enhancing the effectiveness of the push (**Fig. 18.12**).

scales

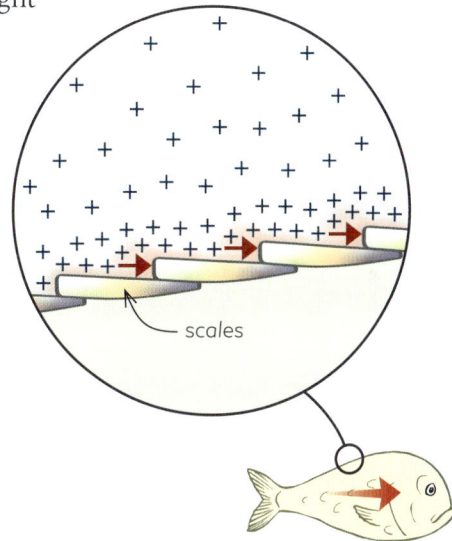

Figure 18.11. Overlapping fish scales. Thick edges always point backward.[w4]

Figure 18.12. Water flow pushes effectively on rear face of each scale, driving fish forward.

The gel-based propulsion mechanism outlined above has the advantage of adaptability. Imagine a fish moving calmly at low speed. Suddenly, a predator appears. To escape, the fish immediately flicks its tail to get going, simultaneously secreting gel-producing mucin from all over its flanks. As the mucin hydrates — explosively, according to evidence[13] — massive numbers of hydronium ions begin appearing within milliseconds, building from the body outward. The fish quickly gains speed. Meanwhile, the hydronium-ion trail moves rearward, amplifying the effectiveness of the propulsive force to the point that the fish may even leap from the water.

The mechanism may also explain why frightened trout consistently surge upstream, a paradox noted by the aforementioned Viktor Schauberger. A trout will often hover practically motionless, facing upstream. When scared, the fish should logically escape in the direction they can travel fastest, downstream. However, the trout invariably darts upstream, surging ahead with impressive speed. The proton-propulsion mechanism might help explain why: Because of the swift current, the abundantly released hydronium ions should flow almost directly downstream. While nearby ions will attract the fish, the rest of those many hydronium ions should create a repulsion-based upstream force, pushing the trout ahead. With that mechanism in place, the upstream escape is anything but illogical.

It's worth mentioning that the gel-based propulsion mechanism just described bears functional similarity to the gill-based propulsion mechanism. Both systems expel protons/hydronium ions to the rear, thereby propelling the fish forward. Those mechanisms may work together to produce the brisk forward motion that is characteristic of most fish.

We are not quite done with propulsive forces. Another proton-based variant exists, similar in principle to the first two. But first, let us detour briefly to consider a notable feature of fish behavior: schooling.

Schooling

Fish often swim in groups (**Fig. 18.13**). Like birds flocking, fish schooling is commonly thought of as a social phenomenon — *i.e.*, the fish appear to enjoy one another's company. The collective structure is thought to

Figure 18.13. A school of Bigeye Snappers swimming around the coral reef, South Andaman Sea, off the coast of Thailand.

confer existential advantage: It could seem outsized enough to scare off predators. Both interpretations ring true. But another explanation may also carry some gravity: The charges surrounding each fish, if modest enough, could create a like-likes-like attraction, drawing the fish together whether they prefer it or not (**Fig. 18.14**).

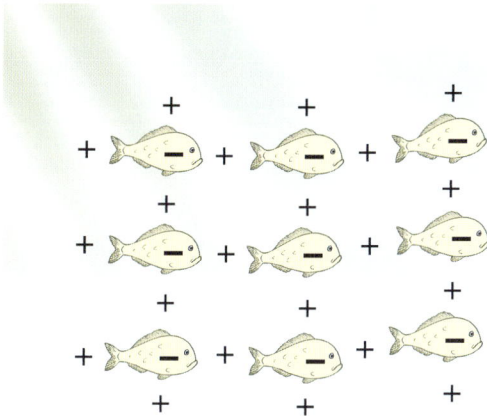

Figure 18.14. Possible mechanism of schooling. Positive charges lying between negatively charged fish may create attractive forces by the like-likes-like mechanism.

To understand how this attraction could occur, consider an individual fish's charge. All cells bear net negative charge (Chapter 16), and since animals are comprised mainly of cells, the animal, as a whole, should theoretically also bear net negative charge. This appears confirmed: In our laboratory, we have measured net negativity of multiple plants and humans — even our own bodies bear a net negativity — offering preliminary confirmation of the presence of this feature in diverse species, which would logically include fish.

Fish appear to concentrate that negativity in a front-heavy manner — measurements show the strongest negativity on the head and the gills.[14] Such charge distribution should hardly be surprising: The vast arrays of hydrophilic filaments running through the gills imply a substantial EZ water presence and hence plenty of negativity in that anterior region.

Thus, the fish can be envisioned as a kind of dipole: negative charge concentrated up front, and positive charge emitted behind. That dipole would create a strong electric field. The downside of that field is the enabling of predators to detect the fish through use of electro-receptive organs, a well-studied example being the shark's *Ampullae de Lorenzini*, an exquisitely sensitive field-detection organ that makes sharks particularly adept at tracking prey through electrical sensing. Divers might want to leave their electronics in their boats!

A consequence of that dipolar charge may be schooling (**Fig. 18.14**). The trailing positivity could attract the negative head of the fish to the rear, keeping those fishes in line. Charge forces could also act side to side, with positive hydronium ions potentially attracting the neighboring negatively charged fish and creating the three-dimensional ensemble.

The ensemble may confuse predators by presenting one large electric field instead of many small ones. Any such advantage, however, would stand as an incidental byproduct of the charge distribution around fish. The survival aspect may be instinctive, but that instinct could well be rooted in a charge-based physical attraction.

Pulling Forward

The schooling discussion hints at a third source of fish propulsion: pulling. The positive charges trailing from a swimming fish ought to attract the negative head of a fish just behind. One fish's "exhaust" becomes another fish's "fuel." That pull could be substantial: The positive charge ahead would induce still more negativity on the head of the trailing fish, enhancing the electrostatic pulling force.

Such a pulling force could occur even in the absence of a fish ahead (**Fig. 18.15**). Any negative-charge concentration in the fish's front region

Figure 18.15. Negative charge in the fish's head region induces positive charge ahead, which pulls the fish forward.

should induce positive charge in the water just ahead, drawing the fish forward. Ordinarily, that pulling force should be modest; otherwise, fish could never stand virtually idle. However, when a fish begins moving ahead, its front region should acquire additional negative charge as water flows rapidly through the gills and builds increased EZ negativity there. That higher front-based negative charge should induce additional positive charge in the water ahead, thereby amplifying the pulling force. By this inductive mechanism (similar to the mechanism drawing hurricane clouds forward; Chapter 10), swimming fish could naturally accelerate to higher speeds.

This latter mechanism could work effectively in mammalian species such as the dolphins and whales that we touched on earlier. Those deep-diving species contain unusually high concentrations of the oxygen-binding protein, myoglobin,[w5] which carry high negative charge.[w6] Since myoglobin-containing organs such as heart and lung lie in the frontal area of those species, it can be reasonably supposed that the front would bear that negative charge, as in fish. If so, then the inductive mechanism illustrated in **Figure 18.15** could operate not only in fish but also in mammalian swimmers, perhaps even more effectively. Those swimmers could be pulled by the induced positive charge ahead, darting forward at high speed and possibly explaining those species' heroic feats.

This pulling mechanism may seem fundamentally different from the pushing mechanisms we've considered so far; however, they share a common feature: electrical charge. In the pulling mechanism, positive charges pull from ahead, while in the pushing mechanisms, positive charges, repelling one another, push from behind. Thus, for aquatic propulsion, charge forces are put forth as central protagonists. I don't suggest that the more traditional mechanisms are necessarily ruled out but do suggest charge-based mechanisms may stand at least as adjuncts to those more familiar mechanisms.

Discussing the mechanisms underlying fish propulsion returns us to the concept of reverse engineering: exploiting successful engineering designs to deduce nature's inner workings. To wit: Ships propel themselves by pushing water backward; hence, we might expect fish propulsion to operate similarly. The gill mechanism certainly pushes water backward; and so does the gel-derived proton mechanism, wherein backward-driven hydronium ions create forward push. Hence, pushing water backward stands as a common mechanistic feature characterizing both fish propulsion and ship propulsion — a satisfying philosophical correspondence.

Indeed, the features of fish propulsion suggested here, if proven valid, could conceivably hold lessons for optimizing ship propulsion. Designers of bulky ships might have something to learn from the great speed and maneuverability that aquatic life demonstrates. One day perhaps, ships may move with the agility of sea creatures.

Summary

Based on plainly seen motions, scientists and naturalists have surmised that fish swim mainly by flapping their fins and/or undulating their bodies. Anyone can witness those movements. On the other hand, good reasons exist for questioning the adequacy of those two mechanisms. Not the least of concerns is the observation that fish can sometimes swim without exhibiting any such motions at all. In those circumstances at least, some other force would seem necessary for propelling the fish forward.

This chapter addressed the evidence that points to charge-based forces — please recall their strength — operating in several complementary modes.

The first mode is the jet-like backflow emerging from the fish's gill slits. From water taken in through the mouth, gills should facilitate the buildup of negatively charged EZ water around each constituent filament. Complementary hydronium ions should then build beyond those interfacial zones. Repelling one another, those hydronium ions ought to create pressure, ultimately driving the water out of the gill slits. Brisk rearward expulsion should push the fish forward, much like the action of a motorboat.

The second propulsive mode originates from the fish's mucus-gel coating. As the gel precursor hydrates to form the gel, free hydronium ions are released. Those positively charged ions repel one another, pushing away from the fish while simultaneously pushing the fish away. Because of the fish's body taper, a component of that pushing force should be forward directed. The faster the fish moves, the more the hydronium-ion trail falls behind and the larger that forward-directed component becomes. Scale architecture also facilitates: With their flat rear surfaces, scales help hydronium-ion forces push effectively in the forward direction. That augmentation feature may explain why all fish have scales.

The third propulsive mode stems from the greater negativity of the fish's head region. That negativity should induce positive charge in the water ahead, which, in turn, should draw the negatively charged fish forward. The negative head region could also help explain the phenomenon of schooling: With the head region of a rear fish attracted to the ejected protons of the forward fish, members of the ensemble could keep together.

All three propulsive modes operate on the basis of separated charges: opposite charges pulling from ahead, and like-charges pushing from behind. Having all three of these mechanisms available to choose from should lend versatility: The gill-slit mechanism ought to work well for maneuverability and short spurts, while the two other mechanisms could work well for steady high-speed swimming over the longer haul. Together, they lend a measure of versatility to aquatic locomotion that resembles the versatility of avian flight mechanisms. Nature's majesty at work.

In the next and final chapter, I review the five foundational principles that run pervasively through this book. They weave through the proposed fabric of understanding. I then conclude with some philosophical principles that have evolved throughout these pages, and that characterize the scientific pursuit more generally.

If everyone is thinking alike, then no one is thinking.

— Benjamin Franklin

The formulation of a problem is often more essential than its solution, which may be merely a matter of mathematical or experimental skill. To raise new questions, new possibilities, to regard old problems from a new angle requires creative imagination and marks real advances in science.

— Albert Einstein

SECTION VII

Summing Up: Unlocking Nature's Mysteries

The previous 18 chapters contain bold proposals on the workings of nature, proposals that build principally on the unsuspected role of electrical charge forces. Those forces can rise to enormous magnitudes. They can accomplish massive amounts of work, often attributed to other causes.

In arriving at those proposals, we stumbled upon natural processes well recognized but not widely applied. Hence, we classify those processes as "secret" rules of Nature. Arguably, those secret rules likely have broad application, and for that reason it seems useful to highlight them.

We then move on to consider the current process of doing science. Scientific knowledge has certainly increased over recent years, but the kinds of scientific revolutions that bring major advances have been few and far between. We consider why. I suggest some inadvertent structural issues that get in the way of major scientific advances, ending the book, finally, with a hopeful outlook for the future.

CHAPTER 19

The Secret Rules of Nature

Readers who have made their way to this final chapter have surely earned themselves some wrap-up, and I will gladly oblige. I will first point to several fundamentals — five scientific principles that have emerged as seminal for an understanding of the natural world around us. Even scientists who know and work with some of these key ideas may not have recognized their full explanatory potential.

I will then shift gears, turning to more general issues confronting today's science. Deep, systemic problems afflict the way we do science today. Those problems pose unintended consequences, working against the emergence of the kinds of world-shattering breakthroughs that had been common in earlier eras. Indeed, the very design of today's scientific enterprise, I argue, discourages revolutionary science.

Finally, I will propose a more fruitful path forward — a way in which the scientific enterprise can fulfill its potential to produce the kinds of thrilling advances that can change the world.

Five Foundational Principles Deserving Attention

In reflecting back on the contents of this book, I have identified five scientific principles that have appeared again and again. All of them involve electrical charge. Allow me to elaborate — I believe those principles may be critical for an understanding of the world around us.

(1) In nature, separated charges exist everywhere. To create charge-based forces, positive and negative charges must remain separated. It appears that they can. Thanks largely to the sun's relentless production of energy, we can find separated charges all over the face of the earth as well as in its atmosphere, and likely beyond.

Consider the earth's charge. Although its presence has largely escaped widespread awareness, the earth's net negative charge has been recognized for over a century. We also know that negatively charged objects repel. Hence, any negatively charged object situated above the earth should experience an upward-directed force, resulting potentially in some kind of levitation. While that seems clear, the principle has only rarely been applied. Few researchers have considered the possibility that charge-based repulsion might account, even in part, for the lift experienced by birds, planes, and even clouds.

The same applies to the earth's positively charged counterparts. Freed from the earth's surface by the energy of the sun, positive charges fill the atmosphere. Again, well-recognized but only sketchily considered, those atmospheric charges could underlie myriad natural features. Examples: (a) Positively charged atmospheric molecules should attract to the negative earth, keeping our atmosphere in place; (b) By condensing negatively charged water droplets into clouds, those positively charged atmospheric molecules could create weather; (c) Since sunlight only strikes one face of the earth at a time, the rising, solar energy-driven positive charges ought to create lateral charge gradients, and those gradients, in turn, could propel the prevailing winds, and perhaps even drive the earth's continued rotation.

Thus, positive and negative charges may underlie many natural phenomena. Their separation creates a battery-like potential, which may drive physical processes for which no genuinely satisfying explanations currently exist (**Fig. 19.1**).

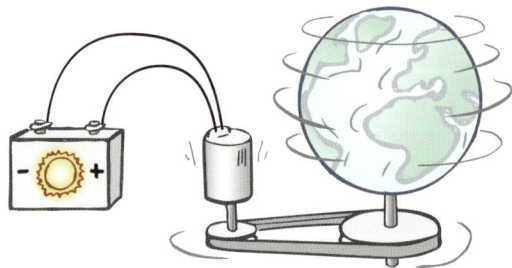

Figure 19.1. Spinning the earth requires energy of some kind.

(2) Movement requires force. According to Newton, producing motion requires force. We also know that it requires expenditure of energy. Increasingly, physicists prefer to focus on the latter rather than the former. Thus, you may hear, "What *energy* creates the observed motion?" While theoretically sound, focusing on energy sidesteps the burden of having to specify the nature of the underlying force. Sometimes those forces can be difficult to pin down, often exposing a deeper problem: an unsound theoretical framework. In such a circumstance, identifying the driving force may become impossible — in which case it may make sense to begin again, employing a completely different theoretical framework.

Let me illustrate the problem with the centuries-old phenomenon of "orbital epicycles."

Long ago, everyone believed that the center of the solar system was the earth, not the sun. That framework seemed perfectly natural: Anyone could observe the sun traversing the sky, so the earth just *had* to be its orbital center. Logically, everything revolved around that center point, including the planets.

But something happened to dispel that notion.

As the increasing power of mathematics began to inform the field of astronomy, it became possible to compute the orbits of those planets, supposedly revolving around the earth. As you can imagine, those orbits were anything but simple. They consisted not only of cycles, but also epicycles superimposed on those cycles (**Fig. 19.2**). Postulating some kind of obscure celestial energy for driving those intricate orbital motions might offer provisional satisfaction for some. But, imagine the more concrete challenge of identifying the set of forces responsible for creating those mind-bogglingly complex orbits. What genius would try?

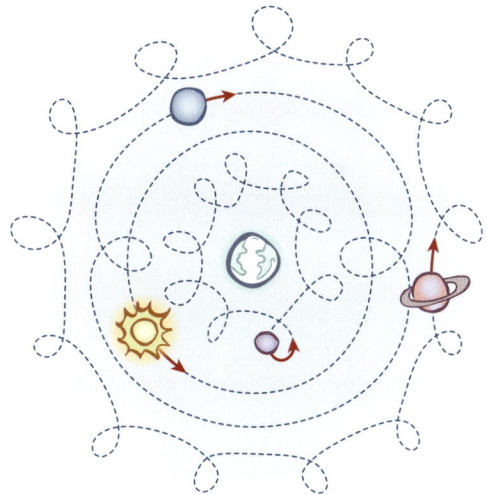

Figure 19.2. For an Earth-centered solar system, planetary orbits would exhibit complex epicycles upon cycles.

Given such a daunting challenge, it didn't take long for the paradigm of an earth-centered solar system to get cast into the dustbin of misbegotten notions. Surely no set of forces could be that complicated. Something had to be wrong — dead wrong.

By insisting that all theoretical developments must specify the relevant forces, we affirm the value of cause-and-effect simplicity. Movement requires force. The nature of the force needs to be specified to determine whether it satisfies the simplicity condition. This book has endeavored to follow along that pathway by identifying attractions and repulsions as pervasive agents of force production. We insisted on that crucial step — not only specifying the nature of the underlying forces, but also tacitly demanding that those forces be simple and direct.

I do appreciate that my quest for simplicity may expose me to criticism. I like to think that simplicity implies elegance, and elegance, in turn, may imply scientific truth. That logic may prove valid sometimes — but not necessarily always. When we get down to the nitty gritty, it's evidence that counts, not philosophy. Thus, while I may defend the simplicity principle, I welcome challenges to that philosophy and to the mechanisms deduced from following that philosophy, for rigorous examination of proffered arguments can only serve to refine our understanding. And that, after all, is our ultimate goal.

Nevertheless, it's my philosophical bent toward cause-and-effect simplicity that drives me to suggest that all forces must be specified. Motion requires force. If we are to succeed in identifying nature's mechanisms, all of which involve one kind of motion or another, then I argue that a necessary step is to identify — or at least *try* to identify — the underlying force. Paradoxically, any failure to identify such force could prove instructive, for it may redirect us to look elsewhere for mechanistic understanding.

(3) Faraday induction. Positioning an isolated charge near an object induces opposite charge on that object — or more specifically on the object's nearest face. The opposing charges then create an attraction. This scenario defines the concept of inductive force, a phenomenon underlying several of this book's proposed mechanisms. A possible example lies in gravitation. The negative charge of the earth should induce positive charge

on the lower surface of any object above the earth's surface (**Fig. 19.3**). That creates an attractive force, which pulls said object toward the earth, creating a gravitational attraction.

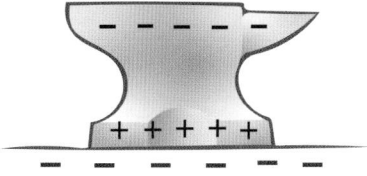

Figure 19.3. The earth's negative charge draws positive charge of any above-surface object downward. Meanwhile, negative charges get pushed upward.

This principle applies similarly for weather. When clouds lodge at sufficiently low altitude, their negatively charged droplets should induce earth positivity immediately beneath. If sufficiently strong, that positivity ought to pull the clouds' negatively charged water droplets downward. The result is familiar: rain.

Physicists understand induction, perhaps because it's simple and intuitively sensible. On the other hand, that principle could play a more critical role in natural phenomena than we may currently appreciate.

(4) Triboelectricity. When one substance rubs another, surface charges transfer; one substance becomes positive, the other negative. Known as "triboelectricity," the principle remains rather obscure among most lay people — I only learned of its subtleties several years ago when my friend Hiro Ishiwatari, a scientist from Japan, informed me of the potential significance of the phenomenon. Hiro pointed to a salient feature that's apparent from the position held by air in the so-called triboelectric series: air flowing past *any* material leaves that material with excess negative charge. The material apparently captures some negative charges from the flowing air, conferring negativity on the material and leaving the air with net positive charge.

Whether the air flows past the object or the object moves through the air doesn't matter. The principle remains the same: The object acquires negative charge. Relevant objects might range from birds and planes all the way to sails pressing against the wind. Any acquired negativity could potentially account for the objects' counter-intuitive dynamics, including both forward motion and lift (**Fig. 19.4**). Even beyond those two features, triboelectricity might matter more than generally presumed (if presumed at all), for charges may transfer whenever something moves. Dealing with those transfers and their potentially widespread impacts would seem obligatory if we are to understand how the world works.

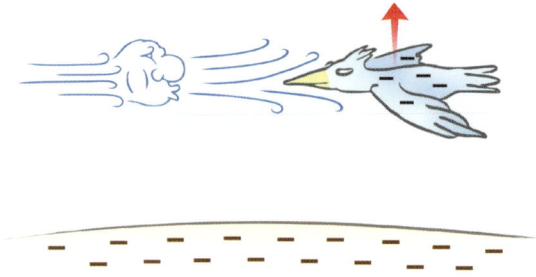

Figure 19.4. Wind passing over any object induces negative charge on that object. That negative charge repels the earth's negative charge, creating a levitational force.

(5) Like-likes-like. Like-charged objects ordinarily repel. However, if a small number of opposite charges manage to sneak in between those like-charged objects, then the objects will draw together — the objects appearing to "like" each other. Thus, Feynman seemed justified in titling this attraction "like-likes-like," adding the explanatory suffix, "because of an intermediate of unlikes." It's the unlike charges lying in between that create the apparent attraction (**Fig. 19.5**).

Figure 19.5. While like-charges repel, placing an unlike charge in between them draws those negative charges together.

Thus creates the hazard: By reflexively presuming that like-charged objects must always repel, we can be seriously misled, for oppositely charged substances lying in between can easily pull those objects together. The attraction is counter-intuitive but confirmed. Throughout this book, I have employed the principle to account for multiple phenomena, ranging from the moon's attraction to the earth to atmospheric droplet condensation into clouds. Though underappreciated (except perhaps by Feynman devotees), the like-like-likes phenomenon may prove critical for a full understanding of nature.

Does Electrical Charge Really Matter?

In one way or another, the five principles outlined above all involve electrical charge. They are essentially simple; anyone can understand them. They remain largely unchallenged, though arguably less appreciated than warranted. Yet, notwithstanding their relative obscurity, those charge-centered concepts may occupy a central position in the arena of natural function.

I opened this book by presenting some enigmas surrounding charge — namely the colossal amounts of force that can be developed by even small amounts of electrical charge. I've posed those riddles time and again to many scientists and nonscientists alike. Only a handful ever came close to guessing the staggering force magnitudes involved, magnitudes whose revelation left them consistently flabbergasted.

In our collective failure to appreciate the strength of those charge forces, we may have overlooked a potentially powerful driver of natural phenomena. Such disregard has not been without consequence: Scientists have naturally sought alternative explanations, some of which have spawned frameworks of ever-growing complexity. Indeed, the proliferation of densely intricate "explanations" and their apparent distance from the tangible natural world may partly explain why many young people have become fearful of physics. Physics today seems increasingly remote from common experience, almost fully disconnected from traditional cause-and-effect reasoning. It's become insanely complex (**Fig. 19.6**).

Figure 19.6. Building science on an unsound foundation results in mechanisms that are both overly complex and likely wrong.

Veering off from the trend toward complexity, I have endeavored to demonstrate something straightforward: how electrical charge could matter. While no incontrovertible proof has been offered for any of my explanations, I do suggest that those explanations amount to simple, ultimately testable hypotheses. Indeed, electrical charge may be the key required to unlock many stubbornly resistant doors, and I can only hope others will be tempted to explore whether those doors can be pried open further, or whether perhaps they may need to be pushed permanently shut.

As you might surmise, I would not have written this book if I had thought that charges didn't matter. My take is that they may matter a lot more than we might think.

Going a Step Further: Way Out on a Limb

In my relentless passion for simplicity, I will hazard a guess that will surely put me even farther out on a limb. I will suggest that charge forces may be universal — *i.e.,* that conceivably, *a single type of force may uniquely account for numerous phenomena throughout nature*. I have already invoked electrical charge as the agent responsible for explaining a substantial number of phenomena, but obviously not all. The ability to reasonably interpret those phenomena in terms of simple attractions and repulsions leads me to wonder whether such forces might turn out to be — gasp! — universal, *i.e.,* whether charge forces may serve as the *chief* mediator of the many diverse phenomena occurring throughout nature.

Initially, I wanted to say *sole* mediator, but hesitated. I wavered because I've come to appreciate that there may be more to nature than the conventional forces we commonly recognize. I refer here to so-called "subtle forces." I will venture a few words about those subtle forces and energies in a moment. Notwithstanding, I do suggest that charge forces constitute a pervasive force operating throughout more of nature than we may think.

I can already sense the negative response coming from some of my academic colleagues — from

physicists dealing with the putative forces operating at the sub-atomic level, all the way to researchers dealing with more spiritually oriented phenomena. Both will likely demur. The former group may view as arrogant my seeming reluctance to assign deepest significance to the many sub-atomic forces they regard as fundamental and may even go so far as to use such hesitancy to tar the entire book as nonsense. Any such blanket dismissal may help preserve their sense of intellectual security, but I hope those scientists will give my claims serious consideration first.

As for the latter group, many serious scientists are coming to recognize that certain spiritual phenomena involve energies different from anything we currently recognize. Perhaps they do. Evidence implies they might, but in this book, I stop short of considering those so-called subtle energies, restricting my consideration to those energies and forces we currently recognize as conventional.

Regarding the possible existence of those subtle energies, I feel obliged to invoke the oft-cited aphorism: "We don't know what we don't know." Such profound uncertainty makes it difficult to lightly dismiss the possibility of forces beyond those we recognize today. On the other hand, my inclination toward simplicity leads me to first seek explanations that build from widely recognized forces, leaving the immediate pursuit of the more esoteric forces for exploration by others — and perhaps soon by my own scientific group as well.

Meanwhile, permit me to go way out on the limb by suggesting that among known conventional forces, simple attractions and repulsions may well dominate scientific understanding. This suggestion should not completely surprise, for the charges in question constitute the very essence of nature's most fundamental building blocks. Here, I suggest that those charges may hold significance beyond their presence inside the atom. I suggest that they can make things happen throughout nature.

Why Has This Seemingly Foundational Paradigm Remained Secret?

If electrical charge matters as much as suggested, then why have others failed to generate paradigms similar to what have been offered here? My pursuit required nothing unique or special. Anyone with a smattering of physics and a modicum of determination could easily have done the same. Why then has nobody done so?

Many observers assert that if an idea is any good, surely someone would have proposed it earlier. Some members of the Electric Universe community[wl] have ventured in directions similar to mine, but to my knowledge, nobody has dealt with the full set of subjects offered in this book. Surely, among the huge cadre of scientists working today (a number that exceeds the aggregate number working throughout all of history), someone would have come forth with a similar set of ideas. Why have they (apparently) not?

One explanation is that the ideas simply don't wash; they fail to provide a convincing story, and if so, then who might be foolish enough to tread along a pathway with a dead end? Not many. Alternatively, the reason why others have failed to advance a similar set of ideas may lie in the restrictive culture of today's science — and here I begin my philosophical rant.

We like to think of science as an unfettered truth-seeking endeavor, set in a largely unconstrained environment. To find out how accurate that description is, go ahead and ask a few working scientists. Many will vociferously deny any such freedom. More likely, they will admit to policy constraints that seriously influence the course of their work, limiting the scope of their investigations to "acceptable" subjects, and feeling pressured by the peer-review system to carry out research along well-established, "safe" lines. Veering off that narrow path can pose serious dangers (**Fig. 19.7**).

Figure 19.7. For career safety, scientists keep radical ideas to themselves.

This issue has long held my attention, and I hope you will indulge this aside. For the moment, let me just suggest that science's constraining culture may help explain why others have hesitated to put forth any radical proposals such as the one offered here. While some academic scientists assuredly recognize the immense power of electrical forces, few would dare to advance any idea that deeply challenges conventional scientific thinking. It's risky — as I'll illustrate with several examples.

Risk gets amplified when scientists venture beyond their immediate specialty, for any such incursions invite summary dismissal. "What are your credentials?", they may ask. And, "How do you justify your intrusion into foreign territory, territory in which you're (supposedly) not well steeped?" You're speechless. Sticking within one's zone of comfort, on the other hand, confers security — like mingling within one's circle of friends. Most scientists prefer to remain in such safe environments, avoiding the risks attendant with foreign travel.

One particularly risky territory is physics. That discipline, especially, seems to hold an elevated and virtually untouchable position. After all, physicists have (supposedly) managed to unearth the foundational principles of the universe. That's profound. Who, outside the immediate field of physics, would have the chutzpah to challenge the likes of Einstein, Planck, or Bohr — or even their more modern counterparts?

While we remain humbled by the contributions of those towering figures, hero worship can smack of religious belief. We like to think those luminaries must have had it right. But any researcher unfortunate enough to have identified inconsistencies in the thinking of any of those scientific icons, and bold enough to bring challenge, will become quickly acquainted with the pressures that dissuade any such open critique. The sin of merely suggesting that an important scientist may have erred opens oneself to outsider disdain. Lacking the required audacity, and perhaps also the required credentials, most scientists will simply acquiesce to the principles set forth by those lofty figures, enshrining their theories as permanent monuments — theories that *must* remain inviolate. Those theories may then take on the aura of ground truth — the supposedly "solid" foundations upon which all else must build.

With a set of belief systems virtually unchallengeable, lying within a scientific culture that represses challenge, it's no surprise that any proposal similar to the one advanced in this book has never apparently seen the light of day. The risk of advancing any such paradigm is high. Others may have conceived a set of similarly unconventional ideas, but the act of swimming against the scientific tide may entail an unacceptable professional gamble, which few are reckless enough — or ahem! foolish enough — to advance.

Burned in the Fires of Orthodoxy

Figure 19.8. Galileo Galilei (1564–1642).

Having philosophized on the constraining culture of science, let me now illustrate by offering a few examples of the fates of scientists willing to challenge that pervasive culture.

- First, consider Galileo (**Fig. 19.8**), a scientist not especially predisposed to the complexity of earth-centric epicycles. In his case, the reigning orthodoxy was the church, which, at the time, held sway over much scientific thinking. For challenging the church's dictum that the earth lies at the center of the universe, the Inquisition placed Galileo under house arrest. His writings, as well as any writings that he might create in the future, were banned from publication.

Though certainly an extreme case, the social isolation of Galileo and the quarantining of his works is fundamentally no different from what can happen to contemporary scientists who dare to challenge the prevailing beliefs of their respective communities. The main difference involves the type and severity of the consequences. House arrest may now be uncommon; but banning publication is not — the peer-review system has become legendary for dispatching serious irreverence.

Figure 19.9. Ignaz Semmelweis (1818–1865).

- Next, consider Ignaz Semmelweis (**Fig. 19.9**). A bright young Hungarian doctor working in Vienna at one of Europe's most prominent hospitals in the mid-19th century, Semmelweis observed a curious situation in the hospital's obstetrics ward.

Incoming women who were triaged to midwife delivery generally fared well; those directed to physician delivery feared for their lives, as so many of those women developed post-delivery infections, frequently leaving the hospital in coffins.

Semmelweis quickly found the reason. Pre-parturient women were examined by doctors just after they had conducted autopsies. That was the hospital's practice. Physicians would start their day by conducting autopsies on the women who had died the day prior; then came the internal exams on the expectant mothers. Imagine! This routine took effect before Pasteur's work on the relationship between microbes and disease, so little was known about infections. Suspecting a connection between poor hygiene and the preponderance of mysterious deaths, Semmelweis began washing his hands before conducting pelvic examinations, with impressive results. The mothers usually survived childbirth.

The hand-washing paradigm seemed self-evident to Semmelweis, but his colleagues failed to see it that way. Those doctors' professional standings were threatened by this upstart. Instead of adopting his regimen, they sought retribution, ultimately exiling him back to Hungary. Semmelweis declined rapidly in an asylum, cursing his detractors, and, in the course of his erratic behaviors, acquired a wound that (in a sad bit of irony) festered gangrenous and brought about his death. Hand washing survived, Semmelweis not.

- Finally, consider Jacques Benveniste (**Fig. 19.10**), a more contemporary example of what can happen when someone challenges the prevailing scientific wisdom. Once a world-class French immunologist, Benveniste inadvertently turned in mid-career from orthodoxy to controversy. He didn't plan it that way; but the consequences, nonetheless, amounted to nothing short of scientific excommunication.

Figure 19.10. Jacques Benveniste (1935–2004).

The story began in Benveniste's laboratory. While studying a type of cell that secretes histamine when exposed to a particular antibody, someone stumbled upon a curious result. Even after diluting the antibody suspension so extensively that, in theory, not even a single antibody could remain, exposure of that extremely dilute suspension to those cells produced the same highly specific response: histamine secretion. No antibodies were present, only water; yet the response was the same. Whoa! Immunology was not supposed to work that way.

For Benveniste, the tragedy began when he dubbed the phenomenon "water memory." The antibody-free water appeared to retain a memory of the antibody molecules with which it had previously been in contact — otherwise how could that diluted water have elicited so specific a response? But the common-sense understanding didn't fit: Individual water molecules supposedly jittered about randomly at breakneck speed, with no stopping to catch their respective breaths. How could those actively jitterbugging molecules, occupying unpredictable orientations at any given time, possibly retain information? It should be a no brainer: They cannot. Yet, Benveniste stubbornly maintained that, apparently, they could. By so asserting, he butted heads with leaders of the scientific establishment.

Before recounting the fate that befell Benveniste for his irreverence, I should mention that his results have now been confirmed in multiple laboratories, all following the same, original, experimental protocol.[1] Furthermore, exposition of water memory has been confirmed in the laboratory of Luc Montagnier,[2,3] recipient of a Nobel Prize for his earlier work on HIV. If confirmation in other laboratories stands as a hallmark of truth, then the phenomenon of water memory would appear real, as much as we might prefer to think otherwise.

(Benveniste did, nevertheless, make a strategic error, as he himself conceded to me one day. He employed the provocative term, "water memory." A softer challenge might have saved him from the ignominious fate that eventually befell him.)

Further to the water-memory issue, studies in my own laboratory have identified a possible basis by which water could retain information. As discussed earlier, we uncovered water's fourth phase, a rather extensive phase built from hydrogen and oxygen atoms arranged in an ordered assembly.[4] Little dancing, or jitter — only liquid-crystalline order resembling the ordering in standard digital memories.

Score one feature necessary for possible information storage in water.

Equally relevant, the oxygen atoms contained in that ordered array can occupy distinct states, just like the "0" and "1" of digital memories. Chemistry books affirm those states. In fact, they tell us that individual oxygen atoms can assume not just two, but *five* different oxidation states, ranging from -2 all the way to +2. So, at least in theory, water's fourth phase constitutes a substrate with all the features necessary for storing information: ordered molecules, each one capable of occupying different states — the five states implying even higher information density than present in today's digital memories.

With those advances in understanding, the concept of water memory has begun shifting among at least some scientists, from a scientific joke to a phenomenon ripe for exploration. Benveniste would appear validated — just like Galileo and Semmelweis.

In 1989, however, "water memory" was heresy. Benveniste's attempts to publish his lab's findings in the respected journal, *Nature*, were thwarted multiple times by the editor, Sir John Maddox. Finally, under pressure to publish a collective submission by Benveniste plus several independent groups reporting the same result, Maddox relented. He would publish the submission under one condition: He'd be sending a committee of "peers" to look over the shoulders of those French scientists as they performed their experiments. The peers would then report back their findings to the readers of *Nature*.

Seeing vindication on the horizon, Benveniste accepted Maddox's offer to visit.

Several weeks later the team arrived in Paris. The visitors could hardly qualify as "peers." The team comprised three members: Maddox himself, a journal editor with limited hands-on experience in biology; Walter Stewart, a professional fraud sleuth from the National Institutes of Health; and the world-class magician, James Randi, otherwise known as "The Amazing Randi." Randi's considerable fame came partly from his own genius at magic, and partly from his ability to uncover how charlatans used stage magic to manipulate their victims.

The committee's makeup signaled Maddox's intent: Since "water memory" seemed inconceivable to him, clearly, the French group must have obtained their results in some dubious manner, and this committee seemed ideally suited to his purpose — to expose what he believed to be fraud.

Although the committee never identified any obvious trick, it found cause to discredit Benveniste's results as coming from unintentional bias, arising in part from sloppy note keeping. The subsequent report to *Nature*'s readers dubbed water memory a "delusion." The impact was practically instantaneous. Water memory became a scientific joke, and Benveniste soon became the scientific community's laughingstock: "Having trouble recalling facts? Why not drink some of Benveniste's memory water?"

Benveniste never recovered from this blow to his credibility. While his continued experiments showed that the stored information was robust enough that it could be retained even after having been transmitted over the Internet, he found himself unable to secure funds to support his work, and his once-huge laboratory quickly collapsed. Benveniste died not long after, following a routine surgical procedure. The extent to which his demoralized status contributed to his premature death is something we will probably never know.

Preserving the Status Quo: A Consequence of Human Nature?

Benveniste, Semmelweis, and Galileo typify scientists who dared challenge the prevailing wisdom of their times. As a result, they suffered harsh consequences. The retributions' triumphs seem easy to understand: Champions of the prevailing paradigm always have the weight of numbers on their side, so they can handily suppress the idea of a lone "revolutionary" daring to challenge their vaunted authority. The tactics, as we have seen, are quite simple: Merely tar the perpetrator as a crackpot. Responding effectively to any such indictment can be a lonely, uphill endeavor, in which the dominant faction almost always prevails.

But why must the orthodoxy persist in this way? In the supposedly open-minded endeavor that is science, why does the orthodoxy consistently act to repress challenge?

The rationale, I believe, lies in the depths of human nature. We tend to protect our home turf, which includes our belief systems. We cherish those belief systems, protecting them against incursion as we might protect our home from burglary. In so doing, we feel secure — security being a foundational need of all living creatures. All of us crave that refuge, even though, in the domain of science, any such proclivity serves only to safeguard the *status quo*.

One would presume that, by now, any such obstacle to scientific progress might have been exposed, and that today's science ought to be enlightened enough to have overcome any such impediments to progress. But human nature remains stubbornly unchanged. In the end, we are all human beings, replete with the foibles and frailties that make us human, even if those attributes inadvertently serve to thwart major scientific advances.

Protecting the *status quo* is obviously not the way to encourage scientific revolutions. Perhaps that's why, of late, such revolutions have been few and far between. Technological revolutions abound, but fundamental scientific revolutions that have succeeded in changing the world during the last decades can be counted on one's fingers. If the ultimate goal is major scientific advancement, then we need to pay heed to the obstacles that limit the emergence of new science, many of which arise directly out of human nature.

The system of science needs to deal with those obstacles. If it does, then perhaps we will no longer need to bear witness to the kinds of fates experienced by the likes of those three scientists whose brave pursuits of truth led to the shattering of their lives. One can only hope.

Toward the Future

We live in a world in peril. The many issues facing our planet today call into question humanity's future, some observers fearing imminent meltdown. While political and socioeconomic approaches may help solve some of those problems, others could resolve through science.

By "science," I don't mean the science as currently practiced; I mean the kind of science recognized by Aristotle and Newton — the seeking of truth through simple observation and deduction. Those approaches have produced many scientific revolutions in the past, which have in turn brought useful technologies that nobody could even have dreamed of beforehand — *e.g.*, smartphones, laptop computers, and artificial intelligence arising from the seemingly obscure discovery of semiconduction. Every reason exists to think that similar translations will continue. New, fundamental discoveries should bring radical new technologies, some of which may help sculpt the world into a better place.

The critical issue is the way we do science. The natural sciences of modern academia not only reward intellectual timidity — no revolutions please! — but focus increasingly on ever-narrower goals. We identify problems and then throw scientific resources at them, each scientist tackling a chunk of said problem. Once those chunks are filled in, the edifice of understanding is presumed to be complete. But that approach doesn't always bring the desired solutions. It was more than a half century ago that President Nixon declared "war on cancer." After huge expenditures of money, time, and intellectual resources, how close are we to winning that war? Goal orientation may work well for technology development — e.g., for amassing the resources required for traveling to the moon — but not necessarily for fundamental science.

On the contrary, it's the curiosity-driven approaches that often uncover fresh ground. Years ago, in a book entitled *Retrospectroscope*,[5] the respected cardiovascular physiologist Julius Comroe traced the scientific origins of the practical therapies of modern cardiology. There have been many. All of them, Comroe argues, arose from curiosity-driven research in fields utterly remote from medicine. (X-ray imaging is a good example.) Predicting the outcome of fundamental scientific endeavors is like predicting the course of a blown leaf. Nobody knows where it will land — but it will always land somewhere.

If science is to help overcome the world's ever-mounting challenges, it may behoove us to turn away from goal-oriented research and return to the long-held principles of truth-seeking science. That's where the major advances have come; and that's where they are likely to continue to come. In so doing, we need to pay homage to Occam's razor —untangling, as did our predecessors, the complexity that makes today's science virtually unapproachable, and seeking simple answers instead. Merely building on what others have reflexively presumed to be true seems akin to adding plumbing fixtures to a house of cards. If the entire edifice may soon collapse, then why bother?

Toward the goal of redirecting today's scientific enterprise, my colleagues and I have made some progress. We set up the *Institute for Venture Science*.[w2] The Institute seeks to support scientists in diverse fields, ready to mount promising experimental challenges to any aspect of conventional thinking that has outlived its usefulness. Examples abound. By offering financial backing, the Institute nurtures those challenges.

The key to success: simultaneous funding of *multiple* groups pursuing the same challenge theme. The resulting critical mass should elevate the theme promptly to center stage, in direct competition with the orthodox view. If the multiply funded research groups find ample evidence to support the theme's validity, then quick community acceptance should follow. The process circumvents the thumb of conformity now pressing heavily on the scale of science. Revolutions could rapidly emerge, as never before.

Any such burgeoning scientific renaissance ought to make a real difference for humanity. The breakthrough technologies that will inevitably emerge from the new science should lend fresh optimism that we can reverse the long-standing decline of our planet. Our amazing Earth comes replete with wonder and mystery — mysteries that humans have always sought to uncover. That quest has not vanished. It merely sits on the sidelines, waiting until we can extricate ourselves from the yoke of today's constraining system of doing science.

Until such a time, let us hope that those who find the allure of truth-seeking science irresistible will carry the torch, even in the face of the obstacles working against their success. Their sacrifice is critical if we are ever to uncover the secret rules of nature.

Acknowledgments

First, I wish to acknowledge several people who have profoundly influenced my way of thinking.

At the top of the list sits my (late) friend, Tatsuo Iwazumi. Apart from creating stunning technologies, Iwazumi devoted his life to an unending search for truth. In the field of muscle contraction, my former research area, he taught me that simple logic could upend even long-standing textbook views. His arguments stand firm to this day — although in this particular field the views of the towering Nobel laureate Sir Andrew Huxley remain preeminent, notwithstanding conceptual flaws that Iwazumi and others could identify. The gutsy message Iwazumi conveyed is one I continue to cherish: simple logic can reign supreme.

A similar lesson came from Gilbert Ling, who recently passed just shy of age 100, By arguing that the water inside the cell was not like the water inside a glass — its molecules were ordered — Gilbert piqued my interest in water science, which became my "second" career. His contributions inspired a fresh appreciation that widely held views should not be considered sacred; they could be dead wrong. For that enduring lesson, I extend Gilbert my heartfelt gratitude.

Deep appreciation also extends to my dear friend Pedro Verdugo. In his enduringly colorful way, Pedro taught me about the "church" of science: how doctrine can stunt scientific progress. Paying homage to the lofty bishops of scientific clergy, he has maintained, can inhibit the controversy that is the very lifeblood of science. Pedro's message continues to resonate.

It is not just those colleagues who have influenced my approach. Similar appreciation extends to various friends and associates who have lent inspiration by devoting their lives to searching for scientific truths, irrespective of potential consequences. Their guiding principle: "Damn the torpedoes, full speed ahead." For practical reasons I will mention only a few.

In the realm of clinical science, a standout is the pioneering physician Dietrich Klinghardt, who cures diseases that others cannot. For achieving this, he employs techniques that actually work, rather than those constrained by mainstream medicine. In the field of human consciousness, I cite Rupert Sheldrake, Dean Radin, and the late Bob Jahn, inspirational scientists whose innate brilliance should have propelled them into commanding positions within more conventional investigatory realms. Instead, they bravely ventured along hazardous routes, penetrating foundational cracks in orthodox science — to the exasperation of their critics. Their risky incursions into the unknown continue to lend inspiration. We can only feel relieved that society no longer burns witches.

For the actual preparation of this book, I warmly acknowledge the contributions of several people who have made all the difference. Without their contributions, this book would not have been possible.

First and foremost is my son, Ethan Pollack, who conceived and produced the artwork. Already sketching at the age of four, Ethan went on to study sculpture at Syracuse University, assisting thereafter the world-class artist Jeff Koons, and finally returning home to settle in Seattle to start his young family. Ethan's contribution to my previous book earned well-deserved accolades. His uncommon level of sensitivity, scrupulous attention to detail, gift for artistic wit, devotion to understanding, and unrelenting dedication to the project's overall success, have made him a pure joy to work with. If you find the concepts attractively illustrated, Ethan is the person to thank.

I likewise thank my elder son, Seth Pollack, who generously read through every sentence of the manuscript. His uncanny ability to spot faulty logic and unsupported claims helped me avert potentially serious blunders. Blunders may well be present still, but their number is certainly diminished by dint of Seth's uncanny eye.

I also thank my late wife, Emily, for her unrelenting support. She understood my proclivity to look for hidden truths even in the darkest of corners, as well as my need for ample time to search for those truths. Without an ounce of self-pity, she put up with those digressive passions. On the other hand, she was equally willing to temper my oft-runaway enthusiasm for simplicity with needed doses of reality. That was most welcome. I thank Emily also for commenting on early drafts of some chapters, and for alerting me to any potentially offensive verbiage. She understood the devotion required to produce a book of this nature. Thank you, dear Emi, for your patience, and for your unending love.

And thanks also to Laura Colton, surely one of the world's most talented facilitators. Whenever a problem arose, Laura was there to fix it. That included not only finding the right editors at the right time, but also solving various logistical problems associated with the book's production. Laura always knew the best solution. Her insightfulness, coupled with bountiful wisdom and thoughtfulness, made a huge difference not only in the ultimate production of the book but also in dealing with the author's foibles.

Most helpful, also, was my Israeli friend, Alex Bronstein. Alex commented on every single paragraph of the manuscript, penning red check marks if the verbiage flowed smoothly and the ideas made sense, or scrawling comments in the white space when issues were identified. He then airmailed me the marked up manuscript, printed at 1/4 size.. Although I required an illuminating magnifier to decipher some of his red scrawls, Alex's careful reading and extensive commentary really helped, especially with presentation nuances.

As for the book's overall tone, I thank my friends David Kaplan and Don Scott for setting me straight. It has been said that "Only artists love their models more than scientists." I confess — I might have succumbed to that predilection, for I am admittedly given to some fondness for my own scientific models. Thus, early drafts of this book tended to present those models as revelations of ground truth, delivered by an oracle (me). I came to understand that others might not see it that way, and when David risked our friendship by suggesting in no uncertain terms that such heavy-handed presentation might come across as offensive, I did get the message. Don, my trustworthy, long-term editor, concurred. For those important course corrections, I thank both of them. I do appreciate that humans can err, and as Thomas Henry Huxley put it, that "Even the most beautiful theory can be slayed by a single ugly fact." I share that view. I therefore thank David and Don for helping me to recognize, and communicate, the need for an extra measure of humility.

In similar vein, I also thank my editors, (the same) Don Scott, along with Gary Wilson and Lauren Russel. Don's way with words is practically unparalleled. If occasional phrases strike you as flowing like silken streams of water, he is likely the agent responsible. Don also set me straight on some foundational scientific concepts that I initially thought were valid but were not. So did Gary. Gary was most helpful in promoting presentation clarity. His edits reminded me that many of my readers might not have very much scientific training, and that some nurturing might be needed along the way. I tried to comply. Lauren was particularly helpful in ensuring the smooth flow from one paragraph to the next and from one section to the next. A naturopathic physician with an editing background, Lauren instilled confidence that the ultimate text might be digestible, perhaps even interesting, even to non-scientists. To all three, I offer my hearty thanks and sincere gratitude.

I thank Elizabeth Mullaly, who took great care weaving Ethan's illustrations together with my text to solve the many challenges attendant with layout. Effective layout involves much more than meets the eye.

I also thank Diellza Mehmeti, who conscientiously executed the miscellaneous tasks involved with the production of the book.

Finally, I wish to thank those who have critiqued the manuscript, or portions thereof, along the way. This list includes scientists, research fellows, students, and even some lay people. Their level of helpfulness was frequently out of proportion to their academic status; hence, I list them alphabetically, and if someone's name has been inadvertently omitted, I surely apologize. That was not my intent.

My thanks to the following people, who have commented on all or a substantial number of sections of the manuscript: Sebastian Alagon, GunWoong Bahng, Alex Bronstein, Charlie Bohlin, Gina Bria, Cadie Buckley, Laura Colton, Barry Craig, Charles Cushing, Simon deWeerdt, Sujata Dube, Michael Ebstyne, Neil Eisman, Peter Espie, Herb Fleschner, Emily Freedman, Jason Gillen, Achim Hoffmann, Jianzhi Huang, Hiromasa Ishiwatari, Alexis Kaplan, David Kaplan, Magdalena Kowacz, Johan Kronholm, Kurt Kung, Sally Landefeld, Zheng Li, Sheldon Lu, Josh Mitteldorf, Len Murray, Amar Neogi, Greg Nigh, Arazi Pinhas, Seth Pollack, Arjun Raman, Brandon Reines, Robert Rotella, Rose Rotella, Valery Shalatonin, Abha Sharma, Rupert Sheldrake, Kevin Shelton, Nathan Siles, Henk Ter Keurs, Tony Thomson, Brent Townsend, Ben Tyers, Anqi Wang, and Tao Ye.

For the combined efforts of the large community of thoughtful people who helped shape this book, I'm deeply grateful. Of course, any remaining errors are mine alone.

Lastly, I thank you, the reader, for taking the time to read through and consider the challenging ideas offered in this book. If they provided you with food for thought, I would be well pleased.

References

CHAPTER 1

1. Scott DE (2006). *The Electric Sky*. Mikamar Publ, Portland, OR.

2. Pollack GH (2013). *The Fourth Phase of Water: Beyond Solid, Liquid, and Vapor*. Ebner and Sons, Seattle, WA. www.ebnerandsons.com.

3. Klimov A and Pollack GH (2007). Visualization of charge-carrier propagation in water. *Langmuir* **23** (23): 11890-11895.

4. Ovchinnikova K and Pollack GH (2009). Can water store charge? *Langmuir* **25** (1): 542-547.

5. Vonnegut B, Moore CB, Semonin RG, Bullock JW, Staggs DW, Bradley WE (1962). Effect of Atmospheric Space Charge on Initial Electrification of Cumulus Clouds. *J Geophys Res* **67** (10).

WEB REFERENCES

w1. https://youtu.be/oY1eyLEo8_A

CHAPTER 2

1. Feynman RP, Leighton RB, and Sands M (1964). *The Feynman Lectures in Physics*. Addison-Wesley, Boston, MA.

2. Harrison RG (2011). Fair weather atmospheric electricity. *Journal of Physics: Conference Series* **301** (1): 012001. DOI: 10.1088/1742-6596/301/1/012001. ISSN: 1742-6596.

3. Aplin KL, Harrison RG and Rycroft MJ (2018). Investigating Earth's Atmospheric Electricity: A Role Model for Planetary Studies. *Space Science Reviews* **137** (1-4): 11-27. DOI: 10.1007/s11214-008-9372-x.

4. Pollack GH (2013). *The Fourth Phase of Water: Beyond Solid, Liquid, and Vapor*. Ebner and Sons, Seattle, WA. www.ebnerandsons.com.

5. de Ninno A (2017). Dynamics of formation of the Exclusion Zone near hydrophilic surfaces. *Chem Phys Lett* **667**: 322-326.

6. Cheng Y and Moraru CI (2018). Long-range interactions keep bacterial cells from liquid-solid interfaces: evidence of a bacterial exclusion zone near Nafion surfaces and possible implications for bacterial attachment. *Colloids and Surfaces B: Biointerfaces* **162**: 16-24.

7. He X, Y Zhou X Wen A (2018). Shpilman and Q. Ren (2018). Effect of Spin Polarization on the Exclusion Zone of Water. *J Phys Chem B* **122** (36): 8493-8502. DOI: 10.1021/acs.jpcb.8b04118.

8. Elia V, Oliva R, Napoli E, Germano R, Pinto G, Lista L, Niccoli M, Toso D, Vitiello G, Trifuoggi M, Giarra A, Yinnon TA (2018). Experimental study of physicochemical changes in water by iterative contact with hydrophilic polymers: A comparison between Cellulose and Nafion. *J Mol Liquids* **268**: 598-609.

9. Chai B, Yoo H and Pollack GH (2009): Effect of Radiant Energy on Near-Surface Water. *J Phys Chem B* **113**: 13953-13958.

10. Sedlak M (2006). Large-Scale Supramolecular Structure in Solutions of Low Molar Mass Compounds and Mixtures of Liquids. III. Correlation with Molecular Properties and Interactions. *J Phys Chem B* **110** (28): 13976-13984.

11. Chai B, Zheng JM, Zhao Q, and Pollack GH (2008). Spectroscopic studies of solutes in aqueous solution. *Phys Chem A* **112**: 2242-2247.

12. Pollack GH. (2014): Cell electrical properties: reconsidering the origin of the electrical potential. *Cell Biol Int* **39**: 237-242. ISSN: 1065-6995 DOI: 10.1002/cbin.10382.

13. Tennant JL (2008). *Healing Is Voltage: The Handbook.* Createspace Independent Publishing Platform.

14. Zheng J-M, Chin W-C, Khijniak E, Khijniak E, Jr, and Pollack GH (2006). Surfaces and interfacial water: Evidence that hydrophilic surfaces have long-range impact. *Adv Colloid Interface Sci* **127**: 19-27.

15. Ypma RE and Pollack GH (2015). Effect of Hyperbaric Oxygen Conditions on the Ordering of Interfacial Water. *Undersea and Hyperbaric Medicine* **42** (3): 257-264.

16. Yoo H, Baker DR, Pirie CM, Hovakeemian B, and Pollack GH (2011). Characteristics of water adjacent to hydrophilic interfaces. In: *Water: the Forgotten Molecule*, ed. Denis LeBihan and Hidenao Fukuyama, Pan Stanford, pp. 123-136.

17. Tschauner O, Huang S, Greenberg E, Prakapenka VB, Ma C, Rossman GR, Shen AH, Zhang D, Newville M, Lanzirotti A, Tait K (2018). Ice-VII inclusions in diamonds: Evidence for aqueous fluid in Earth's deep mantle. *Science* **359**: 1136 (2018) DOI: 10.1126/science.aao3030.

18. Rodebush WH and Fiock EF (1925). The Measurement of the Absolute Charge on the Earth's Surface. *PNAS* **11** (7): 402-404.

19. Rakov VA and Uman, MA (2003). *Lightning: Physics and Effects.* Cambridge University Press, Cambridge.

20. Chevalier G, Sinatra ST, Oschman JL, Sokal K, and Sokal P (2012). Earthing: Health Implications of Reconnecting the Human Body to the Earth's Surface Electrons. *Journal of Environmental and Public Health* 2012: 291541. DOI:10.1155/2012/291541.

21. Pollack GH (2001). *Cells, Gels and the Engines of Life*. Ebner and Sons, Seattle, WA. www.ebnerandsons.com.

22. Aull F (1967). Measurement of the electrical potential difference across the membrane of the Ehrlich mouse ascites tumor cells. *J Cell Physiol* **69**: 21-32.

23. Borle A, Loveday J (1968). Effects of temperature, potassium and calcium on the HeLa cells. *Cancer Res* **28**: 2401-5.

24. Sharma A, Adams C, Cashdollar B, Li Z, Nguyen NV, Sai H, Shi J, Velchuru G, Zhu,KZ, and Pollack GH (2018): Effect of Health-Promoting Agents on Exclusion-Zone Size. *Dose-Response* **16** (3): DOI: 10.1177/1559325818796937.

PATENTS

p1. Plauson, H. Conversion of Atmospheric Electric Energy, Patent #1,540,998, June 9, 1925.

WEB REFERENCES

w1. https://en.wikipedia.org/wiki/Distribution_of_lightning

w2. https://en.wikipedia.org/wiki/Lightning

w3. https://digital.lib.washington.edu/researchworks/handle/1773/33812

w4. https://www.newscientist.com/article/dn25723-massive-ocean-discovered-towards-earths-core/

w5. https://www.primarywaterinstitute.org/

w6. https://primarywater.org/

w7. https://www.scientificamerican.com/article/rare-diamond-confirms-that-earths-mantle-holds-an-oceans-worth-of-water/

CHAPTER 3

1. Mohammadi AH, Tohidi B, and Burgass RW (2003). Equilibrium Data and Thermodynamic Modeling of Nitrogen, Oxygen, and Air Clathrate Hydrates. *J Chem Eng Data* **48** (3): 612-616.

2. Pollack GH (2013). *The Fourth Phase of Water: Beyond Solid, Liquid, and Vapor*. Ebner and Sons, Seattle, WA, p 80; pp. 271-275 www.ebnerandsons.com.

3. Klimov A and Pollack GH (2007). Visualization of charge-carrier propagation in water. *Langmuir* **23** (23): 11890-11895.

4. Ishidoya S, Sugawara S, Hashida G, Morimoto S, Aoki S, Nakazawa T, and Yamanouchi T (2006). Vertical Profiles of the O2/N2 Ratio in the Stratosphere over Japan and Antarctica. *Geophysical Research Letters* **33**: L13701, DOI:10.1029/2006GL025886.

WEB REFERENCES

w1. https://en.wikipedia.org/wiki/Protonosphere

w2. http://www.engineeringtoolbox.com/gas-density-d_158.html

w3. https://commons.wikimedia.org/wiki/File:Msis_atmospheric_composition_by_height.svg

w4. https://en.wikipedia.org/wiki/Clathrate_hydrate

CHAPTER 4

1. Roble RG and Tzur I (1986). The Global Atmospheric-Electrical Circuit. National Research Council 1986. *The Earth's Electrical Environment*. Washington, DC: The National Academies Press. pp. 206-231. DOI: 10.17226/898.

2. Mozer FS (1971). Balloon Measurement of Vertical and Horizontal Electric Fields. *Pure and Applied Geophysics* **84** (1): 32-45.

3. Falconer RE (1953). A Correlation between Atmospheric Electrical Activity and the Jet Stream. GE Technical Research Laboratory Report RL-900, #53-512.

4. Hull AW, Bonnell J, Mozer F, Scudder JD, Chaston CC (2003). Large parallel electric fields in the upward current region of the aurora: Evidence for ambipolar effects. *Journal of Geophysical Research* **108**: DOI: 10.1029/2002JA009682.

WEB REFERENCES

w1. https://en.wikipedia.org/wiki/Hadley_cell

w2. https://en.wikipedia.org/wiki/Kristian_Birkeland

w3. https://en.wikipedia.org/wiki/Aurora

w4. https://www.youtube.com/watch?v=ieILUnkdD90

CHAPTER 5

1. Bar-Zohar D (2010). *From the Sun's Energy Source to the Formation of the Solar System*. E-book. pp. 131-135. https://books.google.com/books?id=8CtfDwAAQBAJ&lpg.

2. Newitt LR, Chulliat A, Orgeval JJ (2009). Location of the North Magnetic Pole in April 2007, *Earth Planets & Space* **61**: 703-710.

WEB REFERENCES

w1. https://www.usgs.gov/educational-resources/magnetic-declination-varies-considerably-across-united-states

w2. https://www.universetoday.com/29811/new-finding-shows-super-huge-space-tornados-power-the-auroras/

w3. https://www.nasa.gov/topics/earth/features/2012-poleReversal.html

w4. https://en.wikipedia.org/wiki/Geomagnetic_reversal

w5. http://www.geomag.bgs.ac.uk/education/earthmag.html#_Toc2075560

w6. http://lifeng.lamost.org/courses/astrotoday/CHAISSON/AT313/HTML/AT31305.HTM

w7. https://www.csmonitor.com/Science/2013/0517/Why-do-planets-farthest-from-sun-have-highest-winds-Team-closes-in-on-answer

w8. https://www.uu.edu/dept/physics/
    scienceguys/2004Sept.cfm

CHAPTER 6

1. Hide R and Dickey JO (1991). The Earth's variable rotation, *Science* **253**: 629-637.

2. Dickey JO, Gegout P and Marcus SL (1999). Earth-Atmosphere Angular Momentum Exchange and ENSO: The Rotational Signature of the 1997-98 Event. NASA report.

3. Lee RB (1995). Long-term total solar irradiance variability during sunspot cycle 22. *J Geophys Res* **100** (A2): 1667-1675.

4. White RH, Wallace JM, Battisti DS (2021). Revisiting the role of mountains in the Northern Hemisphere winter atmospheric circulation. *JAS* **78** (7), 2221-2235.

5. White RH, Battisti DS, Roe GH (2017). Mongolian Mountains Matter Most: Impacts of the Latitude and Height of Asian Orography on Pacific Wintertime Atmospheric Circulation. *J Climate* **30** (11): 4065-4082. DOI: 10.1175/JCLI-D-16-0401.1

6. Chao BF (1989). Length-of-Day Variations Caused by El Niño-Southern Oscillation and Quasi-Biennial Oscillation *Science* **17**: 923-925.

7. Halekas JS, Poppe AR, Harada Y, Bonnell JW, Ergun RE, McFadden JP (2018). A Tenuous Lunar Ionosphere in the Geomagnetic Tail. *Geophys Res Lett* **45** (18): 9450-9459.

WEB REFERENCES

w1. https://www.livescience.com/178-spin-earth-rotation.html

w2. https://www.britannica.com/science/trade-wind

w3. https://trs.jpl.nasa.gov/bitstream/handle/2014/17186/99-0613.pdf?

w4. https://astrobiology.nasa.gov/news/cold-clouds-and-water-in-space/, *and* https://www.jpl.nasa.gov/news/astronomers-find-largest-most-distant-reservoir-of-water/

w5. https://www.space.com/21157-uranus-neptune-winds-revealed.html

CHAPTER 7

1. Vonnegut B, Moore CB, Semonin RG, Bullock JW, Staggs DW, Bradley WE (1962). Effect of Atmospheric Space Charge on Initial Electrification of Cumulus Clouds. *J Geophys Res* **67** (10).

2. Pollack GH (2013). *The Fourth Phase of Water: Beyond Solid, Liquid, and Vapor*. Ebner and Sons, Seattle, WA. www.ebnerandsons.com.

3. Wright H L (1939). Atmospheric opacity: A study of visibility observations in the British Isles, *Quart J Royal Meteorol Soc, London* **65**: 411.

4. Lundgren DA and Cooper DW (1969). Effect of Humidity on Light-Scattering Methods of Measuring Particle Concentration. *Journal of the Air Pollution Control Assoc* **19** (4): 243-247. DOI: 10.1080/00022470.1969.10466482.

5. Ienna F, Yoo H and Pollack GH (2012). Spatially Resolved Evaporative Patterns from Water. *Soft Matter* **8** (47): 11850-11856.

6. Feynman RP, Leighton RB, Sands M (1964). *The Feynman Lectures on Physics*. Addison-Wesley, Reading, MA; Chapter 2, p 2.

7. Nagornyak E, Yoo H and Pollack GH (2009). Mechanism of attraction between like-charged particles in aqueous solution. *Soft Matter* **5**: 3850-3857.

8. Ito K, Yoshida H, Ise N (1994). Void Structure in Colloidal Dispersions. *Science* **263** (5413): 66-68.

9. Ise N (1986). Ordering of Ionic Solutes in Dilute Solutions through Attraction of Similarly Charged Solutes — A Change of Paradigm in Colloid and Polymer Chemistry. *Angew Chem* **25**: 323-334.

10. Fujii S, Ryay AJ and Armes SP (2006). Long-Range Structural Order, Moiré Patterns, and Iridescence in Latex-Stabilized Foams. *J Am Chem Soc* **128** (24): 7882-7886.

WEB REFERENCES

w1. https://headsup.boyslife.org/how-much-does-a-cloud-weigh/

w2. https://www.mentalfloss.com/article/49786/how-much-does-cloud-weigh

w3. https://en.wikipedia.org/wiki/Circumhorizontal_arc

CHAPTER 8

1. Pollack GH (2013). *The Fourth Phase of Water; Beyond Solid, Liquid, and Vapor*. Ebner and Sons, Seattle, WA, Chapter 8. www.ebnerandsons.com.

2. Ise N (2007). When, why, and how does like like like?—Electrostatic attraction between similarly charged species. *Proc Jpn Acad Ser B* **83** (7): 192-198.

3. Warner J (1969). The Microstructure of Cumulus Cloud. Part I. General Features of the Droplet Spectrum. *Journal of the Atmospheric Sciences* **26** (5): 1049-1059. Bibcode:1969JAtS...26.1049W. DOI:10.1175/1520-0469(1969)026<1049: TMOCCP>2.0.CO; 2. pp. 1056, 1058.

4. Ito K, Yoshida H and Ise N (1994). Void Structure in Colloidal Dispersions. *Science* **263** (5413): 66-68.

5. Velev OD and Kaler EW (2000). Structured Porous Materials via Colloidal Crystal Templating: From Inorganic Oxides to Metals. *Adv Mater* **12** (7): 531-534.

6. Montero-Martínez G, Alexander B. Kostinski, Raymond A. Shaw, and Fernando García-García1 (2009). Do all raindrops fall at terminal speed? *Geophysical Research Letters* **36**: L11818, DOI:10.1029/2008GL037111.

7. Soon W, Connolly R and Connolly M (2015). Re-evaluating the role of solar variability on Northern Hemisphere temperature trends since the 19th century, *Earth-Science Reviews* **150**: 409-452.

8. Kirby J (2008). Cosmic Rays and Climate. arXiv:0804.1938v1 [physics.ao-ph] 11 Apr.

WEB REFERENCES

w1. https://weather.org/singer/book.htm

w2. https://www.britannica.com/video/process-cloud-formation-factors-surface-air-water/-203821

w3. https://www.abc.net.au/news/2020-01-23/brown-rain-falls-in-victoria-as-dust-storm-and-rain-collide/11892080

w4. https://www.3aw.com.au/my-car-is-like-a-coffee-cup-melburnians-wake-to-dirty-dumping-of-rain/

w5. https://www.news.com.au/sport/tennis/australian-open/melbourne-weather-horrifies-tennis-world-at-australian-open/news-story/75bc1359a2b17e7849e51bf80084e561

w6. https://en.wikipedia.org/wiki/Cloud_condensation_nuclei

w7. https://mega.nz/file/XYg2VZ4I#h5322MbsE-uCQ4BFax3sfCXXu2R4Xt2faV_yn9MPWio

w8. http://www.nasa.gov/centers/goddard/news/topstory/2007/twilightzone_particles.html

w9. https://www.weather-climate.org.uk/09.php

w10. https://en.wikipedia.org/wiki/Density_of_air

w11. https://www.researchgate.net/figure/Electric-charges-in-a-cumulonimbus-cloud-and-Intra-cloud-inter-cloud-cloud-to-ground_fig2_266289213

w12. https://www.quora.com/When-it-s-raining-why-do-the-clouds-come-down-so-low-and-seem-to-swallow-the-mountains w13. https://www.downtoearth.org.in/news/climate-change/scientists-find-evidence-cosmic-rays-influence-earth-s-climate-65436

w14. https://www.nature.com/news/2011/110824/full/news.2011.504.html

CHAPTER 9

1. Pollack GH (2013). *The Fourth Phase of Water: Beyond Solid, Liquid, and Vapor*. Ebner and Sons, Seattle, WA, pp. 221-252; pp. 307-327.

2. Ehre D, Lavert E, Lahav M, Lubomirsky I (2010): Water Freezes Differently on Positively and Negatively Charged Surfaces of Pyroelectric Materials. *Science* **327**: 672-675.

WEB REFERENCES

w1. https://en.wikipedia.org/wiki/Thunderstorm

w2. https://www.britannica.com/science/thunderstorm

w3. https://www.ncei.noaa.gov/access/data/coastal-water-temperature-guide/npac.html

w4. https://climate.ncsu.edu/edu/Thunderstorm

CHAPTER 10

1. Vonnegut B (1960). Electrical theory of tornadoes. *J Geophysical Research*. Online ISSN: 2156-2202. DOI: 10.1029/JZ065i001p00203.

2. Bolonkin A (2013). *Global Journal of Research in Engineering, F: Electrical & Electronic* **13** (14, v1.0) Type: Double Blind Peer Reviewed International Research Journal, Publisher: Global Journals Inc. (USA) Online ISSN: 2249-4596, Print ISSN: 0975-5861

3. Pollack GH (2013): *The Fourth Phase of Water: Beyond Solid, Liquid, and Vapor*. Ebner and Sons, Seattle, WA, pp. 221-252; pp. 307-327.

WEB REFERENCES

w1. https://www.npr.org/sections/thetwo-way/2017/09/26/552063244/long-after-the-hurricanes-have-passed-hard-work-and-hazards-remain

w2. https://www.thunderbolts.info/wp/2017/06/13/tornado-the-electric-model

w3. http://charles-chandler.org/Geophysics/Tornadoes.php?text=full

w4. https://kids.britannica.com/students/assembly/view/139631

w5. https://www.accuweather.com/en/severe-weather/upside-down-lightning-experts-break-down-insane-viral-video/1165263

w6. https://www.nssl.noaa.gov/education/svrwx101/tornadoes/

CHAPTER 11

1. Sheldrake R (2012). *Science Set Free: 10 Paths to New Discovery*. Random House, NY.

2. Gershteyn ML, Gershteyn LI, Gershteyn A, and Karagioz OV (2002). Experimental evidence that the gravitational constant varies with orientation. http://Arxiv.org/pdf/physics/0202058v2.

3. Streicher C (1993). Weight changes of biological and chemical material in a thermodynamically closed system. PhD Thesis, Maharishi University of Management. http://search.proquest.com/docview/304111790?accountid=37967.

4. Volkamer K, Streicher C, Walton KG, Fagan J, Schenkluhn H, Marlow H (1994). Experimental re-examination of the law of conservation of mass in chemical reactions. *J Sci Exploration* **8** (2): 217-250.

5. Volkamer K (2017). *Discovery of Subtle Matter*. Brosowski, Berlin, ISBN: 978-3-946533-01-6.

6. Volkamer K (2009). Gravitational Spacecraft Anomalies as Well as the at Present Relatively Large Uncertainty of Newton's Gravity Constant Are Explained on the Basis of Force Effects Due to So-Far Unknown Form of Space-like Matter. In DB Cline, ed., *Sources of Detection of Dark Matter and Dark Energy in the Universe*, Am. Inst. Of Phys.

7. Kuusela T, Jäykkä J, Kiukas J, Multamäki T, Ropo M, and Vilja I (2006). Gravitation experiments during the total solar eclipse. *Phys Rev D* **74**: 122004.

8. Vogt DB (1996). *Gravitational Mystery Spots of the United States*. Vector Associates, Bellevue WA.

9. Feynman RP, Leighton RB, and Sands M (1964). *The Feynman Lectures in Physics*. Addison-Wesley, Reading, MA.

10. Laviolette P (2008). *Secrets of Antigravity Propulsion*. Bear & Co.

WEB REFERENCES

w1. http://news.softpedia.com/news/WIMP-Discovery-Claim-Contested-84111.shtml

w2. http://earthobservatory.nasa.gov/Features/GRACE/page3.php

w3. www.thunderbolts.info

w4. https://www.youtube.com/watch?v=YkWiBxWieQU

w5. http://www.youtube.com/watch?v=5QhqEoRvpEA

w6. https://www.youtube.com/watch?v=1Q0yxatzGAg

CHAPTER 12

1. Alfvén H (1948). *The New Astronomy*. Chapter 2, Section III, page 74-79. 2nd ed. 1955, Simon and Schuster, A Scientific American book.

2. Janhunen P (2008). The electric sail — a new propulsion method which may enable fast missions to the outer solar system. *J British Interplanetary Soc* **61**: 322-325.

3. Kuusela T, Jäykkä J, Kiukas J, Multamäki T, Ropo M, and Vilja I (2006). Gravitation experiments during the total solar eclipse. *Phys Rev D* **74**: 122004

4. Scott DE (2006). *The Electric Sky*. Mikamar.

5. Thornhill W and Talbott D (2007). *The Electric Universe*, Mikamar.

WEB REFERENCES

w1. https://teslaresearch.jimdo.com/dynamic-theory-of-gravity/

w2. https://peswiki.com/powerpedia:teslas-dynamic-theory-of-gravity

w3. https://en.wikipedia.org/wiki/Heliospheric_current_sheet

w4. https://www.swpc.noaa.gov/phenomena/solar-wind

w5. https://en.wikipedia.org/wiki/Solar_wind

w6. https://www.space.com/32644-cosmic-rays.html

w7. https://science.nasa.gov/science-news/news-articles/effects-of-the-solar-wind

w8. https://science.nasa.gov/universe/whats-inside-a-dead-star/

w9. http://www.electric-sailing.fi/

w10. https://en.wikipedia.org/wiki/Axial_tilt

w11. https://www.thunderbolts.info/wp/

w12. https://en.wikipedia.org/wiki/Gravitational-wave_observatory

w13. https://en.wikipedia.org/wiki/Kerr_effect

CHAPTER 13

1. Lorenz RD (2018). Lightning detection on Venus: a critical review. *Prog Earth Planet Sci 5*: **34** . DOI: 10.1186/s40645-018-0181-x.

2. Dyudina UA, Del Genio AD, Ingersoll AP, Porco CC, West RA, Vasavada AR, Barbara JM. (2004). Lightning on Jupiter observed in the H[alpha] line by the Cassini imaging science subsystem. *Icarus* **172**: 24-36.

3. Delory GT (2010). Electrical Phenomena on the Moon and Mars. Proc *ESA Annual Meeting on Electrostatics*, Paper A.

4. Arp H (1998). Seeing Red: Redshifts, Cosmology and Academic Science. Apeiron, Montreal http://redshift.vif.com.

5. Thornhill W and Talbot D (2007). *The Electric Universe*. Mikamar.

WEB REFERENCES

w1. https://solarsystem.nasa.gov/missions/cassini/science/saturn/

w2. https://cosmosmagazine.com/space/water-water-everywhere-in-our-solar-system-but-what-does-that-mean-for-life/

w3. https://en.wikipedia.org/wiki/Retrograde_and_prograde_motion

w4. https://www.youtube.com/watch?v=9NbCzbDdd-g

w5. https://www.thunderbolts.info/wp/

w6. https://en.wikipedia.org/wiki/Far_side_of_the_Moon

w7. https://europa.nasa.gov/news/18/hubble-finds-oxygen-atmosphere-on-jupiters-moon-europa/

w8. https://www.space.com/20577-rhea-saturn-s-dirty-snowball-moon.html

w9. https://www.nationalgeographic.com/science/2018/09/news-full-moon-electric-ionosphere-nasa-artemis-space/

w10. https://oceanservice.noaa.gov/facts/highesttide.html

w11. https://www.nhc.noaa.gov/surge/

CHAPTER 14

1. Shinbrot T and Hermann HJ (2008). Static in Motion. *Nature* **45** (14): 773-774.

2. Zheng X, He L and Zhou Y (2004) . Theoretical model of the electric field produced by charged particles in windblown sand flux. *J Geophys Res* **108**: D10.

3. Pollack GH (2013). *The Fourth Phase of Water: Beyond Solid, Liquid, and Vapor*. Ebner and Sons, Seattle, WA. www.ebnerandsons.com.

4. LaViolette PA (2008). *Secrets of Antigravity Propulsion*. Bear & Co., Rochester VT.

5. Xu C, Zi Y, Wang AC, Zou H, Dai Y, He X, Wang P, Yi-Cheng W, Feng P, Li D, Wang ZL (2018). On the Electron-Transfer Mechanism in the Contact-Electrification Effect. *Advanced Materials* **30** (15): e1706790. DOI:10.1002/adma.201706790. PMID 29508454.

6. Jackson TL and Farrell WM (2006). Electrostatic Fields in Dust Devils: An Analog to Mars. *IEEE Transactions On Geoscience and Remote Sensing* **44** (10): 2942-2949.

7. Gorham PW (2013). Ballooning Spiders: The Case for Electrostatic Flight arXiv:1309.4731v2 [physics. bio-ph].

8. Morley EL and Robert D (2018). Electric fields elicit ballooning in spiders. *Current Biology* **28**: 2324-2330.

9. Grebennikov VS (1997). "Chapter 5: The Natural Phenomena of AntiGravitation and Invisibility in Insects due to the Grebennikov Cavity Structure Effect (CSE)." *My World*. English translation with introductory comment: https://archive.org/details/my-world-by-viktor-grebennikov/mode/2up.

WEB REFERENCES

w1. https://en.wikipedia.org/wiki/Dust

w2. http://vimeo.com/26045314

w3. https://www.youtube.com/watch?v=vzZy1Aqleno

w4. https://www.nature.com/articles/d41586-018-07477-9

w5. https://www.keyence.com/ss/products/static/resource/feature/property.jsp

w6. http://www.infoplease.com/ce6/weather/A0843423.html

w7. https://www.nationalgeographic.com/magazine/2019/05/see-how-spiders-fly-with-electricity-video/

w8. https://www.theatlantic.com/science/archive/2018/07/the-electric-flight-of-spiders/564437/

w9. https://www.youtube.com/watch?v=GRrUxi6d7so

w10. https://www.youtube.com/watch?v=SPeZinCynS8

w11. https://www.youtube.com/watch?time_continue=346&v=hYJXE4FCm7Q

w12. http://www.keelynet.com/greb/greb.htm

w13. https://blogs.scientificamerican.com/but-notsimpler/files/2013/08/IMG_3871-cr.jpg

w14. http://blogs.scientificamerican.com/but-not-simpler/the-beautiful-science-of-helicopter-halos

w15. http://www.teachersource.com/ElectricityAndMagnetism/Electricity/FunFlyStick.aspx

w16. http://cst.mos.org/sln/toe/kite.html

w17. https://www.youtube.com/watch?v=MHlAJ7vySC8

w18. https://www.youtube.com/watch?v=GeyDf4ooPdo

w19. https://www.youtube.com/watch?v=tLMpdBjA2SU

w20. https://journals.aps.org/prresearch/abstract/10.1103/PhysRevResearch.4.023131

w21. https://interestingengineering.com/goodyears-a-new-tire-design-can-charge-your-car-as-you-drive

w22. https://en.wikipedia.org/wiki/Lift_(force)#

CHAPTER 15

1. Regis E (2020). The Enigma of Aerodynamic Lift. *Scientific American* **322**: Issue 2.

2. Vedel Tanning (1931). A. Ravens flying up-side-down. *Nature* **127**: 856.

3. Craig Gail (1997). *Stop Abusing Bernoulli! How Airplanes Really Fly*. Regenerative Press, Indiana.

4. Gunn R (1948). Electric Field Intensity Inside of Natural Clouds. *J Appl Phys* **19**: 481-484. DOI: 10.1063/1.1698159.

PATENT

p1. Pal; Anadish Kumar. Triboelectric treatment of wing and blade surfaces to reduce wake and BVI/HSS noise, Patent #US-20070246611-A1, October 25, 2007.

WEB REFERENCES

w1. http://www.dailymail.co.uk/news/article-1265891/
Hold-think-youre-going-Skydiver-grabs-gliders-
tail-fin-fly-2-100-metres-100mph.html

w2. https://www.youtube.com/watch?v=eo_
MGJ5NusY

w3. https://www.youtube.com/
watch?v=d2mITQ53kr8

w4. http://www.centennialofflight.gov/essay/
Evolution_of_Technology/supercritical/
Tech12G1.htm

w5. http://www.flyingfoam.com/Airfoil-Help.html

w6. https://youtu.be/e0l31p6RIaY

w7. http://inflight.squarespace.com/featured/
2011/3/14/the-secret-lives-of-gliders.html

w8. http://www.stuff.co.nz/travel/themes/adven-
ture/68515467/soaring-over-the-south-island

w9. https://www.youtube.com/watch?v=pVPWe-
d52ovI

w10. https://www.youtube.com/watch?v=o6vRbP-
fLBmg

CHAPTER 16

1. Warnke U (2008). Bees, Birds and Mankind: Destroy-
ing Nature by Electrosmog. In: *Effects of Mobile Radio
and Wireless Communication*. Ed K Hecht, M Kem, K
Richter, H-C Scheiner, Bürgerwelle e.V., Dachverband
der Bürger und Initiativen zum Schutz vor Elektros-
mog, Lindenweg 10, D-95643 Tirschenreuth, Germany ISBN: 978-3-00-023124-7 www.buergerwelle.de/.

2. Warnke U (1973). Bees, Birds and Mankind. in:
Herbert König: *Unsichtbare Umwelt*. Heinz Moos
Verlag, München.

3. Warnke U (1989). Information Transmission by
Means of Electrical Biofields. In: *Electromagnetic
Bioinformation* ed: Popp FA, Warnke U, Konig HL,
Peschka, W. Urban and Schwartzenberg, München.

4. Gan-Mor S, Schwartz Y, Bechar A, Eisikowitch D,
Manor G (1995). Relevance of electrostatic forces in
natural and artificial pollination. *Canad Agric Engng*
37: 189-195.

5. Pollack GH (2013). *The Fourth Phase of Water: Beyond
Solid, Liquid, and Vapor*. Ebner and Sons, Seattle,
WA. www.ebnerandsons.com.

6. Pollack GH (2014). Cell electrical properties: recon-
sidering the origin of the electrical potential. *Cell
Biology International* **38**: 237-42. ISSN: 1065-6995
DOI: 10.1002/cbin.10382.

7. Schwalfenberg GK (2012). The Alkaline Diet: Is
There Evidence That an Alkaline pH Diet Ben-
efits Health? *J Environ Public Health*. 727630,
DOI: 10.1155/2012/727630.

8. Grosso Michael (2015). *The Man Who Could Fly:
St. Joseph of Cupertino and the Mystery of Levitation*.
Rowman & Littlefield Publishers.

9. Wikelski M, Tarlow EM, Raim A, Diehl RH, Larkin
RP, Visser GH (2003). Costs of migration in free-fly-
ing songbirds. *Nature* **423**: 704.

10. Swan LW (1961). The Ecology of the High Himalayas. *Sci Am* **205**: 68-78.

11. Laybourne RC (1974). Collision between a vulture
and an aircraft at an altitude of 37,000 feet. *Wilson
Bull* **86**: 461-462.

WEB REFERENCES

w1. http://old.post-gazette.com/sports/
outdoors/20020428shal0428p5.asp

w2. http://ngm.nationalgeographic.com/
2011/02/feathers/zimmer-text/1

w3. https://en.wikipedia.org/wiki/Guano

w4. http://www.nytimes.com/2010/05/25/science/25migrate.html

w5. http://www.nytimes.com/2007/10/23/science/23migr.html

w6. https://spectrum.ieee.org/energywise/tech-history/dawn-of-electronics/bird-poop-can-cripple-power-grids

w7. https://youtu.be/FPRswRWZ23Q

CHAPTER 17

1. Kimball John (2010). *The Physics of Sailing*. CRC Press, Boca Raton, FL. www.crcpress.com.

WEB REFERENCES

w1. https://en.wikipedia.org/wiki/High-performance_sailing

w2. http://news.bbc.co.uk/2/hi/technology/7968860.stm

w3. http://en.wikipedia.org/wiki/Blackbird_(land_yacht)

w4. https://forums.sailboatowners.com/index.php?threads/static-or-electric-shock.53694/

w5. https://en.wikipedia.org/wiki/Atom

w6. https://www.youtube.com/watch?v=GTpfk3BhrLg

w7. https://www.youtube.com/watch?v=WBG1g8s3BT0

w8. https://buildingspeed.org/2012/06/15/the-myth-of-the-200-mph-lift-off-speed

CHAPTER 18

1. Bartholomew A (2014). *Hidden Nature: The Startling Insights of Viktor Schauberger*. Floris Books.

2. Coats Callum (2001). *Living Energies: An Exposition of Concepts Related to the Theories of Viktor Schauberger*. Gateway. ISBN:-13: 978-0717133079.

3. Yu A, Carlson P, Pollack GH (2013). Unexpected axial flow through hydrophilic tubes: Implication for energetics of water. *Eur Physical J Special Topics* **223**: 947-958. DOI: 10.1140/epjst/e2013-01837-8.

4. Rohani M and Pollack GH (2013). Flow through horizontal tubes submerged in water in the absence of a pressure gradient: Mechanistic considerations. *Langmuir* **29** (22):6556-61. DOI: 10.1021/la4001945.

5. Li Z and Pollack GH (2020). Surface-induced flow: A natural microscopic engine using infrared energy as fuel. *Science Advances* **6** (19): eaba0941 DOI: 10.1126/sciadv.aba0941.

6. Pollack GH (2013). *The Fourth Phase of Water: Beyond Solid, Liquid, and Vapor*. Ebner and Sons, Seattle, WA. www.ebnerandsons.com.

7. Klyuzhin I, Symonds A, Magula J, Pollack GH (2008). A new method of water purification based on the particle exclusion phenomenon. *Environ Sci and Techn* **42**(16): 6160-6166.

8. Ye T and Pollack GH (2022). Do Aqueous Solutions Contain Net Charge? *PLOS One* **17** (10). DOI: 10.1371/journal.pone.0275953.

9. Oplatka A (1998). Do the bacterial flagellar motor and ATP synthase operate as water turbines? *Biochem Biophys Res Commun* **249**: 573-578.

10. Tirosh R and Oplatka A (1982). Active streaming against gravity in glass microcapillaries of solutions containing acto-heavy meromyosin and native tropomyosin. *J Biochem* **91**: 1435-1440.

11. Wright PA, Hemming T, Randall D (1986). Downstream pH changes in water flowing over the gills of rainbow trout. *J Exp Biol* **126**: 499-512.

12. Mitsumata T, Ikeda K, Gong JP, Osada Y (2000). Controlled Motion of Solvent-Driven Gel Motor and Its Application as a Generator. *Langmuir* **16**: 307-312.

13. Verdugo P, Deyrup-Olsen I, Martin AW, Luchtel DL (1992). Polymer gel phase transition: The molecular mechanism of product release in mucin secretion? In *Swelling of polymer networks, NATO ASI Series H* (ed. Karalis E), **64**, pp. 671-681 Springer-Verlag, Heidelberg, Germany.

14. Wilkens LA and Hofmann M H (2005). Behavior of animals with passive, low-frequency electrosensory systems. In T. H. Bullock, C. D. Hopkins, A. R. Popper, & R. R. Fay (Eds.), *Electroreception*. New York: Springer, pp. 229-263.

WEB REFERENCES

w1. https://en.wikipedia.org/wiki/Fish_locomotion

w2. https://www.youtube.com/watch?v=HQYhbISCDeM

w3. https://en.wikipedia.org/wiki/Aquatic_locomotion

w4. https://animals.mom.me/what-are-fish-scales-used-for-3010814.html

w5. https://blogs.voanews.com/science-world/2013/06/14/scientists-discover-how-marine-mammals-hold-their-breath-for-long-periods/

w6. https://www.businessinsider.com/how-the-sperm-whale-holds-its-breath-2013-6

2. Montagnier L, Aïssa J, Ferris S, Montagnier J-L, Lavallee C (2009). Electromagnetic signals are produced by aqueous nanostructures derived from bacterial DNA sequences. *Interdiscip Sci Comput Life Sci* **1**: 81-90. DOI: 10.1007/s12539-009-0036-7.

3. Montagnier L, Del Giudice E, Aïssa J, Lavallee C, Motschwiller S, Capolupo A, Polcari A, Romano P, Tedeschi A, Vitiello G (2015). Transduction of DNA information through water and electromagnetic waves. *Electromagn Biol Med* **34** (2):106-12. DOI: 10.3109/15368378.2015.1036072.

4. Pollack GH (2013). *The Fourth Phase of Water: Beyond Solid, Liquid, and Vapor*. Ebner and Sons, Seattle, WA. www.ebnerandsons.com.

5. Comroe JH, Jr (1977). *Retrospectroscope*. Von Gehr Press.

WEB REFERENCES

w1. www.thunderbolts.info

w2. www.ivscience.org

CHAPTER 19

1. Belon P, Cumps J, Ennis M, Mannaioni PF, Roberfroid M, Sainte-Laudy J, Wiegant FAC (2004). Histamine dilutions modulate basophil activation. *Inflamm Res* **53**: 181. DOI: 10.1007/s00011-003-1242-0.

Image Credits

CHAPTER 1 Fig. 1.5 – Pollack Lab   **CHAPTER 2** Fig. 2.1 – courtesy Andrey Klimov; Fig. 2.2 – Floyd Clark / courtesy of Caltech Archives and Special Collections   **CHAPTER 3** Fig. 3.3 – NASA   **CHAPTER 7** Fig. 7.1 – Ethan Pollack; Fig. 7.2 – Gerald Pollack; sunset photo page 112 – Mr. Brendel / GNU Free Documentation License; Fig. 7.3 – Pollack Lab; Fig. 7.4 (*left*) – Pollack Lab; Fig. 7.4 (*right*) – Ethan Pollack; Fig. 7.5 – Pollack Lab; Fig. 7.6 – Pollack Lab; Fig. 7.8 – Ethan Pollack; Fig. 7.9 – Pollack Lab; Fig. 7.10 – Pollack Lab; Spouting Bowl image page 126 – source unavailable   **CHAPTER 8** Fig. 8.1 – source unavailable; Fig. 8.2 – Gerald Pollack; Cloubuster apparatus page 146 – courtesy Madjid Abdelaziz; Fig. 8.4 – Gerald Pollack; Fig. 8.6 – Kenneth Dwain Harrelson / CC-BY-SA 3.0; clouds as art photo (*a*) page 151 – Gerald Pollack; clouds as art photo (*b*) page 151 – Gerald Pollack; clouds as art photo (*c*) page 151 – Ari Pentilla; Fig. 8.13 – checubus / stock.adobe.com; Fitzroy's Retort photos page 164 – Gerald Pollack   **CHAPTER 9** Fig. 9.1 – source unavailable; Fig. 9.2 – source unavailable; Fig. 9.5 – source unavailable   **CHAPTER 10** Fig. 10.1 – NASA; Fig. 10.4 – oporkka / stock.adobe.com; Fig. 10.6 – AP Photo / Lori Mehmen; Fig. 10.9 – NZP Chasers / stock.adobe.com; dust devil photo page 204 – NASA; waterspout photo page 204 – Evgeny Drokov   **CHAPTER 11** Fig. 11.2 – NASA; Mystery Spot images (*a*) page 222 – scanned from Mystery Spot postcard; Mystery Spot photos (*b*) – Gerald Pollack; Fig. 11.7 – public domain; Fig. 11.10 – Kevin Shelton / Pollack Lab   **CHAPTER 12** Fig. 12.1 – Getty Images; Fig. 12.6 – David Phan / CC BY-SA 2.0   **CHAPTER 13** Fig. 13.6 – NASA   **CHAPTER 14** Fig. 14.1 – Eva Kali - stock.adobe.com; Fig. 14.2 – imageBROKER / stock.adobe.com; Fig. 14.3 – Kevin Shelton / Pollack Lab; Fig. 14.5 – Sean McGrath / CC BY-SA 2.0; Viktor Grebennikovs flying machine photo page 290 – source unavailable; Fig. 14.6 – source unavailable; Fig. 14.7 – source unavailable; gyroscope photo page 294 – Tom Nevesely / stock.adobe.com; Fig. 14.8 – DLeonis / stock.adobe.com   **CHAPTER 15** Fig. 15.1 – Staff Sgt. Chris Willis / Public Domain; Fig. 15.2 – 1988, Turkish Gendarmerie; Fig. 15.4 – courtesy Hollister Soaring Center   **CHAPTER 16** Fig. 16.1 – source unavailable; Fig. 16.4 – source unavailable; Fig. 16.13 – Aungsumol / stock.adobe.com   **CHAPTER 17** Fig. 17.1 – source unavailable; Fig. 17.2 – George Grantham Bain Collection; Fig. 17.3 – Eric Gevaert / stock.adobe.com; Fig. 17.4 – Rick Cavallaro; Fig. 17.7 – public domain   **CHAPTER 18** Fig. 18.1 – Ethan Pollack; Fig. 18.3 – Guy Sagi / stock.adobe.com; Fig. 18.7 – Blue Whale Blow © Tim Stenton; Fig. 18.11 – Orlando Florin Rosu / stock.adobe.com; Fig. 18.13 – maemanee / stock.adobe.com   **CHAPTER 19** Fig. 19.6 – public domain; Fig. 19.7 – public domain; Fig. 19.8 – with permission from Dr Jérôme Benveniste, PhD
ABOUT THE AUTHOR Photo of G. Pollack by Ethan Pollack; Photo of E. Pollack by Julia Pollack

Unless otherwise listed above, drawings and illustrations are by Ethan Pollack.
Attributions listed as *source unavailable* could not be located despite considerable effort.
Please alert the publisher with information regarding licensing and usage rights. Licensor will
be remunerated and an updated attribution will appear in subsequent printings.

Index

Page numbers in *italic* denote figures.

About the Author

Scientist, speaker, and author, GERALD H. POLLACK (Jerry) is recognized worldwide for identifying water's fourth phase — H_3O_2. His fundamental belief that all natural phenomena can be explained by identifying simple, straightforward mechanisms often flies in the face of scientific orthodoxy. Despite stepping on many toes along the way, he is known for his open-minded approach — with ideas and with people.

Jerry maintains an active laboratory at the University of Washington, concentrating on varied topics including muscle contraction, water and biology, the origin of life, and the role of electrical charge in nature. His work with water earned him the inaugural Emoto Peace Prize as well as the Prigogine Medal for Thermodynamics.

He is the founding Editor-in-chief of the research journal *WATER*, founding Executive Director of the Institute for Venture Science, and organizer of the annual International Conference on the Physics, Chemistry, and Biology of Water. Further, he has received an honorary doctorate from Ural State University, and was more recently named Honorary Professor of the Russian Academy of Sciences. His university has bestowed on him their highest faculty distinction, the Faculty Lecturer Award.

When prodded, Jerry likes to bring up his involvement with the Travis Rice snowboard movie, The Fourth Phase, which was inspired by his popular book, *The Fourth Phase of Water*.

These days, Jerry can be found laptop in hand, studying water, trees, birds, and clouds outside his window. He is currently writing three books, whose subjects include: the (revised) structure of the atom, the origin of volcanoes and earthquakes, and biology reconsidered.

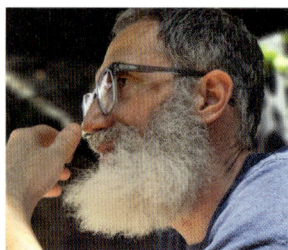

ETHAN POLLACK, illustrator and son of Gerald Pollack, studied art at Syracuse University. Later in NYC, he constructed giant balloon sculptures for renowned artist, Jeff Koons. Ethan then co-founded Conner Pollack Productions, a Brooklyn based specialty fabrication shop creating one-of-a-kind sculptures, prototypes, and models including one for the winning design of the Freedom Tower.

Ethan now lives in Seattle with his lovely wife and two wonderful kids. He enjoys binge-watching Star Wars shows with his family, although they much prefer watching something else. Illustrating his dad's far-out ideas has been an honor and a joy.